T0344856

Glancing Angle Deposition of Thin Films

Wiley Series in Materials for Electronic and Optoelectronic Applications

www.wiley.com/go/meoa

Series Editors

Professor Arthur Willoughby, *University of Southampton, Southampton, UK*
Dr Peter Capper, *SELEX Galileo Infrared Ltd, Southampton, UK*
Professor Safa Kasap, *University of Saskatchewan, Saskatoon, Canada*

Published Titles

Bulk Crystal Growth of Electronic, Optical and Optoelectronic Materials, Edited by P. Capper

Properties of Group-IV, III–V and II–VI Semiconductors, S. Adachi

Charge Transport in Disordered Solids with Applications in Electronics, Edited by S. Baranovski

Optical Properties of Condensed Matter and Applications, Edited by J. Singh

Thin Film Solar Cells: Fabrication, Characterization and Applications, Edited by J. Poortmans and V. Arkhipov

Dielectric Films for Advanced Microelectronics, Edited by M. R. Baklanov, M. Green and K. Maex

Liquid Phase Epitaxy of Electronic, Optical and Optoelectronic Materials, Edited by P. Capper and M. Mauk

Molecular Electronics: From Principles to Practice, M. Petty

CVD Diamond for Electronic Devices and Sensors, Edited by R. S. Sussmann

Properties of Semiconductor Alloys: Group-IV, III–V and II–VI Semiconductors, S. Adachi

Mercury Cadmium Telluride, Edited by P. Capper and J. Garland

Zinc Oxide Materials for Electronic and Optoelectronic Device Applications, Edited by C. Litton, D. C. Reynolds and T. C. Collins

Lead-Free Solders: Materials Reliability for Electronics, Edited by K. N. Subramanian

Silicon Photonics: Fundamentals and Devices, M. Jamal Deen and P. K. Basu

Nanostructured and Subwavelength Waveguides: Fundamentals and Applications, M. Skorobogatiy

Photovoltaic Materials: From Crystalline Silicon to Third-Generation Approaches, Edited by G. Conibeer and A. Willoughby

Glancing Angle Deposition of Thin Films

Engineering the Nanoscale

Matthew M. Hawkeye
Department of Electrical and Computer Engineering
University of Alberta
Canada

Michael T. Taschuk
Department of Electrical and Computer Engineering
University of Alberta
Canada

Michael J. Brett
Department of Electrical and Computer Engineering
University of Alberta
Canada
NRC National Institute for Nanotechnology
Canada

Library of Congress Cataloging-in-Publication Data

Hawkeye, Matthew M.
  Glancing angle deposition of thin films : engineering the nanoscale / Matthew M. Hawkeye, Michael T. Taschuk, Michael J. Brett, NRC National Institute for Nanotechnology.
      pages cm
  Includes bibliographical references and index.
  ISBN 978-1-118-84756-5 (cloth)
  1. Nanotechnology.   2. Chemical vapor deposition.   3. Nanoelectromechanical systems–Design and construction.   I. Taschuk, Michael T.   II. Brett, Michael J.   III. Title.
  T174.7.H39 2014
  621.3815′2–dc23
                                                                                          2014013307

A catalogue record for this book is available from the British Library.

ISBN: 9781118847565

Set in 10/12pt Times by Aptara Inc., New Delhi, India
Printed and bound in Malaysia by Vivar Printing Sdn Bhd

1   2014

Contents

Series Preface

WILEY SERIES IN MATERIALS FOR ELECTRONIC AND OPTOELECTRONIC APPLICATIONS

This book series is devoted to the rapidly developing class of materials used for electronic and optoelectronic applications. It is designed to provide much-needed information on the fundamental scientific principles of these materials, together with how these are employed in technological applications. The books are aimed at (postgraduate) students, researchers and technologists, engaged in research, development and the study of materials in electronics and photonics, and industrial scientists developing new materials, devices and circuits for the electronic, optoelectronic and communications industries.

The development of new electronic and optoelectronic materials depends not only on materials engineering at a practical level, but also on a clear understanding of the properties of materials, and the fundamental science behind these properties. It is the properties of a material that eventually determine its usefulness in an application. The series therefore also includes such titles as electrical conduction in solids, optical properties, thermal properties, and so on, all with applications and examples of materials in electronics and optoelectronics. The characterization of materials is also covered within the series in as much as it is impossible to develop new materials without the proper characterization of their structure and properties. Structure-property relationships have always been fundamentally and intrinsically important to materials science and engineering.

Materials science is well known for being one of the most interdisciplinary sciences. It is the interdisciplinary aspect of materials science that has led to many exciting discoveries, new materials and new applications. It is not unusual to find scientists with a chemical engineering background working on materials projects with applications in electronics. In selecting titles for the series, we have tried to maintain the interdisciplinary aspect of the field, and hence its excitement to researchers in this field.

Peter Capper
Safa Kasap
Arthur Willoughby

Preface

The design and fabrication of thin films have been central to technological advances in microelectronics and information processing, optics and displays, energy generation and storage, and in many other devices pervasive in society. The capability to further engineer film properties through physical structuring on the sub-micrometre scale and by control of porosity enable new opportunities for performance improvements. Glancing angle deposition (GLAD) provides a simple method to fabricate nanostructured columnar architectures using existing and prevalent physical vapour deposition processes such as sputtering or evaporation, while requiring only minor modifications to enable carefully controlled substrate tilt and rotation. GLAD is remarkable in that substrate motions on the macroscale translate over six orders of magnitude to create three-dimensional precision in fabrication of nanometre-scale arrays of isolated structures.

We are encouraged by the popularity of GLAD. Over 120 groups have adopted the GLAD technique for fabricating films and have published nearly 1000 papers detailing new nanostructure designs or describing applications such as photonics, sensing, photovoltaics, chemical analysis, batteries and more. Perhaps this interest is due to the ready accessibility of the GLAD process, its compatibility with existing deposition systems, the fine level of structural control and the novelty of the structures that can be fabricated for specific needs. For us, there is a captivation with the symmetry and physical elegance of arrays of nanocolumnar structures, and a belief that these new architectures could lead to significant improvements in specific physical or chemical properties.

It is our hope that this book can provide guidance so that current and new researchers can achieve the best precision and performance in designing and fabricating GLAD nanostructures. As much as possible, we have attempted to adopt a 'how to' style, describing what is necessary for fabrication of a particular architecture. Although we do discuss some specific applications, our focus is on the technical detail of procedures to make the ideal helix (or periodic array, or sinusoidal index gradient, etc.), enabling the reader to best utilize the strong foundation that has been laid over the last 20 years.

Our intended audience is anyone familiar, even to a limited extent, with thin-film deposition and vacuum technology. Those who are not familiar should consult one of the many standard textbooks available in the field. *Glancing Angle Deposition of Thin Films: Engineering the Nanoscale* is structured with an introduction to GLAD in Chapters 1 and 2, and then successive Chapters 3–6 present progressively more refined and complex structural control techniques. Chapter 7 concludes with practical information such as uniformity, motor control and vacuum requirements. There is not nearly enough room to properly survey all device applications of GLAD films; however, Chapter 1 includes a brief overview.

The book provides extensive references to the works of others in this field, and in particular has attempted to include all critical works advancing the GLAD process. Some very worthy reports may have been omitted owing to lack of space, or because the focus was on device applications, or perhaps through unfortunate oversight. We would like to thank the many authors whose work we have profiled – a large proportion of the book is a review of the research of others, and we apologize for any errors that may have inadvertently slipped into the manuscript. We would especially like to thank the many members of the GLAD research group at the University of Alberta who, since 1994, have worked to improve the technique and develop applications of nanostructured coatings. We additionally thank the current group members, who have acted as early manuscript readers, and particularly Ryan Tucker and Steven Jim, who created new figures.

We hope that you find this book a useful guide to nanostructuring of porous thin-film coatings, and that the information herein helps you reduce the many cycles of film optimization through successive rounds of fabrication and characterization. We welcome your comments and look forward to reading in the literature about your advances and applications.

Matthew Hawkeye, Michael Taschuk and Michael Brett

Canada

January, 2014

1 Introduction: Glancing Angle Deposition Technology

1.1 NANOSCALE ENGINEERING AND GLANCING ANGLE DEPOSITION

One key area of technological development since the 1950s has focused on fulfilling the possibilities described by Richard Feynman in his 1959 talk 'Plenty of Room at the Bottom', where he describes how engineering material structure on a progressively smaller and smaller scale enables enormous technological potential. Novel material properties emerge as this structural control approaches the nanometre scale, leading to the development of nanotechnologies based on exploiting these innovative engineering capabilities. Two major approaches have been used to achieve nanoscale engineering: *top-down* methods, where a process or instrument is used to impose patterning or functionality on a small scale, and *bottom-up* approaches, where individual atoms or molecules are coaxed into self-assembly to produce materials with desirable functionality. The power of top-down approaches is demonstrated by the success of microelectronics fabrication techniques, such as photolithography. Leading microprocessor companies are engaged in a race to the nanoscale, developing advanced fabrication techniques and tools which drive the integrated circuit industry and contribute to the rapid technological development we have experienced over the last half century. Although bottom-up approaches are less mature than top-down methods, they have been the focus of substantial research effort in recent years and are promising routes to low-cost nanoscale technologies. Researchers in bottom-up nanofabrication hope to achieve functionality reminiscent of that found in living organisms, which exploit directed self-assembly techniques polished by billions of years of evolution.

Glancing angle deposition (GLAD) is a thin-film deposition process that may be viewed as a combination of top-down and bottom-up nanofabrication approaches. During conventional film deposition, a stream of vapour-phase atoms strikes and condenses upon a perpendicular substrate to form a dense, solid film. In GLAD, the substrate is tilted to a glancing angle, thus creating an oblique deposition geometry. As the atoms condense on the substrate they spontaneously form microscopic nuclei (Figure 1.1a). Ballistic (i.e. line-of-sight) shadowing prevents incoming vapour from condensing into regions behind the nuclei, causing the nuclei to develop into columns that tilt towards the vapour source (Figure 1.1b). Thus, we see the bottom-up aspect of the GLAD process: the shadowing mechanism in an oblique deposition geometry causes the condensing vapour to self-assemble into oriented columnar nanostructures. The top-down component of GLAD involves using macroscopic substrate

Glancing Angle Deposition of Thin Films: Engineering the Nanoscale, First Edition.
Matthew M. Hawkeye, Michael T. Taschuk and Michael J. Brett.
© 2014 John Wiley & Sons, Ltd. Published 2014 by John Wiley & Sons, Ltd.

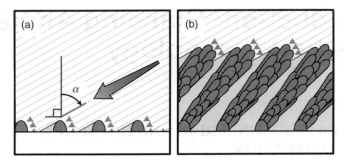

Figure 1.1 A conceptual view of the GLAD process. (a) The incident vapour atoms, arriving at an angle α with respect to the surface normal, are intercepted by the growing nanocolumns. (b) Shadowed regions are thus created where no growth occurs, and vapour deposition is restricted to the nanocolumns that grow towards the vapour source. The basic principle of GLAD is to manipulate this shadowing effect and guide the directional column growth by dynamically controlling the substrate orientation.

rotations during deposition to change the apparent vapour direction and manipulate the column growth direction. By combining shadowing and substrate rotation, the evolving film serves as its own mask, which can be dynamically reconfigured during deposition to steer columns into the desired geometry. The art of GLAD is found in combining the top-down substrate motions with the specific material's bottom-up tendencies. A sense of the challenge in performing GLAD can be understood by an analogy to a field of growing grass. Imagine a square field 200 m on a side, covered by 2 cm tall grass that grows towards a sunlamp carried by a helicopter flying at a distance of 3 km. As the helicopter moves around, the grass grows towards it and is thereby coaxed into desired shapes. The GLAD process controls the nanostructure of thin films at a distance greater than 10 million times the minimum morphological feature size. With the techniques and understanding presented in this book, producing extremely high quality nanostructures is possible.

A schematic view of the local shadowing environment controlled by GLAD's substrate motions is given in Figure 1.2. In this film, a helical nanostructure is produced by slowly rotating the substrate around its normal axis. In Figure 1.2a, one full turn has been completed. By this time, the nucleation layer shown in Figure 1.1a has already developed into individual, neighbouring columns. The black vapour flux particles are those captured by the black helix in the centre of the field, slowly growing towards the vapour source, thereby recording the substrate orientation in the growing nanostructure. As the substrate continues to rotate, the film completes additional turns, shown at 1.33 rotations in Figure 1.2b, and 1.66 turns in Figure 1.2c.

Many different GLAD processes have been designed and developed, based on manipulating the substrate position to control the ballistic shadowing effect and guide the columnar growth process. The substrate position in a GLAD process is specified by two angles (Figure 1.3a). The deposition angle α is defined as the angle between the substrate normal and the incident vapour flux direction. The angle φ measures the substrate rotation about its normal, thus providing the azimuthal position of the substrate. Controlling α and φ provides sufficient rotational freedom to set the apparent vapour source position at any desired altitudinal (controlled by α) and azimuthal (controlled by φ) coordinates, as viewed from the

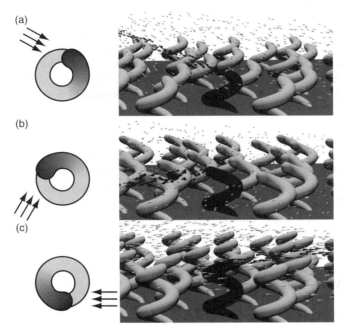

Figure 1.2 A helical GLAD film is built up by rotating the substrate around its axis, shown at (a) one full turn, (b) one and a third turns and (c) one and two-thirds turns. Neighbouring columns shadow each other, concentrating incident vapour flux on the tip of the growing column. Varying the incident flux vector determines a GLAD film's nanostructure.

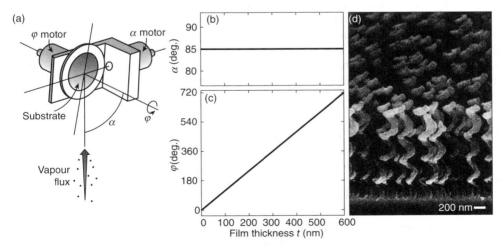

Figure 1.3 (a) A prototypical set-up for GLAD motion control wherein the substrate is mounted on a rotating stage with two degrees of freedom. The α motor controls the substrate tilt, while the φ motor rotates the substrate about its normal. The substrate is positioned to intercept the vapour flux and a deposition rate monitor (not shown) measures the growth rate and provides feedback for motion control. A GLAD process is specified by how the (b) α and (c) φ positions are changed during film growth. (d) GLAD-fabricated Si helical nanostructures.

substrate surface. The α angle can vary between $0°$ (normal incidence) and $90°$ (parallel to the substrate), whereas the φ angle can assume any value and is periodic.

In this book, we present the many GLAD processes in terms of substrate motion algorithms that describe how the α and φ angles change over the course of the deposition. As an instructive example, Figure 1.3b and c specifies a simple motion algorithm that would be used to fabricate a basic helical columnar microstructure (Section 2.5). In this example, α is fixed at $85°$ to create a GLAD geometry, and φ is continuously rotated at $1.2°/nm$, thus completing one revolution every 300 nm of growth. The φ rotation continuously changes the column growth direction and steers the columns into helical structures, as shown in Figure 1.3d. The α and φ positions are specified as a function of the film thickness because motion algorithms are generally designed relative to the desired column microstructure. Note that we define film thickness as the *vertical* thickness of the film (i.e. measured along the substrate normal), as opposed to the length of a column element, which may be different than the vertical thickness depending on the column shape/orientation. In practice, accurately implementing a complex substrate motion algorithm generally requires using a calibrated deposition rate monitor and motion feedback system.

1.2 GLAD-VANTAGES

1.2.1 Nanoscale morphology control

There is a pressing need in nanotechnology to develop fabrication methods that are simultaneously capable of nanoscale precision while also providing flexibility in the range of nanostructures that can be achieved. Nanofabrication techniques that satisfy both these requirements provide superior design freedom to scientists and engineers seeking to create and optimize technological devices based on nanostructured materials. GLAD provides such freedom as it directly and accurately controls material structure across a range of size scales and allows creation of a diverse array of columnar morphologies through modifications of the same basic technique. A small cross-section of possible structures is showcased by the selected columnar nanostructures shown in Figure 1.4. These six examples are all deposited using different GLAD processes designed to manipulate the ballistic shadowing conditions during growth in different ways; specifically:

- **Zigzag structures (Figure 1.4a):** Depositing from alternating directions at high α realizes a film where the column orientation switches back and forth, creating a film with a zigzag column morphology (Section 2.5.3).
- **Helical structures (Figure 1.4b):** Helical columns are created by high-α deposition coupled with slow substrate rotation, a motion that continuously redirects the column growth orientation and shapes the columns into a helical structures (Section 2.5.1).
- **Vertical columns (Figure 1.4c):** Depositing at high α and with fast substrate rotation produces a vertical column growth direction and yields a vertically oriented columnar morphology (Section 2.5.2).
- **Graded-density structures (Figure 1.4d):** Periodically modulating α between high and low values induces a corresponding modulation in the column diameter and film density (Section 5.4).

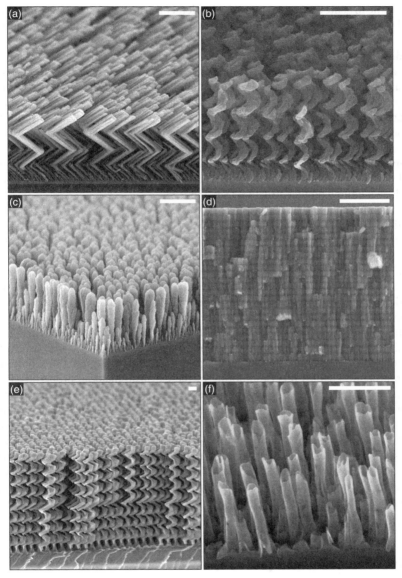

Figure 1.4 A wide variety of nanostructured columnar films can be realized by modifying the basic GLAD fabrication procedure, as showcased by these selected examples. Films with (a) zigzag, (b) helical, and (c) vertical columnar morphologies are created by implementing different substrate rotations during growth, while (d) changing the α angle during deposition introduces controlled porosity gradients to the film. Pre- and post-deposition processes can also be used to further control film structure: (e)depositing onto pre-patterned substrates with periodic 'seed' arrays allows fabrication of high-uniformity column arrays, and (f) as-deposited GLAD structures can be used as templates for subsequent coating and etch processing, used to create hollow-core nanotube structures. All scale bars indicate 1 μm.

It also possible to achieve greater control by implementing pre- and post-deposition processes designed to further tailor the growth mechanisms and/or modify the final film nanostructure.

- **Periodic structure arrays (Figure 1.4e):** Normal GLAD structures are arranged randomly across the substrate due to the stochastic nucleation phase of film growth. However, depositing onto a preprocessed substrate with periodic surface features controls the nucleation phase and enables production of highly uniform, periodic arrays of nanocolumn structures. For optimal growth control and structural uniformity, the substrate features must be properly designed and the GLAD process optimized (Chapter 3).
- **Hollow-core structures (Figure 1.4f):** The as-deposited GLAD structures can be used as templates for further processing. Here, the original GLAD structures were conformally coated with a second material and then the inner material was removed via an etch process to produce a hollow tubular structure (Section 6.8).

As mentioned previously, these examples represent a limited selection of the overall microstructural diversity that can be achieved using GLAD, and processes for realizing the many possible morphologies are explored throughout this book.

An additional feature of GLAD nanofabrication is that it enables structural control across multiple size scales. At the macroscale, the GLAD-structured thin film is simultaneously deposited over the entire wafer, allowing massively parallel fabrication of nanostructures over large areas. Furthermore, standard lithographic techniques can be subsequently used to pattern the as-deposited film and define specific microdevice geometries. Moving down to the submicrometre scale, the column structure (on the order of 10–100 nm) can be flexibly manipulated and shaped into diverse geometries, as shown in Figure 1.4. Controlling the columnar structure enables precise tailoring of the extrinsic material properties associated with nanoscale engineering. Moving down even further, the internal columnar morphology (on the order of 1–10 nm) can also be controlled using the GLAD process. These sub-10 nm features are critical to many applications as they provide the most significant contribution to the total surface area enhancement of the porous film and directly impact the interfacial properties of the engineered material. As an example, the mesopore structure can be controlled by altering the deposition angle (Section 4.5.2). The ability to quickly realize a wide variety of nanostructures and to control the film morphology across several size scales provides a significant advantage for application design and device optimization, especially when coupled with the broad material compatibility discussed below.

1.2.2 Broad material compatibility

Because GLAD is based upon physical vapour deposition (PVD; see Section 2.2.1 for discussion) techniques it inherits much of the broad material compatibility provided by PVD thin-film deposition. This flexibility is a major scientific and technological advantage as GLAD can therefore be used to fabricate diverse nanostructural morphologies in materials with a wide range of physicochemical properties. This advantage provides further design freedom and has led to high-surface-area, nanoengineered GLAD films being developed for numerous applications that exploit specific intrinsic material characteristics (see Section 1.6 for a brief summary). Table 1.1 presents a selection of materials that have been successfully deposited using GLAD. As can be seen, the material compatibility of the GLAD

Table 1.1 Materials successfully deposited using GLAD and the corresponding PVD methods employed. IBS: ion-beam sputtering; IP: ion plating; MBE: molecular beam epitaxy; PLD: pulsed laser deposition.

Material	PVD method	Ref.	Material	PVD method	Ref.
Elements			HfO_2	E-beam	[1]
Ag	E-beam	[2, 3]	InN	IP	[4]
Al	Thermal, e-beam	[5, 6]	In_2S_3	Thermal	[7]
Au	Sputtering	[8]	ITO	E-beam	[9–11]
C	IBS, PLD, E-beam	[12–14]	MgF_2	Thermal	[15, 16]
Cr	E-beam, sputtering	[17, 18]	MgO	Effusion	[19]
Co	Thermal, e-beam	[20–22]	MoO_3	E-beam	[23]
Cu	E-beam, thermal, sputtering	[21, 24, 25]	Nb_2O_5	E-beam	[26, 27]
Ge	E-beam	[28, 29]	RuO_2	Sputtering	[8]
Fe	Thermal, MBE	[30–32]	SiO	Thermal	[17, 33]
Mg	E-beam	[34, 35]	SiO_2	E-beam	[10, 16]
Mn	Evaporation	[36]	Ta_3N_5	Evaporation	[37]
Nb	Sputtering	[38, 39]	Ta_2O_5	E-beam	[40–42]
Ni	E-beam	[43]	TiAlN	Sputtering	[44]
Pd	Effusion	[45]	TiC	E-beam	[46, 47]
Pt	E-beam, sputtering	[48–50]	TiO_2	E-beam	[51–53]
Ru	Sputtering	[50, 54]	TiZrV	Sputtering	[55]
Se	Thermal	[56]	WO_3	Sputtering, thermal	[57–59]
Si	E-beam	[60, 61]	$W_xSi_yO_z$	Sputtering	[59, 62]
Ta	E-beam, sputtering	[63, 64]	Y_2O_3:Eu	E-beam	[65, 66]
Te	Thermal	[56]	YSZ	E-beam	[67, 68]
Ti	Sputtering, e-beam	[69–71]	ZnO	PLD, e-beam, sputtering	[72, 73]
W	Sputtering	[21, 58, 74]	ZnS	E-beam	[75]
Inorganic			ZrO_2	E-beam	[41, 76]
Al_2O_3	E-beam	[77, 78]			
As_2O_3	Thermal	[79, 80]	*Organic*		
ATO	E-beam	[81]	Alq_3	Thermal	[82, 83]
$BiVO_4$	E-beam	[84]	C_{60}	Thermal	[85]
CaF_2	Thermal	[15, 86]	CuPc	Thermal	[87]
CeO_2	E-beam	[88]	α-NPD	Thermal	[89]
CrN	Sputtering	[90]	Parylene C	Nozzle	[91]
Fe_2O_3	E-beam	[92]	Pentacene	Thermal	[85]
GeSbSn	Thermal	[93]	PPX	Nozzle	[94]
$GeSe_2$	Thermal	[95]			

process covers several classes of materials: elements; compounds with complex stoichiometry, including carbides, fluorides, nitrides, oxides and sulfides; metals, semiconductors, and dielectrics; binary and ternary chalcogenide glasses; organic small-molecules and polymers. The various deposition methods used to prepare the GLAD films are also listed, with several materials affording multiple fabrication options. Across all these methods and materials, the GLAD process remains bottom-up and catalyst free, with several techniques involving reactive processing. Note that although Table 1.1 presents an extensive list of materials and references, the survey is not comprehensive, and more examples can be found by exploring the literature. Additional levels of material flexibility are introduced by using co-deposition processes (Section 2.8.2), which enable fabrication of mixed-material and doped structures, and via post-deposition processing such as templating (Section 6.8) and conformal surface coating (Section 6.2.5).

In general, GLAD films exhibit similar columnar morphologies across the various materials that can be deposited. Figure 1.5 presents examples of this general columnar structure arising in three different materials: (a, b) Al$_2$O$_3$, (c, d) indium tin oxide (ITO) and (e, f) MgF$_2$. Each film was deposited at $\alpha = 85°$ and without substrate rotation. The SEM images in the left column correspond to side views of the film structure, whereas the right column

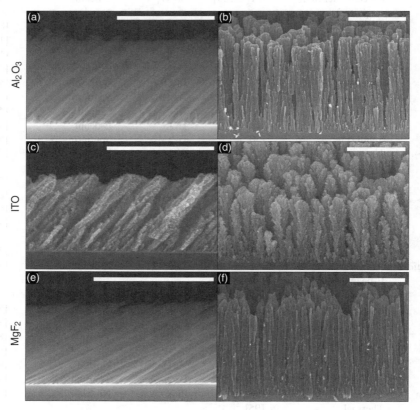

Figure 1.5 Tilted columnar films of (a, b) Al$_2$O$_3$, (c, d) ITO, and (e, f) MgF$_2$, each deposited at $\alpha = 85°$ and without substrate rotation. The scanning electron microscope (SEM) images in the left column show side views of the columnar structure, while the right column shows the structures viewed from above but at right angles to the column axes. To a first-order approximation, the films all possess the same general tilted columnar morphology, as the ballistic deposition process underlying GLAD is similar for each material. However, the specific nanoscale structure of the columnar features (e.g. column tilt angles and column surface roughness) are different owing to higher order material-specific growth properties. All scale bars indicate 1 μm. Reproduced from [96] with permission of J.B. Sorge.

shows the films imaged from above but at right angles to the column axis. In each material, the deposited film consists of randomly arranged columns inclined towards the incident vapour source. This columnar morphology is generally consistent between different materials and is a consequence of the ballistic growth process, which is material independent to a first-order approximation. However, beyond the general structural similarities, each film exhibits material-specific structural properties that reflect differences in higher order growth characteristics, such as atomic diffusion, nucleation rates, vapour flux distributions, and so on. For instance, even though each film is deposited at the same α, the column tilt angle is different for each material, with the MgF_2 columns being considerably more inclined than the Al_2O_3 and ITO structures. The material dependence makes column tilt angle prediction difficult without prior information (Section 2.4.2). Comparing the SEM images also reveals differences in the nanoscale structure of the columnar features. Subtle differences can be discerned between the Al_2O_3 and MgF_2 columns, primarily in the column bundling geometry, the sharpness of the column tips and the level of nanoscale surface roughness along the columns. However, the ITO columns show a remarkably different nanostructure, with the columns exhibiting substantially greater surface roughness and nonuniformity. Thus, it is important to recognize that, although similar GLAD structures can be readily fabricated in many materials using the same general techniques, the specific deposition and growth behaviour of every material is different.

Perhaps the most striking example of a material-specific structure is in GLAD-fabricated tris-(8-hydroxyquinoline)aluminium (Alq_3) films. First deposited in a GLAD process by Hrudey et al. [82], Alq_3 shows several unique structural characteristics (Figure 1.6). First, the surface of Alq_3 columns is extremely smooth in comparison with typical GLAD structures, which often exhibit substantial nanoscale roughness. Second, the columnar structures are significantly more uniform than other GLAD columns and show minimal column-to-column structural variation. Third, the Alq_3 columns are evenly spaced on the substrate, following a nearly hexagonally close-packed arrangement. Although the specific growth properties are not understood, these structural characteristics arise spontaneously during the Alq_3 GLAD process, whereas achieving similar levels of order and uniformity in other materials requires additional preprocessing techniques to seed column growth (Chapter 3).

Figure 1.6 (a) Side view and (b) top-down SEM images of helical Alq_3 columns deposited at $\alpha = 85°$, revealing the unique structures produced by GLAD-fabricated Alq_3 films. The Alq_3 structures are exceptionally uniform and highly ordered in comparison with typical GLAD microstructures. Reproduced with permission from [82]. Copyright © 2006 Wiley–VCH Verlag GmbH Co. KGaA, Weinheim.

1.2.3 Novel thin-film material properties

The properties of nanostructured materials are often significantly different than those of the material in its bulk form, and nanofabrication techniques attempt to control the material's properties (e.g. thermal, mechanical, electrical, chemical, optical) and also create new effects. By using GLAD to engineer the nanoscale structure, film properties can be controlled and tuned over a substantially greater range than typically achieved with conventional film deposition. For example, simply by altering the deposition angle to control the film porosity, the electrical conductivity of the film can be tuned by several orders of magnitude (Section 4.7). In another example, tailoring the porosity allows the film refractive index to be continuously varied between the bulk material index and the index of the void regions (Section 5.3). Such capabilities enable the realization of structured films with physical properties that can be substantially different than observed in the bulk material, thus providing greater freedom to the designing engineer. However, the GLAD process offers even more exciting technological opportunities by further exploiting the intimate link between material properties and nanoscale structure to design structured films with entirely new physical characteristics that are not observed in conventional films. Consider that by using GLAD, any compatible material can be transformed into a highly porous film exhibiting greatly enhanced surface area and a tailored mesopore structure (Section 4.5). This provides a direct route towards improving surface chemical activity and mass transport kinetics, which have enormous implications on catalyst efficiency in a number of technological applications, as discussed in Section 1.6. In another example, the individual submicrometre columns fabricated with GLAD can be used as micromechanical components, such as microsprings and microcantilevers (Section 4.8). Intriguingly, these microstructures behave like scaled-down versions of their macroscopic counterparts, suggesting applications in microelectromechanical system (MEMS) devices. Finally, nanostructured GLAD films have been designed with artificially engineered birefringence properties in order to realize optical coatings with complex, polarization-dependent properties (Section 5.5). In these devices, the polarization anisotropy is entirely created by the columnar morphology, as opposed to being an intrinsic crystalline or stress-related birefringence, thus allowing anisotropic effects to be engineered in ordinary materials.

1.2.4 Compatibility with standard microfabrication processes

While the GLAD process provides many techniques for engineering film properties, the possibilities for technological development do not end with the as-deposited films. Rather, nanostructured GLAD films have been repeatedly demonstrated to be robust to subsequent handling and post-deposition processing, thus providing a useful capability for integrating GLAD films into different microdevice designs (see Chapter 6). Much of this compatibility originates from GLAD being a thin-film process. This facilitates the integration of GLAD fabrication steps into common microfabrication process flows based on standard techniques such as thin-film deposition, photolithography, chemical etching and thermal annealing. Furthermore, the material freedom provided by GLAD allows selection of materials compatible with subsequent processing techniques. Process integration is a key part of transforming functional GLAD films into technological devices. Many demonstrations can be found in the literature, and selected examples are showcased in Figure 1.7. Exploiting the gas-sensing

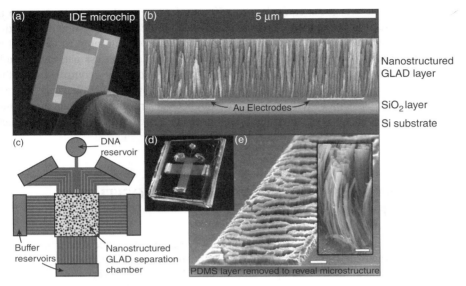

Figure 1.7   Examples of nanostructured GLAD films being integrated into different microdevice geometries. (a) A GLAD layer added to an interdigitated electrode (IDE) microchip, where (b) the functional porous layer is deposited directly onto the electrode surface. Structured GLAD films can also be directly patterned using standard lithographic processes. (c) Here, a polydimethylsiloxane (PDMS) microfluidic device was designed using the GLAD film as a DNA separation medium [97]. (d) A photograph of the completed device. (e) The multistep process patterns the structured film into a microchannel geometry, as shown by this SEM image. (a) and (b) Reproduced from [98]. With permission J.J. Steele. (c) and (d) Reproduced from [99]. With permission from L.W. Bezuidenhout. (e) Reproduced from [97] with permission from The Royal Society of Chemistry.

capabilities of nanostructured GLAD films requires adding electrical contacts to transduce the impedance changes associated with the target analyte. In Figure 1.7a and b, the GLAD film was fabricated on an IDE microchip to enable electrical read-out [100]. Porous GLAD layers have been integrated into PDMS microfluidic devices as shown in Figure 1.7c–e, where they have been used as high-surface-area media for DNA separation [97]. Such integration required developing a multistep sacrificial etch process flow to embed the three-dimensional (3D) GLAD nanostructures into lithographically patterned PDMS microchannels.

1.2.5 Scalable fabrication method

At its core, GLAD is an extension of traditional thin-film vacuum deposition, a well-established manufacturing process that is a critical component of several industries. This aspect of GLAD mitigates many of the scalability and economic concerns typically associated with bottom-up nanofabrication techniques that, while promising, lack a proven route to scalability and process control. Because the vacuum deposition industry is very mature, the necessary hardware is highly developed and readily available, as is expert technical guidance. Thin-film deposition is inherently wafer scale; using our research-grade systems, we routinely deposit over areas equivalent to 250 mm diameter wafers. Cycle time can be reduced and production throughput increased using standard deposition system designs that implement load-locked vacuum chambers, high-performance vacuum pumps, large wafers and even roll-to-roll configurations. Furthermore, the deposition process can be carefully controlled and fully automated, enabling detailed reliability and quality control studies that focus on process and yield optimization. These advantages help to bridge the gap between small-scale prototype demonstrations at the research level and high-throughput produc-tion at the manufacturing level, an important factor in the commercial success of new technologies.

1.3 THE ROOTS OF GLANCING ANGLE DEPOSITION: OBLIQUE DEPOSITION

Although modern interest in the GLAD process has grown rapidly since the technique was initially demonstrated in 1995 [101], GLAD developed out of a long-standing interest in film morphology and studies of oblique deposition. It has long been recognized that films deposited at oblique incidence exhibit many unique properties when compared with films prepared at normal incidence. Beginning with work by Kundt in 1886 [102], early studies established that obliquely deposited films exhibit anisotropy in their electrical [103], magnetic [104] and optical [105] properties. Importantly, it was found that these anisotropic properties were not a by-product of stress or nonuniformity in the film. Instead, these early investigations concluded that anisotropy in these films is a microstructural effect directly associated with the oblique deposition conditions. Subsequent experimental research was greatly accelerated by improvements in electron microscopy techniques that enabled direct inspection of the film morphology at high resolution. A major contribution of these morphology investigations was identifying atomic-scale shadowing as a crucial factor in thin-film growth and the development of columnar morphologies [106, 107]. This prompted many experimental and theoretical (see Section 1.5) investigations examining how the oblique deposition geometry impacts the resulting columnar morphology. Studies through the 1960s and 1970s established several key results regarding the microstructure of obliquely deposited films, summarized by Dirks and Leamy [108]. Smith *et al.* observed that obliquely deposited films tended to be elongated and bundled along the direction perpendicular to the incident flux [107]. This morphological asymmetry, now referred to as column fanning, is responsible for many of the anisotropic physical properties typical of obliquely deposited films and is a consequence of the one-dimensional nature of the shadowing effect (Section 2.6.4). Nieuwenhuizen and Haanstra performed key microfractography studies, where the film cross-section is exposed by fracturing the sample, demonstrating that the columnar structures are tilted towards the source at an angle β with respect to the substrate normal [109]. Their results showed that β increases monotonically with, but always remains less than, the deposition angle α and they proposed the well-known tangent rule (Eq. 2.1) to quantify the empirical β versus α data. The incorporation of shadow-defined void regions in the deposited film leads to a monotonically decreasing film density as α is increased [110].

The fabrication of columnar films with controlled porosity is a central feature of obliquely deposited films, leading to the formation of high-surface-area materials and directly impact-ing many physical properties. In particular, the magnetic properties of obliquely deposited films have been the subject of substantial research interest as the shadowing-induced columnar separation strongly reduces the magnetic exchange coupling between adjacent columns, leading to a dramatic rise in the magnetic coercivity as α is increased [111]. The columnar structures can thus be stably magnetized and the film can be used for data storage applications [112–114], an application that quickly attracted significant commercial attention. Obliquely deposited metals have been used for many years as magnetic layers in metal-evaporated tape storage media, by companies such as Sony, Fujifilm and IBM. Scaling the oblique deposition process to mass production provides an intriguing case study [112]: by using a roll-to-roll coating process with a series of baffles to define the oblique deposition geometry and deposition at a high rate (>0.5 µm/s), several kilometres of tape could be created in a matter of a few hours. Proposals have been put forth to realize GLAD in such a roll-to-roll processing scheme (Section 7.6).

Aside from a single experiment in 1959 by Young and Kowal [115], the majority of oblique deposition studies consider deposition conditions where the substrate is held stationary throughout the deposition process. However, experimental studies by Taga and Motohiro [116] showed that the column orientation changed by rotating the substrate 180° during deposition, thus producing the first zigzag columnar morphology. This work was quickly followed by theoretical proposals to create chiral structures via continuous rotation [117]. However, these reports remained in the $\alpha < 70°$ regime, where the film morphology is highly dense. The invention of GLAD emerged from experimental investigations combining dynamic sub­strate motions with deposition in the $\alpha > 80°$ regime, thus exploiting extreme shadowing and directional column growth to fabricate unusual nanostructured materials using basic thin-film tools [15, 101]. As α is increased beyond ~80°, the geometric shadowing lengths become extremely long, which has a remarkable impact on film morphology: the intercolumn void regions expand and the film transitions to an isolated column morphology (Section 2.4). Rotation about the substrate normal also has a dramatic effect. Such substrate rotation alters the apparent direction of the vapour source from the perspective of the columns, thus changing the local shadowing orientation and reorienting the column growth direction (Section 2.5). The nanoscale structure of the thin film can therefore be engineered by depositing in this *glancing angle* regime and implementing specifically designed substrate motions. GLAD motion algorithms are typically computer controlled and incorporate feedback from a deposition rate monitor in order to accurately execute complex motions and achieve precise structural control.

1.4 THE IMPORTANCE OF EXPERIMENTAL CALIBRATION

As is common in any thin-film fabrication technique, careful control of processing variables and experimental calibration of film properties is crucial towards achieving repeatable GLAD fabrication. While the general GLAD growth mechanisms are sufficiently well understood, thus enabling predictive design of the film microstructure and many physical properties, the exact properties of the film are sensitive to numerous experimental factors. It is the role of this book to present key experimental results and GLAD techniques, thus providing initial parameters for subsequent device design and process development. The relationships between important processing variables (such as α and substrate temperature) and the film microstructure and properties are largely consistent across many reports, making it possible to recreate effects and characteristics reported elsewhere. However, it is very challenging to *precisely* reproduce a reported value without substantial effort and a highly detailed knowledge of the specific processing conditions used. Dedicated experimental efforts are required to reliably achieve desired film properties, and process calibration, control and monitoring are central aspects of GLAD technology.

It is important to identify the critical experimental variables and control these parameters during GLAD process development. Although film properties are influenced by a long list of variables, in our experience there are several factors that should be considered first when designing a GLAD process.

1. **Deposition geometry:** Ensuring reproducible control over the deposition angles α and φ is a primary requirement for repeatable GLAD work. It is important that the motors and gearing systems used to control the substrate position and motion are able to execute the desired motion algorithm with sufficient accuracy (Section 7.5). Position/motion accuracy should be confidently verified using measurement tools such as spirit levels or position encoders. The placement of substrates in the system should be precise and repeatable to avoid run-to-run geometry variations (see Section 7.4 for the relevant calculations).

2. **Chamber pressure and gas composition:** For GLAD, it is important to control the deposition pressure to provide repeatable gas particle scattering conditions (Section 7.2.1) and produce a consistent vapour flux distribution. This reduces run-to-run variations in the vapour collimation that would affect the ballistic shadowing and hinder reproducible film growth. Producing a consistent vacuum environment for deposition also allows control of background gases (e.g. H_2O and O_2), as these can influence the nucleation kinetics and will be incorporated into the film during growth. Note that controlling the gas composition and pressure are critically important in any sputtering, reactive or plasma process. Accurate pressure monitoring and flow control are recommended for such processing, and gas species identification via a residual gas analyser is very useful as well.

3. **Deposition rate:** The deposition rate is always an important processing variable in thin-film fabrication as it impacts the nucleation and condensation process during film growth. Rate calibration is also necessary to accurately deposit columns of specific length or structural pitch. The deposition rate should, therefore, be controlled and monitored during film growth. Note also that, when depositing at an angle α, the incident vapour flux rate is geometrically reduced by a $\cos \alpha$ factor, an effect that must be properly calibrated to ensure deposition of accurate thicknesses (Section 7.3). Deposition rate fluctuations can also indirectly alter the substrate temperature, due to a change in radiative heating and energy transfer via latent heat of condensation.

4. **Substrate temperature:** GLAD is typically carried out at substrate temperatures significantly lower than the melting point of the material to ensure limited adatom mobility conditions and the formation of columnar morphology (Section 2.2.3). During normal processing, these conditions are usually satisfied without direct cooling, and we do not typically employ a substrate temperature control system for most GLAD work. However, it is important to recognize that radiative heating and vapour condensation will heat the substrate during deposition, which may be important for low-melting-point materials/substrates. There are also several advanced GLAD techniques (e.g. VLS-GLAD) that require high-temperature substrate conditions, making substrate heating and temperature control a central aspect of process control.

5. **Initial substrate condition:** Because GLAD is a surface technique, it is highly sensitive to the condition of the substrate surface. Realizing consistent film growth requires

minimizing variation in surface parameters that impact the nucleation and ballistic shadowing mechanisms. For example, the substrate roughness, surface chemistry and defect density should be well controlled, and the effects of any surface treatments (e.g. cleaning) should be considered as well.

Controlling these experimental variables and establishing acceptable processing windows will provide good first-order repeatability and serve as an excellent starting point for further process optimization studies. One should be aware that these parameters are often interrelated, and proper design of experimental techniques should be used to examine any interactions. Additionally, it is important to develop appropriate monitoring procedures to detect any long- or short-term parameter drift. Confident drift detection also necessitates periodic calibration checks of important process control equipment, such as pressure gauges, thermocouples, deposition rate monitors and flow rate controllers. Process calibrations should be verified after any significant change is made to the deposition system, such as modifying pressure gauge placement or major cleanings.

1.5 COMPUTER SIMULATIONS OF GLANCING ANGLE DEPOSITION GROWTH

Although GLAD research is primarily experimental, computer simulations have provided many important contributions towards understanding the critical growth mechanisms and predicting film morphologies. Substantial insight into GLAD and thin-film phenomena has emerged from growth simulations using simple ballistic deposition (BD) models, which employ a Monte Carlo approach to model the diffusion-limited aggregation process. In BD, the incoming vapour flux is represented by a large number of hard particles launched towards a substrate along ballistic paths selected to approximate the angular distribution of the vapour source (incident angle and collimation). These particles, launched one at a time, travel along linear trajectories until colliding with the substrate or a previously deposited particle, at which point the particle is 'deposited' and becomes part of the growing film microstructure. This particle aggregation process reproduces important aspects of low-temperature vapour deposition: vapour particles arrive at the substrate travelling along straight lines, particle mobilities are severely limited after deposition and the sticking coefficient of incident particles is near unity. Because the BD model incorporates these basic physical effects it qualitatively reproduces the columnar morphology characteristic of many vapour-deposited films, and early work with two-dimensional (2D) BD simulations was critical in explaining the microstructure of obliquely deposited films in terms of the ballistic shadowing effect [108, 118]. Subsequent 3D simulations also examined the shadowing effect and investigated the dynamic evolution of obliquely deposited columnar structures [119]. Simply by considering geometric shadowing phenomena, these computer simulations reproduced several key empirical results, such as the increase in column tilt angle β with α, the observation that $\beta < \alpha$ and the monotonic decrease in film density as α increases. These results provided modelling support for experimental efforts in thin-film design and engineering.

To realize more realistic modelling, additional growth factors can be added to the basic BD simulation by introducing physical models for important kinetic processes, such as surface and bulk diffusion. These are added to the BD model by allowing for particle motion after contact. Basic restructuring can be implemented by allowing aggregated particles to relax into the nearest 'pocket' formed by previously deposited particles [108, 118], thereby mimicking a minimum energy configuration. More complex and realistic diffusion kinetics can be simulated by different methods. In one approach, Smy *et al.* allow deposited particles to diffuse short distances according to a curvature-dependent chemical potential [120]. Karabacak *et al.* model the diffusion process by allowing a fixed number of random particle hops after aggregation [121]. Other simulations have implemented a random walk diffusion mechanism with a hopping probability defined by the particle coordination number and the surface energy [122, 123]. Incorporating appropriate physical representations of the growth kinetics is necessary for modelling material differences and sensitivities to experimental conditions, such as substrate temperature. As superior computational resources become available, it may become possible to scale up molecular dynamics simulations to larger, experimentally relevant dimensions, thus providing accurate 3D modelling of more complicated growth phenomena, such as grain formation, crystallographic defects, strain and thermal vibrations [124].

The use of fully 3D BD simulations with integrated kinetic modelling to describe the GLAD process has been growing in recent years, being used for both qualitative and quantitative descriptions of film growth, thin-film microstructure and the resulting physical properties. Figure 1.8 presents several images of simulated GLAD structures [120], including helical columns (Figure 1.8a,b), vertically oriented columns (Figure 1.8c,d) and a zigzag column structure (Figure 1.8e). These structures were created by dynamically altering the angular coordinates of the vapour source to reproduce the corresponding GLAD substrate motion. The film morphology created by the BD simulations provides a realistic reproduction of actual film structures, as is evident when comparing the images in Figure 1.8 with the SEM images of Figure 1.4. The agreement between model and experiment indicates that the simulation realistically reproduces the competitive growth mechanisms characteristic

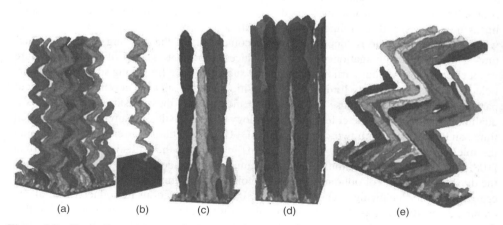

(a)	(b)	(c)	(d)	(e)

Figure 1.8   Three-dimensional simulations of GLAD film microstructure produced by a kinetic Monte Carlo BD model [120]. Using an advanced growth simulator, realistic representations of actual GLAD films can be obtained: (a) helical column structures and (b) a single isolated column; (c) and (d) vertically oriented columns; (e) zigzag columns. Reprinted with permission from [120]. Copyright 2000, American Vacuum Society.

of the GLAD process. An advanced growth simulator can be very useful, enabling quantitative prediction of numerous physical properties, including film density [120, 123, 125], surface roughness [71] and column nonuniformity [121, 126]. The simulator can also be used to design precise growth algorithms that suppress undesired growth effects [125], and to understand the effects of different process variables such as gas pressure [127] and substrate rotation rates [128].

1.6 MAJOR APPLICATION AREAS IN GLANCING ANGLE DEPOSITION TECHNOLOGY

GLAD has been applied in multiple different technology areas over the course of its development as scientists and engineers devise new and innovative means of exploiting and combining the numerous advantages discussed above. The following topical discussion surveys application areas that have been the focus of substantial research effort in recent years: energy and catalysis, sensing and optics. Each of these areas leverages one or more of the unique properties of GLAD films and provides excellent examples of where GLAD technology has been successfully used for nanoscale material control and device engineering. However, the following is by no means a comprehensive list as GLAD has been applied in many other technologies, including thermal barrier coatings [67], nanomechanical actuators [129], photonic crystals [130], magnetic devices [131], nanopropellers [132], microfluidic devices [97], and so on.

1.6.1 Energy and catalysis

The development of alternative energy sources and improved energy storage devices has emerged as a central technological challenge for twenty-first century science, and solving energy problems is a critical topic in the nanotechnology field. GLAD research and development has provided several important contributions to a range of energy-related technologies. The following summarizes recent efforts in this area, highlighting the advantages provided by the GLAD process.

Photovoltaics

The invention of dye-sensitized solar cells (DSSCs) provided a breakthrough in low-cost solar cells and has been a major topic of research in renewable energy ever since [133]. In a DSSC, photoactive charge-transfer dye molecules are adsorbed on a wide bandgap semiconductor, thereby photosensitizing the surface. Effective light harvesting relies on using a high-surface-area semiconductor electrode to increase the device efficiency. The material and microstructural flexibility of GLAD processing provides possibilities for DSSC optimization, and the robustness to post-deposition processing (e.g. annealing) enables further control over material properties and device structure. As a result, multiple reports have successfully investigated GLAD for DSSC fabrication [134–136]. In organic photovoltaic (OPV) cells, the development of bulk heterojunction (BHJ) devices led to high-surface-area interfaces between active layers in the cell, which reduces the exciton recombination loss and provides a considerable improvement in cell efficiency [137]. The success of the BHJ device archi-tecture has since prompted significant research into applying nanofabrication techniques to engineer the interface morphology and yield further efficiency gains. Consequently, multiple research groups have applied GLAD to OPV fabrication since it is simultaneously compat-ible with several common OPV materials and capable of realizing high-surface-area films with precise morphology control. Owing to the processing flexibility provided, GLAD can be applied to any of the OPV device layers, leading to a large parameter space for device and process optimization. Several approaches are thus being pursued. GLAD has been used to directly deposit a nanostructured organic active layer, with demonstrated examples includ-ing chloroaluminium phthalocyanine (ClAlPc) [138], C_{60} [139,140], copper phthalocyanine (CuPc) [87] and zinc phthalocyanine (ZnPc) [141]. GLAD nanostructured TiO_2 layers have also been used as inorganic acceptor layers in hybrid OPV cells [142]. GLAD has also been used to create nanostructured electrode layers (e.g. ITO bottom contacts [143]) or buffer layers (e.g. CuI [144]) that enhance the performance of the resulting OPV cell.

Fuel cells

Because modern fuel cell devices typically use expensive noble metal catalysts such as Pt, limiting the total amount of catalyst required without sacrificing cell performance is a critical goal of fuel cell cost-reduction efforts. A common design strategy is to increase the catalyst surface-to-volume ratio, thus enhancing catalytic activity and enabling more effective use of the catalyst material. High-surface-area catalysts can be realized by either directly structuring the catalyst itself or by grafting it onto a high-surface-area support structure. The GLAD process has been investigated for both of these approaches. For example, GLAD has been successfully used to directly fabricate nanostructured Pt layers for oxygen reduction electrocatalysis in proton exchange membrane fuel cells [145,146]. In another demonstration, multilayered Pt–Ru nanorods were created with a GLAD process and used as methanol oxidation catalysts in direct methanol fuel cells [147]. The material and post-deposition processing flexibility of the GLAD technique provide a large experimental parameter space for realizing support structures that have additional synergistic effects on catalyst activity, as well as on other fuel cell performance factors. The GLAD process has also been used to fabricate high-surface-area support structures, using diverse materials such as C [148], Cr [149], CrN [150], Nb_2O_5 [27] and Ti [70]. Because GLAD provides the ability to simultaneously control catalyst morphology across multiple size scales, from fine columnar features on the order of a few nanometres to intercolumnar spacings greater than 100 nm, the mass transport properties of the catalyst layers can be manipulated as well to reduce any kinetically limiting effects on fuel cell performance [145].

Lithium-ion battery materials

Li-ion battery performance is intimately linked with the Li-ion storage and transport per-formance of the electrode–electrolyte material system, thus offering many opportunities to optimize battery characteristics via nanoscale materials engineering. The use of nanostruc-tured materials for Li-ion battery electrodes provides numerous advantages [151], includ-ing increased Li-ion flux due to high-surface-area interface formation, decreased diffusion lengths and enhanced Li-ion transport characteristics, and improved electrode stability during the Li-ion intercalation–deintercalation process. These device benefits have motivated several investigations into using GLAD to fabricate nanostructured electrodes of Si [152–155], SiCu [153, 155], $Si_{1-x}Ge_x$ [156], and TiO_2 [157], which are integrated into standard coin- or beaker-cell packages. These reports have all demonstrated increased specific capacities compared with normally deposited films, as well as stability over many charge and discharge cycles. Controlling the composition of the GLAD structured electrode via co-deposition, demonstrated in the SiCu and $Si_{1-x}Ge_x$ works cited above, provides the ability to further enhance the electrode stability and optimize charging rate performance.

Photoelectrochemical water splitting

The efficient generation of hydrogen and oxygen from water is an important requirement for the developing hydrogen economy. Of the many approaches being investigated, solar-based techniques (where a photoelectrochemical reaction is used to split water molecules) offer a clean, highly promising method for hydrogen production. Currently, there is a substantial effort to develop semiconductor material systems (typically metal-oxide based) with optimal combinations of spectral absorption, electrochemical activity, long-term stability and cost-effectiveness. Coupled with this research is a desire to further enhance hydrogen production efficiencies by using nanoscale engineering to increase material surface-to-volume ratios, decrease charge transport lengths and reduce recombination losses, optimize crystallinity and morphology, and improve mass transport kinetics. Many researchers have investigated the GLAD nanoengineering process for water splitting applications as it provides the broad material compatibility and microstructural diversity necessary for device optimization research and development. High-surface-area GLAD films of $BiVO_4$ [84], α-Fe_2O_3 [92], Ta_3N_5 [37] and ZnO [72] have been shown to exhibit positive water-splitting characteristics. More complex material systems have been demonstrated as well, such as Si-doped α-Fe_2O_3 [158], Co-doped Ta_3N_5 [37] and core–shell TiO_2–WO_3 nanocolumn structures [159], indicating routes for further material control and optimization.

1.6.2 Sensing applications

The open, porous microstructure associated with GLAD films makes them ideal candidates for sensing applications, as target analytes can readily interact with the high-surface-area microstructure. This advantage leveraged with the broad material compatibility has led to GLAD films being incorporated into a wide range of environmental sensing and analytical chemistry devices that target many different chemical and biological applications.

Gas sensors

The electrical impedance of porous GLAD films is highly sensitive to relative humidity due to the capillary condensation of water vapour in the porous film, and electrical double-layer formation on the internal surfaces. Exploiting this transduction mechanism, preliminary studies demonstrated high-sensitivity, GLAD-fabricated relative-humidity sensors with ~30 ms response times, significantly faster than other state-of-the-art approaches [160]. These results prompted several microstructural and material optimization studies designed to better understand the sensing mechanism and to further improve sensor performance [77, 100, 161].

High-performance relative-humidity sensors based on optical [162] and colorimetric [53] read-out have been developed as well. Moving beyond water vapour as a target analyte, recent work has successfully developed GLAD sensors for other gas species. Hwang *et al.* demonstrated a multi-material (ITO and TiO_2) electronic nose able to discriminate between H_2, CO and NO_2 gases at the 10 ppm level [163]. Beckers *et al.* developed an SiO_2-based device for sensing various alcohols (methanol, ethanol, 1-propanol, 2-propanol and 1-butanol) [164]. While sensitive, these devices use different transduction mechanisms (such as impedance changes due to carrier depletion [163] and catalytic oxidation [164]) that result in slower response times ($\sim$10–100 s) compared with the relative-humidity sensors.

Surface-enhanced Raman scattering sensors

Raman spectroscopy is a powerful tool for analytical chemistry, able to measure the complex molecular fingerprints associated with the specific vibrational characteristics of bonds in the molecule. While these Raman signals are exceedingly weak, they can be dramatically enhanced by placing the analyte in close proximity to nanostructured metal surfaces, thus enabling successful detection of trace molecular signatures (so-called surface-enhanced Raman scattering (SERS)) [165]. A major source of the signal enhancement is hot-spot formation, regions of intense electromagnetic fields created by the metal (typically Ag or Au) nanostructures. Designing metallic substrates with reliable enhancement properties is a major goal of SERS research, and there has been lots of recent work using GLAD to engineer SERS substrates. GLAD provides important advantages for SERS devices: precise nanostructure control (necessary for the design of tailored, high-intensity hot spots [2, 166]) and single-step, wafer-scale nanofabrication (which enables cost-effective processing). Promising GLAD SERS substrates have been developed for biosensor applications, including virus [167], bacteria [168] and toxin [169] detection, and integrated into more complex device geometries such as fibre-optic probes [170] and ultra-thin-layer chromatography plates [171].

1.6.3 Optics

Since its invention, GLAD has been used to engineer novel optical nanomaterials by controlling the columnar morphology, thus exploiting the intimate link between nanoscale structure and a material's optical response. Important optical properties, such as refractive index and absorption, can be accurately tuned over a very wide range and novel optical effects can be realized, such as engineered polarization-sensitive phenomena. GLAD-fabricated optical nanomaterials have been developed for many optical applications, and we discuss a selection of these in the following.

Artificially engineered birefringence

It has long been recognized that the tilted columnar microstructure created during oblique film deposition creates birefringent properties, where the optical characteristics are highly polarization sensitive. This artificial polarization anisotropy can be exploited to fabricate common polarizing optics such as thin-film waveplates [172]. However, because the orientation of the optical axis is parallel to the column axis, which can be dynamically controlled using the GLAD process, it is possible to realize optical materials with complex anisotropies and novel polarization responses not observed in conventional materials. In a helical structure, for example (Figure 1.4b), the columns are wound into a chiral morphology by the continuous substrate rotation, which induces a corresponding rotation of the optical axis. The resulting film preferentially reflects circularly polarized light (CPL) with the same chiral handedness (i.e. right-handed (RH) CPL is reflected by an RH helix) and will rotate the vibration plane of linearly polarized light passing through the film [173]. (This and other associated chiral optical phenomena are discussed at length in Chapter 5.) The ability to carefully control column morphology in GLAD films using advanced substrate rotations has enabled the fabrication of optical coatings with optimal birefringence properties [174, 175] and engineered birefringence profiles [176, 177]. Furthermore, because these effects can be realized in a wide variety of materials it is possible to fabricate multifunctional optical coatings, such as microstructured ITO films that can simultaneously act as transparent conductive and optically retarding layers for liquid-crystal devices [11].

Optical interference coatings

Optical coatings have long been an important part of thin-film applications, with prominent examples being antireflective coatings for glass lenses and narrow bandpass filters for fibre-optic communication technologies. Conventional optical coating fabrication is based on stacking alternating layers of two different materials in a sequence designed to generate the correct constructive/destructive interference conditions and obtain the desired spectral properties. However, GLAD provides an alternative fabrication concept based on tailoring film porosity to introduce controlled refractive index variations in a *single* material. This approach has proven to be very adaptable, and a wide variety of optical filter designs have been successfully demonstrated, including antireflection coatings [178, 179], distributed Bragg reflectors [180], graded-index interference filters [181] and optical microcavity filters [182]. Furthermore, the material flexibility provided by GLAD allows index variations to be designed into materials with additional useful properties, such as high conductivity [183], fluorescence [66] and low acoustic impedance [184]. The combined microstructural and material flexibility afforded by the GLAD process provides numerous design opportunities to the optical device engineer.

1.7 SUMMARY AND OUTLINE OF THE BOOK

The text is developed as follows. Chapter 2 presents the basics of GLAD film engineering, examining the conventional thin-film growth process and how GLAD substrate motions are used to control ballistic shadowing effects and engineer film nanostructure. Chapter 3 presents methods to design special micropatterned substrates to overcome the randomness of film growth and produce high-uniformity GLAD nanostructure arrays, useful for applications requiring minimal structural variations. Chapter 4 provides a comprehensive experimental overview of the novel growth- and structure-related physical properties that emerge in GLAD-fabricated films, as well as characterization and analysis methods for GLAD film metrology. Engineering of optical films with GLAD has been a major focus of GLAD research, and Chapter 5 describes the optical properties of GLAD nanostructures and presents methods to design and fabricate a wide range of GLAD optical films. Chapter 6 presents post-deposition processes, such as thermal annealing, chemical treatments, photolithographic patterning and nanotemplating, that can be used to further control GLAD film structure, modify properties and integrate GLAD films into microdevices. Chapter 7 covers the practical design aspects of GLAD fabrication, including necessary vacuum conditions, thickness calibration, substrate uniformity effects and motor accuracy requirements. Finally, Appendix A tabulates the GLAD-related patent literature, providing a ready reference of innovations in GLAD technology.

REFERENCES

[1] Ni, J., Zhu, Y., Zhou, Q., and Zhang, Z. (2008) Morphology in-design of HfO_2 thin films. *Journal of the American Ceramic Society*, **91**, 3458–3460.

[2] Chaney, S.B., Shanmukh, S., Dluhy, R.A., and Zhao, Y.P. (2005) Aligned silver nanorod arrays produce high sensitivity surface-enhanced Raman spectroscopy substrates. *Applied Physics Letters*, **87**, 031908.

[3] Zhao, Y.P., Chaney, S.B., and Zhang, Z.Y. (2006) Absorbance spectra of aligned Ag nanorod arrays prepared by oblique angle deposition. *Journal of Applied Physics*, **100**, 063527.

[4] Inoue, Y., Tamaguichi, A., Fujihara, T. *et al.* (2007) Biomimetic improvement of electrochromic properties of indium nitride. *Journal of the Electrochemical Society*, **154**, J212–J216.

[5] Dick, B., Brett, M.J., and Smy, T. (2003) Controlled growth of periodic pillars by glancing angle deposition. *Journal of Vacuum Science and Technology B*, **21**, 23–28.

[6] Jen, Y.J. and Yu, C.W. (2007) Metal and dielectric duality for an aligned Al nanorod array. *Applied Physics Letters*, **91**, 021109.

[7] Cansizoglu, M.F., Engelken, R., Seo, H.W., and Karabacak, T. (2010) High optical absorption of indium sulfide nanorod arrays formed by glancing angle deposition. *ACS Nano*, **4**, 733–740.

[8] Deniz, D. and Lad, R.J. (2011) Temperature threshold for nanorod structuring of metal and oxide films grown by glancing angle deposition. *Journal of Vacuum Science and Technology A*, **29**, 011020.

[9] Schubert, M.F., Xi, J.Q., Kim, J.K., and Schubert, E.F. (2007) Distributed Bragg reflector consisting of high- and low-refractive-index thin film layers made of the same material. *Applied Physics Letters*, **90**, 141115.

[10] Poxson, D.J., Mont, F.W., Schubert, M.F. *et al.* (2008) Quantification of porosity and deposition rate of nanoporous films grown by oblique-angle deposition. *Applied Physics Letters*, **93**, 101914.

[11] Harris, K.D., van Popta, A.C., Sit, J.C. *et al.* (2008) A birefringent and transparent electrical conductor. *Advanced Functional Materials*, **18**, 2147–2153.

[12] Cuomo, J.J., Pappas, D.L., Lossy, R. *et al.* (1992) Energetic carbon deposition at oblique angles. *Journal of Vacuum Science and Technology A*, **10**, 3414–3418.

[13] Vick, D., Tsui, Y.Y., Brett, M.J., and Fedosejevs, R. (1999) Production of porous carbon thin films by pulsed laser deposition. *Thin Solid Films*, **350**, 49–52.

[14] Colgan, M.J. and Brett, M.J. (2001) Field emission from carbon and silicon films with pillar microstructure. *Thin Solid Films*, **389**.

[15] Robbie, K., Brett, M.J., and Lakhtakia, A. (1995) First thin film realization of a helicoidal bianisotropic medium. *Journal of Vacuum Science and Technology A*, **13**, 2991–2993.

[16] Gospodyn, J. and Sit, J.C. (2006) Characterization of dielectric columnar thin films by variable angle Mueller matrix and spectroscopic ellipsometry. *Optical Materials*, **29**, 318–325.

[17] Seto, M.W., Dick, B., and Brett, M.J. (2001) Microsprings and microcantilevers: studies of mechanical response. *Journal of Micromechanics and Microengineering*, **11**, 582–588.

[18] Lintymer, J., Gavoille, J., Martin, N., and Takadoum, J. (2003) Glancing angle deposition to modify microstructure and properties of sputter deposited chromium thin films. *Surface and Coatings Technology*, **174–175**, 316–323.

[19] Dohnálek, Z., Kimmel, G.A., McCready, D.E. *et al.* (2002) Structural and chemical characterization of aligned crystalline nanoporous MgO films grown via reactive ballistic deposition. *Journal of Physical Chemistry B*, **106**, 3526–3529.

[20] Tang, F., Liu, D.L., Ye, D.X. *et al.* (2003) Magnetic properties of Co nanocolumns fabricated by oblique-angle deposition. *Journal of Applied Physics*, **93**, 4194–4200.

[21] Karabacak, T., Wang, G.C., and Lu, T.M. (2004) Physical self-assembly and the nucleation of three-dimensional nanostructures by oblique angle deposition. *Journal of Vacuum Science and Technology A*, **22**, 1778–1784.

[22] Umlor, M.T. (2005) Uniaxial magnetic anisotropy in cobalt films induced by oblique deposition of an ultrathin cobalt underlayer. *Applied Physics Letters*, **87**, 082505.

[23] Beydaghyan, G., Doiron, S., Haché, A., and Ashrit, P.V. (2009) Enhanced photochromism in nanostructured molybdenum trioxide films. *Applied Physics Letters*, **95**, 051917.

[24] Alouach, H. and Mankey, G.J. (2004) Texture orientation of glancing angle deposited copper nanowire arrays. *Journal of Vacuum Science and Technology A*, **22**, 1379–1382.

[25] Karabacak, T., DeLuca, J.S., Wang, P.I. *et al.* (2006) Low temperature melting of copper nanorod arrays. *Journal of Applied Physics*, **99**, 064304.

[26] Xiao, X., Dong, G., Xu, C. *et al.* (2008) Structure and optical properties of Nb_2O_5 sculptured thin films by glancing angle deposition. *Applied Surface Science*, **255**, 2192–2195.

[27] Bonakdarpour, A., Tucker, R.T., Fleischauer, M.D. *et al.* (2012) Nanopillar niobium oxides as support structures for oxygen reduction electrocatalysts. *Electrochimica Acta*, **85**, 492–500.

[28] Choi, W.K., Li, L., Chew, H.G., and Zheng, F. (2007) Synthesis and structural characterization of germanium nanowires from glancing angle deposition. *Nanotechnology*, **18**, 385302.

[29] Li, L., Fang, X., Chew, H.G. *et al.* (2008) Crystallinity-controlled germanium nanowire arrays: potential field emitters. *Advanced Functional Materials*, **18**, 1080–1088.

[30] Okamoto, K., Hashimoto, T., Hara, K. *et al.* (1987) Columnar structure and texture of iron films prepared at various evaporation rates. *Thin Solid Films*, **147**, 299–311.

[31] Okamoto, K. and Itoh, K. (2005) Incidence angle dependences of columnar grain structure and texture in obliquely deposited iron films. *Japanese Journal of Applied Physics*, **44** (3), 1382–1388.

[32] Bubendorff, J.L., Garreau, G., Zabrocki, S. *et al.* (2009) Nanostructuring of Fe films by oblique incidence deposition on a $FeSi_2$ template onto Si(111): growth, morphology, structure and faceting. *Surface Science*, **603**, 373–379.

[33] Seto, M.W., Robbie, K., Vick, D. *et al.* (1999) Mechanical response of thin films with helical microstructures. *Journal of Vacuum Science and Technology B*, **17**, 2172–2177.

[34] Tang, F., Parker, T., Li, H.F. *et al.* (2007) Unusual magnesium crystalline nanoblades grown by oblique angle vapor deposition. *Journal of Nanoscience and Nanotechnology*, **7**, 3239–3244.

[35] He, Y., Zhao, Y., and Wu, J. (2008) The effect of Ti doping on the growth of Mg nanostructures by oblique angle codeposition. *Applied Physics Letters*, **92**, 063107.

[36] Broughton, J.N. and Brett, M.J. (2002) Electrochemical capacitance in manganese thin films with chevron microstructure. *Electrochemical and Solid-State Letters*, **5**, A279–A282.

[37] Dang, H.X., Hahn, N.T., Park, H.S. *et al.* (2012) Nanostructured Ta_3N_5 films as visible-light active photoanodes for water oxidation. *Journal of Physical Chemistry C*, **116**, 19225–19232.

[38] Mukherjee, S. and Gall, D. (2009) Anomalous scaling during glancing angle deposition. *Applied Physics Letters*, **95**, 173106.

[39] Mukherjee, S. and Gall, D. (2010) Power law scaling during physical vapor deposition under extreme shadowing conditions. *Journal of Applied Physics*, **107**, 084301.

[40] Hirakata, H., Matsumoto, S., Takemura, M. *et al.* (2007) Anisotropic deformation of thin films comprised of helical nanosprings. *International Journal of Solids and Structures*, **44**, 4030–4038.

[41] Hodgkinson, I., Wu, Q.H., and Collett, S. (2001) Dispersion equations for vacuum-deposited tilted-columnar biaxial media. *Applied Optics*, **40**, 452–457.

[42] Sánchez-Valencia, J.R., Borrás, A., Barranco, A. *et al.* (2008) Preillumination of TiO_2 and Ta_2O_5 photoactive thin films as a tool to tailor the synthesis of composite materials. *Langmuir*, **24**, 9460–9469.

[43] Dick, B., Sit, J.C., Brett, M.J. *et al.* (2001) Embossed polymeric relief structures as a template for the growth of periodic inorganic microstructures. *Nano Letters*, **1**, 71–73.

[44] Shetty, A.R., Karimi, A., and Cantoni, M. (2011) Effect of deposition angle on the structure and properties of pulsed-DC magnetron sputtered TiAlN thin films. *Thin Solid Films*, **519**, 4262–4270.

[45] Kim, J., Dohnálek, Z., and Kay, B.D. (2005) Structural characterization of nanoporous Pd films grown via ballistic deposition. *Surface Science*, **586**, 137–145.

[46] Flaherty, D.W., Hahn, N.T., Ferrer, D. *et al.* (2009) Growth and characterization of high surface area titanium carbide. *Journal of Physical Chemistry C*, **113**, 12742–12752.

[47] Flaherty, D.W., May, R.A., Berglund, S.P. *et al.* (2010) Low temperature synthesis and characterization of nanocrystalline titanium carbide with tunable porous architectures. *Chemistry of Materials*, **22**, 319–329.

[48] Harris, K.D., McBride, J.R., Nietering, K.E., and Brett, M.J. (2001) Fabrication of porous platinum thin films for hydrocarbon sensor applications. *Sensors and Materials*, **13** (4), 225–234.

[49] Dolatshahi-Pirouz, A., Hovgaard, M., Rechendorff, K. *et al.* (2008) Scaling behavior of the surface roughness of platinum films grown by oblique angle deposition. *Physical Review B*, **77**, 115427.

[50] Ye, D.X., Lu, T.M., and Karabacak, T. (2008) Influence of nanotips on the hydrophilicity of metallic nanorod surfaces. *Physical Review Letters*, **100**, 256102.

[51] Van Popta, A.C., Sit, J.C., and Brett, M.J. (2004) Optical properties of porous helical thin films. *Applied Optics*, **43**, 3632–3639.

[52] Taschuk, M.T., Krause, K.M., Steele, J.J. *et al.* (2009) Growth scaling of metal oxide columnar thin films deposited by glancing angle depositions. *Journal of Vacuum Science and Technology B*, **27**, 2106–2111.

[53] Hawkeye, M.M. and Brett, M.J. (2011) Optimized colorimetric photonic-crystal humidity sensor fabricated using glancing angle deposition. *Advanced Functional Materials*, **21**, 3652–3658.

[54] Morrow, P., Tang, F., Karabacak, T. *et al.* (2006) Texture of Ru columns grown by oblique angle sputter deposition. *Journal of Vacuum Science and Technology A*, **24**, 235–245.

[55] Li, C.C., Huang, J.L., Lin, R.J. *et al.* (2008) Microstructures, surface areas, and oxygen absorption of Ti and Ti–Zr–V films grown using glancing-angle sputtering. *Journal of Materials Research*, **23**, 579–587.

[56] Peterson, M.J. and Cocks, F.H. (1980) Selenium and tellurium selective absorber coatings produced by an oblique vacuum deposition technique. *Solar Energy*, **24**, 249–253.

[57] Beydaghyan, G., Bader, G., and Ashrit, P.V. (2008) Electrochromic and morphological investigation of dry-lithiated nanostructured tungsten trioxide thin films. *Thin Solid Films*, **516**, 1646–1650.

[58] Deniz, D., Frankel, D.J., and Lad, R.J. (2010) Nanostructured tungsten and tungsten trioxide films prepared by glancing angle deposition. *Thin Solid Films*, **518**, 4095–4099.

[59] Garcia-Garcia, F.J., Gil-Rostra, J., Yubero, F., and González-Elipe, A.R. (2013) Electrochromism in WO_x and $W_xSi_yO_z$ thin films prepared by magnetron sputtering at glancing angles. *Nanoscience and Nanotechnology Letters*, **5**, 89–93.

[60] Kaminska, K., Amassian, A., Martinu, L., and Robbie, K. (2005) Growth of vacuum evaporated ultraporous silicon studied with spectroscopic ellipsometry and scanning electron microscopy. *Journal of Applied Physics*, **97**, 013511.

[61] Buzea, C., Beydaghyan, G., Elliot, C., and Robbie, K. (2005) Control of power law scaling in the growth of silicon nanocolumn pseudo-regular arrays deposited by glancing angle deposition. *Nanotechnology*, **16**, 1986–1992.

[62] Gil-Rostra, J., Cano, M., Pedrosa, J.M. *et al.* (2012) Electrochromic behaviour of $W_xSi_yO_z$ thin films prepared by reactive magnetron sputtering at normal and glancing angles. *ACS Applied Materials and Interfaces*, **4**, 628–638.

[63] Rechendorff, K., Hovgaard, M.B., Chevallier, J. *et al.* (2005) Tantalum films with well-controlled roughness grown by oblique incidence deposition. *Applied Physics Letters*, **87**, 073105.

[64] Zhou, C.M. and Gall, D. (2006) Branched Ta nanocolumns grown by glancing angle deposition. *Applied Physics Letters*, **88**, 203117.

[65] Hrudey, P.C.P., Taschuk, M., Tsui, Y.Y. *et al.* (2005) Optical properties of porous nanostructured Y_2O_3:Eu thin films. *Journal of Vacuum Science and Technology A*, **23**, 856–861.

[66] Gospodyn, J., Taschuk, M.T., Hrudey, P.C. *et al.* (2008) Photoluminescence emission profiles of Y_2O_3:Eu films composed of high-low density stacks produced by glancing angle deposition. *Applied Optics*, **47**, 2798–2805.

[67] Harris, K.D., Vick, D., Gonzalez, E.J. *et al.* (2001) Porous thin films for thermal barrier coatings. *Surface and Coatings Technology*, **138**, 185–191.

[68] Saraf, L., Matson, D.W., Shutthanandan, V. *et al.* (2005) Ceria incorporation into YSZ columnar nanostructures. *Electrochemical and Solid-State Letters*, **8**, A525–A527.

[69] Sit, J.C., Vick, D., Robbie, K., and Brett, M.J. (1999) Thin film microstructure control using glancing angle deposition by sputtering. *Journal of Materials Research*, **14**, 1197–1199.

[70] Bonakdarpour, A., Fleischauer, M.D., Brett, M.J., and Dahn, J.R. (2008) Columnar support structures for oxygen reduction electrocatalysts prepared by glancing angle deposition. *Applied Catalysis A*, **349**, 110–115.

[71] Vick, D., Smy, T., and Brett, M.J. (2002) Growth behavior of evaporated porous thin films. *Journal of Materials Research*, **17**, 2904–2911.

[72] Wolcott, A., Smith, W.A., Kuykendall, T.R. *et al.* (2009) Photoelectrochemical study of nanostructured ZnO thin films for hydrogen generation from water splitting. *Advanced Functional Materials*, **19**, 1849–1856.

[73] LaForge, J.M., Taschuk, M.T., and Brett, M.J. (2011) Glancing angle deposition of crystalline zinc oxide nanorods. *Thin Solid Films*, **519**, 3530–3537.

[74] Karabacak, T., Mallikarjunan, A., Singh, J.P. *et al.* (2003) β-phase tungsten nanorod formation by oblique-angle sputter deposition. *Applied Physics Letters*, **83**, 3096–3098.

[75] Wang, S., Fu, X., Xia, G. *et al.* (2006) Structure and optical properties of ZnS thin films grown by glancing angle deposition. *Applied Surface Science*, **252**, 8734–8737.

[76] Wang, S., Xia, G., Fu, X. *et al.* (2007) Preparation and characterization of nanostructured ZrO_2 thin films by glancing angle deposition. *Thin Solid Films*, **515**, 3352–3355.

[77] Steele, J.J., Gospodyn, J.P., Sit, J.C., and Brett, M.J. (2006) Impact of morphology on high-speed humidity sensor performance. *IEEE Sensors Journal*, **6**, 24–27.

[78] Steele, J.J., Taschuk, M.T., and Brett, M.J. (2008) Nanostructured metal oxide thin films for humidity sensors. *IEEE Sensors Journal*, **8** (8), 1422–1429.

[79] Starbova, K., Dikova, J., and Starbov, N. (1997) Structure related properties of obliquely deposited amorphous a-As_2S_3 thin films. *Journal of Non-Crystalline Solids*, **210**, 261–266.

[80] Bhardwaj, P., Shishodia, P.K., and Mehra, R.M. (2003) Photo-induced changes in optical properties of As_2S_3 and As_2Se_3 films deposited at normal and oblique incidence. *Journal of Materials Science*, **38**, 937–940.

[81] Xiao, X., Dong, G., Shao, J. *et al.* (2010) Optical and electrical properties of SnO_2:Sb thin films deposited by oblique angle deposition. *Applied Surface Science*, **256**, 1636–1640.

[82] Hrudey, P.C.P., Westra, K.L., and Brett, M.J. (2006) Highly ordered organic Alq_3 chiral luminescent thin films fabricated by glancing-angle deposition. *Advanced Materials*, **18**, 224–228.

[83] Hrudey, P.C.P., Szeto, B., and Brett, M.J. (2006) Strong circular Bragg phenomena in self-ordered porous helical nanorod arrays of Alq_3. *Applied Physics Letters*, **88**, 251106.

[84] Berglund, S.P., Flaherty, D.W., Hahn, N.T. *et al.* (2011) Photoelectrochemical oxidation of water using nanostructured $BiVO_4$ films. *Journal of Physical Chemistry C*, **115**, 3794–3802.

[85] Zhang, J., Salzmann, I., Rogaschewski, S. *et al.* (2007) Arrays of crystalline C_6O and pentacene nanocolumns. *Applied Physics Letters*, **90**, 193117.

[86] Li, H.F., Parker, T., Tang, F. *et al.* (2008) Biaxially oriented CaF_2 films on amorphous substrates. *Journal of Crystal Growth*, **310**, 3610–3614.

[87] Van Dijken, J.G., Fleischauer, M.D., and Brett, M.J. (2011) Controlled nanostructuring of CuPc thin films via glancing angle deposition for idealized organic photovoltaic architectures. *Journal of Materials Chemistry*, **21**, 1013–1019.

[88] Hodgkinson, I., Cloughley, S., Wu, Q.H., and Kassam, S. (1996) Anisotropic scatter patterns and anomalous birefringence of obliquely deposited cerium oxide films. *Applied Optics*, **35** (28), 5563–5568.

[89] Saxena, K., Mehta, D.S., Srivastava, R., and Kamalasanan, M.N. (2008) Effect of oblique angle deposition of α-napthylphenylbiphenyl diamine on the performance of organic light-emitting diodes. *Journal of Physics D*, **41**, 015102.

[90] Frederick, J.R., D'Arcy-Gall, J., and Gall, D. (2006) Growth of epitaxial CrN on MgO(001): role of deposition angle on surface morphological evolution. *Thin Solid Films*, **494**, 330–335.

[91] Pursel, S., Horn, M.W., Demirel, M.C., and Lakhtakia, A. (2005) Growth of sculptured polymer submicronwire assemblies by vapor deposition. *Polymer*, **46**, 9544–9548.

[92] Hahn, N.T., Ye, H., Flaherty, D.W. *et al.* (2010) Reactive ballistic deposition of α-Fe_2O_3 thin films for photoelectrochemical water oxidation. *ACS Nano*, **4**, 1977–1986.

[93] Martín-Palma, R.J., Ryan, J.V., and Pantano, C.G. (2007) Spectral behavior of the optical constants in the visible/near infrared of GeSbSe chalcogenide thin films grown at glancing angle. *Journal of Vacuum Science and Technology A*, **25**, 587–591.

[94] Cetinkaya, M., Malvadkar, N., and Demirel, M.C. (2008) Power-law scaling of structured poly(*p*-xylylene films deposited by oblique angle. *Journal of Polymer Science Part B*, **46**, 640–648.

[95] Bhardwaj, P., Shishodia, P.K., and Mehra, R.M. (2007) Optical and electrical properties of obliquely deposited a-$GeSe_2$ films. *Journal of Materials Science*, **42**, 1196–1201.

[96] Sorge, J.B. (2012) Argon-assisted glancing angle deposition, PhD thesis, University of Alberta.

[97] Bezuidenhout, L.W., Nazemifard, N., Jemere, A.B. *et al.* (2011) Microchannels filled with diverse micro- and nanostructures fabricated by glancing angle deposition. *Lab on a Chip*, **11**, 1671–1678.

[98] Steele, J.J. (2007) Nanostructured thin films for humidity sensing, PhD thesis, University of Alberta.

[99] Bezuidenhout, L.W. (2011) Molecular separations using nanostructured porous thin films fabricated by glancing angle deposition, PhD thesis, University of Alberta.

[100] Steele, J.J., Taschuk, M.T., and Brett, M.J. (2009) Response time of nanostructured relative humidity sensors. *Sensors and Actuators B*, **140**, 610–615.

[101] Robbie, K., Friedrich, L.J., Dew, S.K. *et al.* (1995) Fabrication of thin films with highly porous microstructures. *Journal of Vacuum Science and Technology A*, **13**, 1032–1035.

[102] Kundt, A. (1886) Ueber die electromagnetische drehung der polarisationsebene des lichtes im eisen. *Annalen der Physik*, **263**, 191–202.

[103] Pugh, E.W., Boyd, E.L., and Freedman, J.F. (1960) Angle-of-incidence anisotropy in evaporated nickel-iron films. *IBM Journal of Research and Development*, **4**, 163–171.

[104] Smith, D.O. (1959) Anisotropy in permalloy films. *Journal of Applied Physics*, **30**, S264–S265.

[105] Holland, L. (1953) The effect of vapor incidence on the structure of evaporated aluminum films. *Journal of the Optical Society of America*, **43**, 376–380.

[106] König, H. and Helwig, G. (1950) Über die Struktur schräg aufgedampfter Schichten und ihr Einfluß auf die Entwicklung submikroskopischer Oberflächenrauhigkeiten. *Optik*, **6**, 111–124.

[107] Smith, D.O., Cohen, M.S., and Weiss, G.P. (1960) Oblique-incidence anisotropy in evaporated permalloy films. *Journal of Applied Physics*, **31**, 1755–1762.

[108] Dirks, A.G. and Leamy, H.J. (1977) Columnar microstructure in vapor-deposited thin films. *Thin Solid Films*, **47**, 219–233.

[109] Nieuwenhuizen, J.M. and Haanstra, H.B. (1966) Microfractography of thin films. *Philips Technical Review*, **27**, 87–91.

[110] Nakhodin, N.G. and Shaldervan, A.I. (1972) Effect of vapour incidence angles on profile and properties of condensed films. *Thin Solid Films*, **10**, 109–122.

[111] Okamoto, K., Hashimoto, T., Hara, K., and Tatsumoto, E. (1971) Origin of magnetic anisotropy of iron films evaporated at oblique incidence. *Journal of the Physical Society of Japan*, **31**, 1374–1379.

[112] Feuerstein, A. and Mayr, M. (1984) High vacuum evaporation of ferromagnetic materials – a new production technology for magnetic tapes. *IEEE Transactions on Magnetics*, **20**, 51–56.

[113] Luitjens, S.B., Stupp, S.E., and Lodder, J.C. (1996) Metal evaporated tape: state of the art and prospects. *Journal of Magnetism and Magnetic Materials*, **155**, 261–265.

[114] Lodder, J.C. (2001) Metal evaporated tape, in *Encyclopedia of Materials: Science and Technology* (eds K.J. Buschow, R.W. Cahn, M.C. Flemigns, B. Ilschner, E.J.K. Kramer, S. Mahajan, and P. Veyssiere), Elsevier, pp. 5360–5367.

[115] Young, N.O. and Kowal, J. (1959) Optically active fluorite films. *Nature*, **183**, 104–105.

[116] Taga, Y. and Motohiro, T. (1990) Growth of birefringent films by oblique deposition. *Journal of Crystal Growth*, **99** ((1–4)), 638–642.

[117] Azzam, R.M.A. (1992) Chiral thin solid films: Method of deposition and applications. *Applied Physics Letters*, **61**, 3118–3120.

[118] Henderson, D., Brodsky, M.H., and Chaudhari, P. (1974) Simulation of structural anisotropy and void formation in amorphous thin films. *Applied Physics Letters*, **25**, 641–643.

[119] Meakin, P. and Krug, J. (1992) Three-dimensional ballistic deposition at oblique incidence. *Physical Review A*, **46**, 3390–3399.

[120] Smy, T., Vick, D., Brett, M.J. *et al.* (2000) Three-dimensional simulation of film microstructure produced by glancing angle deposition. *Journal of Vacuum Science and Technology A*, **18**, 2507–2512.

[121] Karabacak, T., Singh, J., Zhao, Y.P. *et al.* (2003) Scaling during shadowing growth of isolated nanocolumns. *Physical Review B*, **68**, 125408.

[122] Müller-Pfeiffer, S., van Kranenburg, H., and Lodder, J.C. (1992) A two-dimensional Monte Carlo model for thin film growth by oblique evaporation: simulation of two-component systems for the example of Co–Cr. *Thin Solid Films*, **213**, 143–153.

[123] Suzuki, M. and Taga, Y. (2001) Numerical study of the effective surface area of obliquely deposited thin films. *Journal of Applied Physics*, **90**, 5599–5605.

[124] Hubartt, B.C., Liu, X., and Amar, J.G. (2013) Large-scale molecular dynamics simulations of glancing angle deposition. *Journal of Applied Physics*, **114**, 083517.

[125] Kaminska, K., Suzuki, M., Kimura, K. *et al.* (2004) Simulating structure and optical response of vacuum evaporated porous rugate filters. *Journal of Applied Physics*, **95**, 3055–3062.

[126] Ye, D.X., Ellison, C.L., Lim, B.K., and Lu, T.M. (2008) Shadowing growth of three-dimensional nanostructures on finite size seeds. *Journal of Applied Physics*, **103**, 103531.

[127] García-Martín, J.M., Alvarez, R., Romero-Gómez, P. *et al.* (2010) Tilt angle control of nanocolumns grown by glancing angle sputtering at variable argon pressures. *Applied Physics Letters*, **97**, 173103.

[128] Dick, B., Brett, M.J., and Smy, T. (2003) Investigation of substrate rotation at glancing-incidence on thin-film morphology. *Journal of Vacuum Science and Technology B*, **21**, 2569–2575.

[129] Dice, G.D., Brett, M.J., Wang, D., and Buriak, J.M. (2007) Fabrication and characterization of an electrically variable, nanospring based interferometer. *Applied Physics Letters*, **90**, 253101.

[130] Kennedy, S.R., Brett, M.J., Toader, O., and John, S. (2002) Fabrication of tetragonal square spiral photonic crystals. *Nano Letters*, **2** (1), 59–62.

[131] Albrecht, O., Zierold, Patzig, C. *et al.* (2010) Tubular magnetic nanostructures based on glancing angle deposited templates and atomic layer deposition. *Physica Status Solidi (b)*, **247**, 1365–1371.

[132] Gibbs, J.G. and Zhao, Y. (2010) Self-organized multiconstituent catalytic nanomotors. *Small*, **6**, 1656–1662.

[133] O'Regan, B. and Grätzel, M. (1991) A low-cost, high-efficiency solar cell based on dye-sensitized colloidal TiO$_2$ films. *Nature*, **353**, 737–740.

[134] Kiema, G.K., Colgan, M.J., and Brett, M.J. (2005) Dye sensitized solar cells incorporating obliquely deposited titanium oxide layers. *Solar Energy Materials and Solar Cells*, **85**, 321–331.

[135] González-García, L., González-Valls, I., Lira-Cantu, M. *et al.* (2011) Aligned TiO$_2$ nanocolumnar layers prepared by PVD–GLAD for transparent dye sensitized solar cells. *Energy and Environmental Science*, **4**, 3426–3435.

[136] González-García, L., Idígoras, J., González-Elipe, A.R. *et al.* (2012) Charge collection properties of dye-sensitized solar cells based on 1-dimensional TiO$_2$ porous nanostructures and ionic-liquid electrolytes. *Journal of Photochemistry and Photobiology A*, **241**, 58–66.

[137] Tang, C.W. (1986) Two-layer organic photovoltaic cell. *Applied Physics Letters*, **48**, 183–185.

[138] Li, N. and Forrest, S.R. (2009) Tilted bulk heterojunction organic photovoltaic cells grown by oblique angle deposition. *Applied Physics Letters*, **95**, 123309.

[139] Thomas, M., Worfolk, B.J., Rider, D.A. *et al.* (2011) C$_{60}$ fullerene nanocolumns-polythiophene heterojunctions for inverted organic photovoltaic cells. *Applied Materials and Interfaces*, **3**, 1887–1894.

[140] Thomas, M., Li, W., Bo, Z.S., and Brett, M.J. (2012) Inverted photovoltaic cells of nanocolumnar C$_{60}$ filled with solution processed small molecule 3-Q. *Organic Electronics*, **13**, 2647–2652.

[141] Van Dijken, J.G., Fleischauer, M.D., and Brett, M.J. (2011) Solvent effects on ZnPc thin films and their role in fabrication of nanostructured organic solar cells. *Organic Electronics*, **12**, 2111–2119.

[142] Gerein, N.J., Fleischauer, M.D., and Brett, M.J. (2010) Effect of TiO$_2$ film porosity and thermal processing on TiO$_2$-P3HT hybrid materials and photovoltaic device performance. *Solar Energy Materials and Solar Cells*, **94**, 2343–2350.

[143] Rider, D.A., Tucker, R.T., Worfolk, B.J. *et al.* (2011) Indium tin oxide nanopillar electrodes in polymer/fullerene solar cells. *Nanotechnology*, **22**, 085706.

[144] Zhou, Y., Taima, T., Miyadera, T. *et al.* (2012) Glancing angle deposition of copper iodide nanocrystals for efficient organic photovoltaics. *Nano Letters*, **12**, 4146–4152.

[145] Gasda, M.D., Teki, R., Lu, T.M. *et al.* (2009) Sputter-deposited Pt PEM fuel cell electrodes: particles vs. layers. *Journal of the Electrochemical Society*, **156**, B614–B619.

[146] Khudhayer, W.J., Kariuki, N.N., Wang, X. *et al.* (2011) Oxygen reduction reaction electrocatalytic activity of glancing angle deposited platinum nanorod arrays. *Journal of the Electrochemical Society*, **158**, B1029–B1041.

[147] Yoo, S.J., Jeon, T.Y., Kim, K.S. *et al.* (2010) Multilayered Pt/Ru nanorods with controllable bimetallic sites as methanol oxidation catalysts. *Physical Chemistry Chemical Physics*, **12**, 15240–15246.

[148] Gasda, M.D., Eisman, G.A., and Gall, D. (2010) Nanorod PEM fuel cell cathodes with controlled porosity. *Journal of the Electrochemical Society*, **157**, B437–B440.

[149] Khudhayer, W.J., Kariuki, N., Myers, D.J. *et al.* (2012) GLAD Cr nanorods coated with SAD Pt thin film for oxygen reduction reaction. *Journal of the Electrochemical Society*, **159**, B729–B736.

[150] Gasda, M.D., Eisman, G.A., and Gall, D. (2010) Sputter-deposited Pt/CrN nanoparticle PEM fuel cell cathodes: limited proton conductivity through electrode dewetting. *Journal of the Electrochemical Society*, **157**, B71–B76.

[151] Bruce, P.G., Scrosati, B., and Tarascon, J.M. (2008) Nanomaterials for rechargeable lithium batteries. *Angewandte Chemie International Edition*, **47**, 2930–2946.

[152] Fleischauer, M.D., Li, J., and Brett, M.J. (2009) Columnar thin films for three-dimensional microbatteries. *Journal of the Electrochemical Society*, **156**, A33–A36.

[153] Au, M., He, Y., Zhao, Y. *et al.* (2011) Silicon and silicon-copper composite nanorods for anodes of Li-ion rechargable batteries. *Journal of Power Sources*, **196**, 9640–9647.

[154] Abel, P.R., Lin, Y.M., Celio, H. *et al.* (2012) Improving the stability of nanostructured silicon thin film lithium-ion battery anodes through their controlled oxidation. *ACS Nano*, **6**, 2506–2516.

[155] He, Y., Yang, B., Yang, K. *et al.* (2012) Designing Si-based nanowall arrays by dynamic shadowing growth to tailor the performance of Li-ion battery anodes. *Journal of Materials Chemistry*, **22**, 8294–8303.

[156] Abel, P.R., Chockla, A.M., Lin, Y.M. *et al.* (2013) Nanostructured $Si_{(1-x)}Ge_x$ for tunable thin film lithium-ion battery anodes. *ACS Nano*, **7**, 2249–2257.

[157] Lin, Y.M., Abel, P.R., Flaherty, D.W. *et al.* (2011) Morphology dependence of the lithium storage capability and rate performance of amorphous TiO_2 electrodes. *Journal of Physical Chemistry C*, **115**, 2585–2591.

[158] Chemelewski, W.D., Hahn, N.T., and Mullins, C.B. (2012) Effect of Si doping and porosity on hematite's (α-Fe_2O_3) photoelectrochemical water oxidation performance. *Journal of Physical Chemistry C*, **116**, 5255–5261.

[159] Smith, W., Wolcott, A., Fitzmorris, R.C. *et al.* (2011) Quasi-core-shell TiO_2/WO_3 and WO_3/TiO_2 nanorod arrays fabricated by glancing angle deposition for solar water splitting. *Journal of Materials Chemistry*, **21**, 10792–10800.

[160] Harris, K.D., Sit, J.C., and Brett, M.J. (2002) Fabrication and optical characterization of template-constructed thin films with chiral nanostructure. *IEEE Transactions on Nanotechnology*, **1** (3), 122–128.

[161] Steele, J.J., Fitzpatrick, G.A., and Brett, M.J. (2007) Capacitive humidity sensors with high sensitivity and subsecond response times. *IEEE Sensors Journal*, **7** (6), 955–956.

[162] Steele, J.J., van Popta, A.C., Hawkeye, M.M. *et al.* (2006) Nanostructured gradient index optical filter for high-speed humidity sensing. *Sensors and Actuators B: Chemical*, **120**, 213–219.

[163] Hwang, S., Kwon, H., Chhajed, S. *et al.* (2013) A near single crystalline TiO_2 nanohelix array: enhanced gas sensing performance and its application as a monolithically integrated electronic nose. *Analyst*, **138**, 443–450.

[164] Beckers, N.A., Taschuk, M.T., and Brett, M.J. (2013) Selective room temperature nanostructured thin film alcohol sensor as a virtual sensor array. *Sensors and Actuators B: Chemical*, **176**, 1096–1102.

[165] Moskovits, M. (2005) Surface-enhanced Raman spectroscopy: a brief retrospective. *Journal of Raman Spectroscopy*, **36**, 485–496.

[166] Zhou, Q., Zhang, X., Huang, Y. *et al.* (2012) Enhanced surface-enhanced Raman scattering performance by folding silver nanorods. *Applied Physics Letters*, **100**, 113101.

[167] Driskell, J.D., Zhu, Y., Kirkwood, C.D. *et al.* (2010) Rapid and sensitive detection of rotavirus molecular signatures using surface enhanced Raman spectroscopy. *PLoS One*, **5**, e10222.

[168] Hennigan, S.L., Driskell, J.D., Dluhy, R.A. *et al.* (2010) Detection of *Mycoplasma pneumoniae* in simulated and true clinical throat swab specimens by nanorod array surface-enhanced Raman spectroscopy. *PLoS One*, **5**, e13633.

[169] Wu, X., Gao, S., Wang, J.S. *et al.* (2012) The surface-enhanced Raman spectra of aflatoxins: spectral analysis, density functional theory calculation, detection and differentiation. *Analyst*, **137**, 4226–4234.

[170] Zhu, Y., Dluhy, R.A., and Zhao, Y. (2011) Development of silver nanorod array based fiber optic probes for SERS detection. *Sensors and Actuators B: Chemical*, **157**, 42–50.

[171] Chen, J., Abell, J., Huang, Y., and Zhao, Y. (2012) On-chip ultra-thin layer chromatography and surface enhanced Raman spectroscopy. *Lab on a Chip*, **12**, 3096–3102.

[172] Motohiro, T. and Taga, Y. (1989) Thin film retardation plate by oblique deposition. *Applied Optics*, **28**, 2466–2482.

[173] Robbie, K., Brett, M.J., and Lakhtakia, A. (1996) Chiral sculptured thin films. *Nature*, **384**, 616.

[174] Wu, Q., Hodgkinson, I.J., and Lakhtakia, A. (2000) Circular polarization filters made of chiral sculptured thin films: experimental and simulation results. *Optical Engineering*, **39**, 1863–1868.

[175] Jen, Y., Lakhtakia, A., Yu, C.W. *et al.* (2011) Biologically inspired achromatic waveplates for visible light. *Nature Communications*, **2**, 363.

[176] Van Popta, A.C., van Popta, K.R., Sit, J.C., and Brett, M.J. (2007) Sidelobe suppression in chiral optical filters by apodization of the local form birefringence. *Journal of the Optical Society of America A*, **24**, 3140–3149.

[177] Leontyev, V., Wakefield, N.G., Tabunshchyk, K. *et al.* (2008) Selective transmittance of linearly polarized light in thin films rationally designed by FDTD and FDFD theories and fabricated by glancing angle deposition. *Journal of Applied Physics*, **104**, 104302.

[178] Kennedy, S.R. and Brett, M.J. (2003) Porous broadband antireflection coating by glancing angle deposition. *Applied Optics*, **42**, 4573–4579.

[179] Xi, J.Q., Schubert, M.F., Kim, J.K. *et al.* (2007) Optical thin-film materials with low refractive index for broadband elimination of Fresnel reflection. *Nature Photonics*, **1**, 176–179.

[180] Kaminska, K., Brown, T., Beydaghyan, G., and Robbie, K. (2003) Vacuum evaporated porous silicon photonic interference filters. *Applied Optics*, **42**, 4212–4219.

[181] Kaminska, K. and Robbie, K. (2004) Birefringent omnidirectional reflector. *Applied Optics*, **43**, 1570–1576.

[182] Hawkeye, M.M. and Brett, M.J. (2006) Narrow bandpass optical filters fabricated with one-dimensionally periodic inhomogeneous thin films. *Journal of Applied Physics*, **100**, 044322.

[183] Yan, X., Mont, F.W., Poxson, D.J. *et al.* (2011) Electrically conductive thin-film color filters made of single-material indium-tin-oxide. *Journal of Applied Physics*, **109**, 103113.

[184] Hajireza, P., Krause, K., Brett, M., and Zemp, R. (2013) Glancing angle deposited nanostructured film Fabry–Perot etalons for optical detection of ultrasound. *Optics Express*, **21**, 6391–6400.

2 Engineering Film Microstructure with Glancing Angle Deposition

2.1 INTRODUCTION

The heart of the GLAD technique is the design of substrate motions to control ballistic shadowing and engineer nanoscale growth during thin-film deposition. This is a conceptually simple yet powerful nanofabrication approach that enables directed self-assembly of a wide variety of micro- and nanostructured morphologies. Although ballistic shadowing is a factor in conventional film growth, it is a dominant effect in GLAD as the oblique deposition geometry amplifies the resulting shadow lengths. In this shadow-dominated growth mode, a highly porous columnar film morphology is developed and substrate rotations can be used to dynamically control the shadowing dynamics and guide the column evolution. To understand GLAD it is necessary to first understand the basics of film growth, particularly the ballistic shadowing and surface diffusion mechanisms and their influence on film morphology. We therefore begin with a discussion of PVD methods and conventional film growth, focusing on achieving the shadowing-dominated, columnar growth conditions critical to successful GLAD. The discussion then turns to the extreme shadowing conditions present in GLAD, examining the design of substrate motions to alter the deposition geometry and manipulate shadowing during growth and the structural characteristics of the resulting films. The substrate tilt controls the deposition angle α and is used to create the porous film microstructure and directional column growth associated with extreme shadow effects. Rotating the substrate about its normal, measured via the azimuthal angle φ, exploits directional growth to steer columns along the desired growth path. Basic rotation schemes can be designed to realize a range of nanoscale column morphologies. Building on these concepts, advanced substrate motion procedures for achieving precise growth control are discussed with an emphasis on utility and practical implementation. These advanced motion algorithms demonstrate how the shadowing mechanics can be carefully engineered to provide even greater control over the growth processes and the resulting microstructure. This chapter presents these approaches as well as additional techniques, such as substrate temperature control and co-deposition processing, that can be used to further control the deposited microstructure.

Glancing Angle Deposition of Thin Films: Engineering the Nanoscale, First Edition.
Matthew M. Hawkeye, Michael T. Taschuk and Michael J. Brett.
© 2014 John Wiley & Sons, Ltd. Published 2014 by John Wiley & Sons, Ltd.

2.2 BASICS OF CONVENTIONAL FILM GROWTH

2.2.1 Physical vapour deposition

PVD techniques are atomistic deposition processes widely used in a number of industries to prepare functional thin-film coatings. Conceptually, PVD is a remarkably simple process: the desired source material is first vaporized in a vacuum environment and then subsequently recondensed on a separate surface (the substrate). This vapour-to-solid condensation is general to all PVD techniques, which mainly differ in the vaporization method and vacuum requirements. PVD processes can be used to deposit films of many different materials, either directly or via reactive techniques where compounds are formed through reaction at the substrate between the source material and a process gas (e.g. O_2 or N_2) added to the vacuum chamber. GLAD inherits much of this material flexibility and is compatible with a diverse range of materials, as seen in Table 1.1. This table also shows that the primary PVD techniques used in GLAD work are evaporation and sputter deposition, with a few cases using the related PLD and ion-plating methods. In the following, we briefly review evaporation and sputter deposition and direct the reader to more comprehensive references if required [1, 2].

Evaporation refers to PVD methods where the source material is thermally vaporized, most often using a resistive heating approach (thermal evaporation) or by high-energy electron beam bombardment (e-beam evaporation). PLD is similar but less common, and involves vaporizing the source material with a high-intensity laser pulse. Evaporation is typically performed in a high-vacuum environment, with gas pressures in the 10^{-5} to 10^{-9} Torr range. These low-pressure conditions ensure inter-particle collisions are rare (Section 7.2.1) and that the vapour particles travel to the substrate along linear trajectories, even when the source-to-substrate throw distance is long. The combination of low pressure and long throw conditions produces a highly directional, collimated vapour flux that enhances the geometric shadowing effect critical to GLAD, leading to improved nanostructures. Evaporation is therefore the most widely used PVD method for GLAD work.

In sputter deposition, the source material is physically vaporized by energetic particle bombardment. These particles, generally gaseous ions accelerated from a plasma or an ion gun, collide with and physically eject surface atoms in a momentum-transferring collision cascade. Sputter deposition is complex and provides many experimental variables that can be tuned to control film properties. The plasma gas composition (typically Ar), pressure and energetics all influence the film deposition process. Many different sputtering geometries exist, such as DC diode sputtering, radio frequency (RF) sputtering, DC and RF magnetron sputtering, hollow cathode sputtering and high-power impulse magnetron sputtering. These configurations employ different electrode and magnet geometries for plasma generation and acceleration, leading to differences in important process parameters such as effective source size, energy and angular distribution of sputtered flux, and deposition rate. Each of these parameters can influence the properties of the deposited film. In general, sputter deposition requires a higher pressure environment (>0.1 mTorr) than evaporation to sustain an ion-generating plasma, thus increasing the vapour flux scattering rate. (The exception here is ion-beam sputtering, which can be carried out at pressures $<10^{-5}$ Torr.) Furthermore, sputtering sources are usually larger than evaporation sources, leading to an increased angular flux distribution. These effects decrease the vapour flux collimation, which has a negative impact on GLAD nanostructures, making it important to carefully consider the working pressure, source size and throw distance when developing sputter-based GLAD processes (Section 7.2.2).

2.2.2 Nucleation and coalescence

Practical film deposition is as much art as science, with individual film properties sensitive to an extensive collection of processing variables. However, film growth occurs via several basic physical phenomena that have long been studied in an attempt to improve process control and reliability. Film growth progresses in several phases (Figure 2.1): vapour condensation and nucleation, island formation and growth, island coalescence, agglomeration and continuous structure formation, and thickness growth. The evolution of film structure through these growth phases is complex, and each phase impacts the properties of the final film. In-depth reviews of these phenomena can be found in several monographs on thin-film growth, such as texts by Chopra [3] and Ohring [2]. In the following discussion we focus on the deposition phenomena most relevant to GLAD, beginning with the atomistic processes that are the basis of early film growth.

The initial vapour condensation and nucleation phase, schematically depicted in Figure 2.1a, is an atomistic process with several concomitant microscopic physical phenomena [2–7]. Incoming vapour atoms or molecules arrive at the substrate and are quickly physisorbed on the surface. Adsorbed particles (termed adatoms) then diffuse around the surface, with a mobility determined by thermal energy and the strength of the adatom–substrate interaction. The diffusing adatom can bind with another adatom or adatom cluster, forming microscopic nuclei that decorate the substrate surface. The adatoms can also re-evaporate from the surface, but the re-evaporation rate is generally very small (except at high substrate temperatures when the thermal energy approaches the adatom binding energy or in certain high vapour pressure materials). The balance between the arrival rate of new adatoms and the adatom detachment probability determines the nucleus growth rate. As a nucleus grows, the rate of adatom detachment from the nucleus decreases and eventually the nucleus reaches a critical size where the adatom attachment and detachment rate are balanced and the nucleus stabilizes. Although these critical nuclei are not much larger than atomic sized, they grow rapidly upon stabilization and quickly become the dominant surface features in this initial growth phase.

Deposition and film growth continue as the critical nuclei capture additional atoms, either through attachment of diffusing adatoms or direct condensation from the vapour phase. The nuclei expand following one of three growth modes: 3D island formation (Volmer–Weber growth), 2D layer-by-layer growth (Frank–van der Merwe growth) or a combined layer plus island growth (Stranski–Krastanov growth). Which growth mode is active is determined by the surface energies of the system [4]. When the deposited atoms are strongly bound to one another and only weakly bound to the substrate, it is energetically favourable for the adatoms to form 3D islands. Conversely, the 2D layer configuration is more stable when the adatom–substrate interaction dominates. Stranski–Krastanov growth results where 3D nucleation becomes preferable after the first few 2D monolayers form. This switch is driven by a change in the binding energetics due to strain build-up, lattice symmetry changes and/or molecular orientation changes, and so on. For GLAD, the most important growth mode is

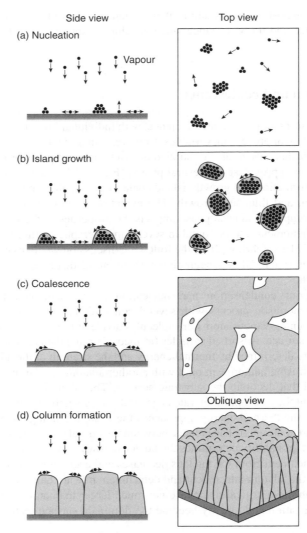

Figure 2.1 A schematic description of the basic film growth process. (a) In the initial nucleation phase, vapour atoms condense on the substrate and bind together to create atomic-scale nuclei at random surface locations. (b) These initial nuclei capture other surface-adsorbed atoms and form larger island-like features that continue to grow. (c) Eventually, at high surface coverages, islands coalesce together to form a continuous, agglomerated network of deposited material that (d) develops into a thin film with a columnar microstructure.

Volmer–Weber as it introduces roughness to an initially featureless substrate (Figure 2.1b). Microscopic surface topographies are thereby spontaneously generated and then amplified by ballistic shadowing, as will be discussed in Section 2.3.

The microscopic islands expand with continued deposition, allowing neighbouring structures to contact and coalesce (Figure 2.1c). Coalescence events are driven by several mass transport mechanisms, including cluster migration, Ostwald ripening and sintering. For small islands with sufficiently mobile grain boundaries, significant crystallization changes may accompany this coalescence, including recrystallization, defect formation/removal and reorientation, driven by a reduction in surface free energy.

Several experimental factors influence the nucleation phase of growth, and control over these parameters is key to repeatable film deposition. A crucial parameter is the substrate temperature, which strongly affects the adatom kinetics and the nucleation process. The deposition rate is another factor, determining the adatom collision probability, and hence the nucleation rate. The substrate condition is also very important, directly impacting the adatom binding energy. Surface defects, such as scratches, foreign surface atoms, cleavage steps or grain boundaries, may act as condensation centres, thus promoting nucleation at the defect sites. Surface contamination, either due to insufficient initial cleaning or the adsorption of residual gases from the vacuum environment (such as water or oxygen), will influence the surface chemistry and affect the binding energy. Improved control over any of these parameters will have a beneficial effect on GLAD film reproducibility.

Finally, it is important to recognize that the nucleation growth process is intrinsically random. The stochastic nature of vapour arrival and adatom diffusion produce non-uniform nuclei with a randomized surface-distribution. This randomness carries over to later growth, priming the competitive growth environment characteristic of GLAD and broadening the statistical distribution of columnar structures.

2.2.3 Column microstructure

Eventually, at high surface coverages, the islands agglomerate together and a continuous structural network is created (Figure 2.1c). After agglomeration, film growth continues via direct vapour nucleation on the previously deposited surface, developing into larger structures oriented in a columnar morphology (Figure 2.1d). The formation of columnar structures in thin films is well known and has long been interpreted in terms of structure zone models (SZMs) [7–10]. While several SZMs exist, the earliest model, proposed by Movchan and Demchishin based on morphological characterization of several vacuum-deposited materials [8], succinctly expresses the development of different columnar morphologies (Figure 2.2). This three-zone SZM relates the film microstructure to the homologous temperature T_s/T_m (where T_s is the substrate temperature and T_m is the melting point of the deposited material, both measured in kelvin), a ratio that effectively captures the magnitude of surface and bulk diffusion processes. Most relevant to GLAD is the low-temperature zone I regime, corresponding to $T_s/T_m < 0.3$, where surface diffusion is limited and unable to fill-in void regions that form in the microstructure. These conditions produce an underdense, fine nanofibrous microstructure that develops into a columnar film microstructure as film growth continues. At higher temperatures ($0.3 < T_s/T_m < 0.5$), zone II films are produced where surface diffusion is the dominant effect and the resulting microstructure consists of tightly packed columnar grains. Diffusive transport fills in void regions and the film density under these conditions is much closer to bulk. Entering zone III ($T_s/T_m > 0.5$) activates bulk diffusion processes, enabling grain boundary diffusion and recrystallization during growth. The resulting equiaxed structures are often single crystalline and the overall films are very dense. Finally, we note that the film microstructure transitions gradually between the zones, and

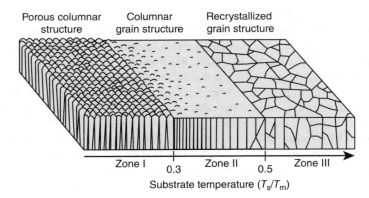

Figure 2.2 The SZM proposed by Movchan and Demchishin [8], describing three basic film morphologies found in films deposited at different temperatures. The boundaries between each zone are based on the ratio between the substrate temperature T_s and the material melting point T_m. In zone I ($T_s/T_m < 0.3$), ballistic growth effects dominate as surface diffusion is highly limited. The result is columnar structures separated by void regions, yielding a frozen-in, underdense microstructure. As the temperature rises the microstructure transitions to zone II ($0.3 < T_s/T_m < 0.5$), where increased surface diffusion produces a film consisting of a tightly packed columnar network with minimal void regions. The zone III microstructure is associated with high-temperature growth ($T_s/T_m > 0.5$) and the activation of bulk diffusion processes. Films in zone III are highly dense and consist of equiaxed columnar grains. Zone I growth is most relevant to GLAD growth.

that other SZMs have been proposed, incorporating additional zones and encompassing the effects of other processing parameters.

As revealed in high-resolution microscopy studies of zone I films by Messier *et al.* [11], an individual column is itself not a homogeneous entity. Rather, the film microstructure consists of a number of basic structural units broadly classed as macrocolumns, microcolumns and nanocolumns, which are arranged in a hierarchical structure (Figure 2.3). The largest features are macrocolumns, which are visible using SEM techniques and are the same columnar structures considered in the previously discussed SZMs. Higher resolution microscopy techniques (TEM and field-ion microscopy) show that these macrocolumns are composed of smaller microcolumn structures that are tightly bundled together and separated by an interstitial void network. Moving down the structural hierarchy, the microcolumn subunits consist of smaller clustered nanocolumns. The hierarchical nature of the film produces a self-similar geometry, where the microstructural details are similar when viewed at different length scales. During growth, transitions between the nano/micro/macrocolumn structure occur as a function of thickness following an evolutionary growth model. Very thin films (~15 nm thick) consist of nanocolumns on the order of 1–3 nm in size. With increasing film thickness, the nanocolumnar features bundle together to form the larger microcolumn structure. This bundling process is repeated as film growth continues to produce ever larger macrocolumn features, with the microstructural transitions appearing to be equally spaced on a logarithmic thickness scale. Consequently, when deposited under low-mobility conditions, the film morphology generally consists of columnar features with an intricate, hierarchical structure that systematically spans a wide range of characteristic dimensions, from 1–3 nm to hundreds of nanometres. Note that, with higher mobility conditions, the resulting columns are often much denser.

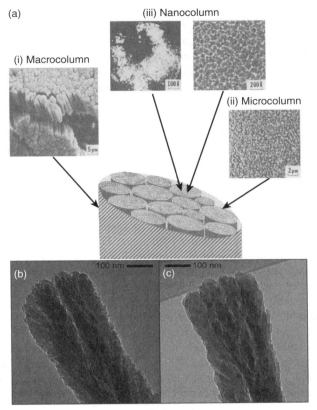

Figure 2.3 (a) The evolutionary columnar growth model developed by Messier *et al.* based on studies of Ge films deposited at low temperature [11]. Film microstructures are hierarchical, with columnar features consisting of bundled units of smaller column subunits. (i) Large macrocolumn features are composed of (ii) smaller microcolumn structures, which are themselves composed of even smaller (iii) nanocolumns. The film microstructure thus spans a wide range of characteristic dimensions and exhibits a self-similar geometry, a phenomenon general to films deposited under low-mobility conditions, as shown here by transmission electron microscope (TEM) images of individual GLAD-fabricated (b) TiO_2 and (c) Al_2O_3 columns. (a) Reprinted with permission from [11]. Copyright 1984, American Vacuum Society. (b) and (c) © 2000 IEEE. Reprinted with permission from [12].

For example, the high surface mobility of Al during deposition promotes the nanocolumn coalescence during column growth, resulting in a visibly denser internal structure (Figure 4.6d).

2.3 GLANCING ANGLE DEPOSITION TECHNOLOGY: MICROSTRUCTURAL CONTROL VIA SUBSTRATE MOTION

Two mechanisms critically impact film growth at all stages. The first is surface diffusion, which, as discussed in the previous section, enables adatom transport and strongly affects nucleation dynamics and column formation. The second mechanism is ballistic shadowing: a purely ballistic effect whereby features on the surface intercept incident vapour particles and cast line-of-sight shadows over the adjacent regions. Even in depositions performed on nominally flat surfaces (for example, atomically smooth Si or mica wafers), the preliminary nucleation events create nanoscale topographies on the substrate, thus providing the initial surface roughness for the ballistic shadowing effect. Vapour is unable to directly condense within the shadow zone immediately behind a surface feature, and material can only reach these areas via adatom diffusion. Shadowing is thus a particularly important effect when the surface mobility is limited as diffusive transport is unable to fill-in the shadowed regions, leading to preferential growth of existing nuclei and islands. As these features grow, the shadowing effect is enhanced, which further amplifies nanoscale surface roughness and promotes the formation of columnar structures [14]. To create sharply delineated shadowed regions, it is important that the incident vapour is highly directional, as depositing with a poorly collimated incident flux allows vapour to directly access shadowed areas. Collimation and shadow definition are maximized when evaporation is performed at sufficiently low pressures (Section 7.2.1) and with small vapour sources (Section 7.3.3).

In conventional film deposition, the vapour flux typically arrives perpendicular to the substrate. The ballistic shadowing lengths are therefore very small and the resulting columnar morphology is tightly packed with an interstitial network of small void regions (for zone I films; Figure 2.2). Although the film is slightly porous, the overall film density remains high, albeit less than the density of the bulk material. However, when the vapour flux arrives off-axis, thereby defining the deposition angle α as the angle between the substrate normal and the incident vapour flux trajectories, the ballistic shadow length increases due to the oblique angle of incidence (Figure 2.4a). Oblique deposition magnifies the shadowing effect and prevents vapour condensation and film growth over a greater area of the substrate. Off-axis deposition has a substantial impact on film growth and morphology, particularly at highly oblique angles where the shadow lengths are very long and a significant fraction of the substrate is shadowed. Figure 2.4b demonstrates these phenomena, showing preliminary film growth onto an array of lithographically defined features designed to act as initial

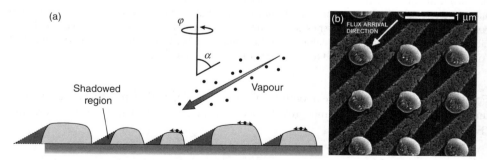

Figure 2.4   The ballistic shadowing effect is amplified when the vapour flux is off-axis as longer geometric shadows are cast at oblique incidence. The deposition angle α is the angle between the substrate normal and the incident vapour, and φ measures rotation of the substrate about its normal (which determines the flux arrival direction). (b) Features on the substrate intercept vapour flux, casting line-of-sight shadows onto adjacent regions of the substrate and strongly impacting film growth. (b) © 2005 IEEE. Reprinted with permission from [13].

nucleation centres (see Chapter 3). These raised features capture the incident vapour and prevent film growth in the immediately adjacent region, leading to the formation of sharply defined shadow zones. The shape and size of the shadow regions are determined by the deposition geometry: the shadow orientations are aligned parallel to the vapour flux arrival direction, and the shadow length is set by α and the feature height. Oblique deposition onto an unpatterned flat surface produces similar effects, although the size and placement of the shadowing features is randomized due to the stochastic nature of the initial nucleation growth phase.

Changing the vapour flux direction to manipulate ballistic shadowing is the key principle of GLAD, providing control over this important film growth mechanism. Implementing this control in practice is straightforward as α can be accurately set by tilting the substrate with respect to the deposition source. Furthermore, a second degree of freedom is obtained by rotating the substrate about its normal axis, thus controlling the substrate azimuthal angle φ (Figure 2.4a). Changing φ alters the apparent direction of the incident vapour flux and provides control over the shadow orientation, thus determining which regions of the substrate are shadowed. Figure 2.5 depicts a typical GLAD apparatus, showing the basic hardware elements required for successful GLAD. The substrate is mounted on a rotatable stage capable of simultaneously and continuously adjusting the α and φ substrate position. The deposition is performed in a PVD vacuum chamber using a collimated vapour flux source to enhance the ballistic shadowing effect. The substrate motions are often computer controlled, with the α and φ positions controlled based on feedback from a deposition-rate monitoring element, such as a quartz crystal microbalance.

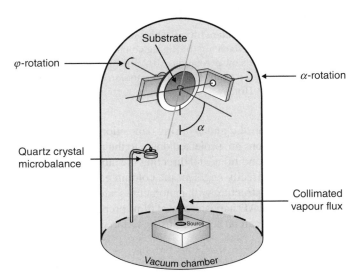

Figure 2.5 A typical GLAD apparatus showing key hardware elements. The substrate is mounted onto a rotatable stage capable of controlling both the α and φ rotation. Deposition rate monitoring, for example using a quartz crystal microbalance, can be used as feedback for α and φ position control. To ensure well-defined ballistic shadows, the vapour source provides a highly collimated flux. This generally requires the deposition be conducted in a high-vacuum environment to prevent loss of collimation due to flux scattering. Reproduced from [15] with permission of M.A. Summers.

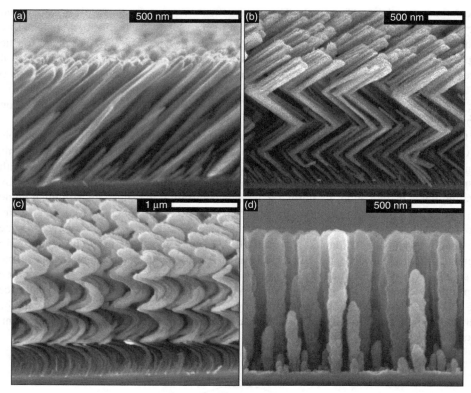

Figure 2.6 Common GLAD microstructured films fabricated using basic substrate motions at high α (>80°). (a) Tilted columns (fixed α, fixed φ). (b) Zigzag columns (fixed α, discrete φ rotations). (c) Helical columns (fixed α, continuous slow φ rotation). (d) Vertical columns (fixed α, continuous rapid φ rotation). (a), (c) and (d) Reproduced from [16] with permission of A.C. van Popta. (b) Reproduced from [17] with permission of P.C.P. Hrudey.

Rotating the substrate to control α and φ during deposition is the crux of GLAD technology: macroscale substrate motions are projected down to the nanoscale growth environment, dictating the local shadowing conditions and directing the atomic growth processes [18, 19]. Using GLAD it is possible to directly engineer the column growth process in highly porous films and control the film microstructure with nanometre-scale precision. Figure 2.6 presents a sample gallery of common GLAD nanostructured films prepared at high α (>80°), showing various film morphologies that can be achieved using tailored α and φ motions:

- Figure 2.6a, tilted columns are produced by deposition at fixed α and fixed φ;
- Figure 2.6b, zigzag columns (also called chevron columns) are fabricated by depositing at fixed α coupled with discrete 180° φ rotations;
- Figure 2.6c, helical columns are realized through deposition at fixed α and slow, continuous φ rotation; and
- Figure 2.6d, vertical columns are created when depositing at fixed α with rapid, continuous φ rotation.

These microstructures represent a small cross-section of the capabilities of the GLAD technique, as complex substrate motions can be designed to realize a diverse range of unique microstructures. However, these basic α and φ motions demonstrate how ballistic shadowing can be controlled to guide column formation and growth. The following sections describe the effect of α and φ in greater detail.

2.4 ENGINEERING FILM MORPHOLOGY WITH α

The deposition angle α has a strong impact on the film microstructure and virtually every physical property of the deposited film, making α one of the most important control variables in GLAD. Changing α affects the microstructure in multiple ways. First, α controls the shadow length and the size of void regions formed in the film, and the microstructure therefore becomes more porous as α is increased. Second, the shadow-dominated growth mode associated with high α conditions leads to the formation of a distinctive, isolated columnar microstructure and highly competitive growth mode (Section 2.6). Third, the column growth is directional and biased towards the source, meaning that the column orientation can be controlled by changing α (and φ – Section 2.5).

2.4.1 Controlling microstructure and porosity

Tuning α via substrate motion provides control over the local deposition geometry and the ballistic shadowing length. Given a surface feature of height h, the geometric shadow length is equal to $h \tan \alpha$, a relationship plotted in Figure 2.7. As α is increased, the shadow length increases, thereby preventing film deposition over a greater fraction of the substrate. Because of the increased shadowing, the intercolumn void region expands, creating more room between columns and correspondingly reducing the density of the film morphology. This trend is monotonic, with shadow-generated void regions comprising an ever greater portion of the film microstructure as α increases. Up to moderately large angles ($\alpha < 75°$), the shadow length remains small and is less than a few times the height of the surface feature. For microscopic islands formed during the nucleation growth phase, the shadow length may be comparable to the surface diffusion length and shadowing may not be the dominant growth mechanism. While films deposited in this oblique regime exhibit porosity and modified properties, the film microstructure remains relatively dense (Figure 2.7A). However, the shadow lengths increase rapidly for $\alpha > 75°$ and become extremely long when moving to glancing angles of incidence ($\alpha > 85°$). When depositing at high α, the shadow length can greatly exceed the surface diffusion length and ballistic shadowing effects dominate film growth.

Entering this extreme shadowing regime dramatically changes the film microstructure, as seen in the SEM images of Figure 2.7 that demonstrate the progressive structural evolution as α is increased. As expected from the increased shadow length, the film becomes more porous as α is increased due to the incorporation of greater amounts of void in the film. The relationship between α and film density has been investigated using theory and experiment, demonstrating that GLAD provides a direct means of controlling the film density (Section 4.4). This control is an important capability as the density directly impacts several physical properties. For $\alpha < 75°$, the film consists of tightly packed columns with

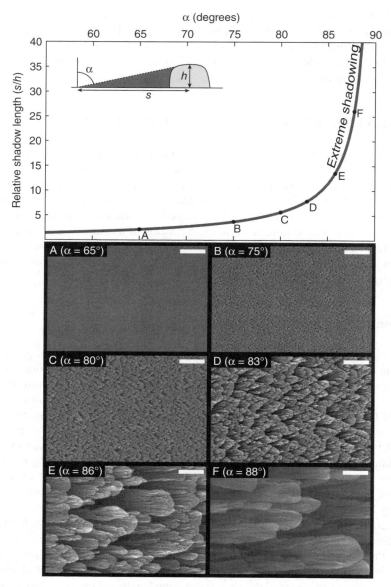

Figure 2.7 As α increases, the geometric shadow length s grows correspondingly longer. At high α, s may be many times greater than the height h of the shadowing surface feature, leading to conditions of extreme shadowing. Striking changes in the film microstructure occur as α increases (A–F). The film microstructure becomes increasingly porous as the shadowed regions increase in size, creating an expanding network of intercolumn voids. In the extreme shadowing regime the microstructure, highly sensitive to small changes in α, consists of completely isolated nanoscale columns, a drastic departure from conventional thin-film morphologies. The scale bars in the top-down SEM images all indicate 100 nm.

sizes difficult to resolve via SEM. As α is increased, a gradual microstructural transition is observed. As can be seen from the SEM images, the increasing shadow length expands the size of the intercolumn void domains. These domains first form a channel-like interstitial pore network that separates the bundled columnar structures. The pore network continues to enlarge with increasing α before an isolated columnar morphology eventually emerges in the extreme shadowing regime. At high α ($>85°$), the film microstructure is open and composed of distinct columns separated by distances on the order of the column diameter. At these highly oblique angles, even the nanoscopic islands of early film growth can cast exceptionally long shadows that exceed several atomic diffusion lengths, thereby suppressing island coalescence and preventing the formation of a continuous film, leading to a drastic departure from conventional film growth and morphology before even a single nanometre of growth [20] (Figure 2.8). The film microstructure is highly sensitive to α in this regime, as even a small change in α produces a large change in the ballistic shadow length, thus impacting the microscopic film growth dynamics.

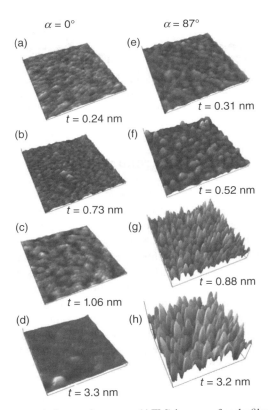

Figure 2.8   Non-contact atomic force microscope (AFM) images of early film growth at (a–d) $\alpha = 0°$ and (e–h) $\alpha = 87°$, taken at different thicknesses. Although film morphology is similar over the initial few ångströms, as seen by comparing (a) and (e), extreme shadowing leads to a rapid roughening of the film surface and a drastic divergence between the two nanoscale structures as deposition progresses. The images correspond to 250 nm × 250 nm film areas. Reprinted with permission from [20]. Copyright 2007, AIP Publishing LLC.

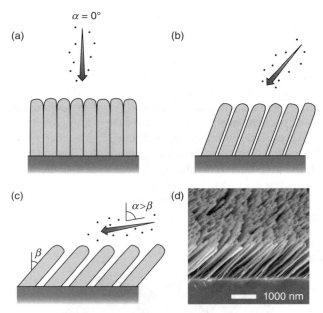

Figure 2.9   Column growth is oriented towards the incident vapour flux, forming an angle β with the substrate normal that increases with (but is always less than) α (a–c). (d) This directional column growth produces a tilted columnar microstructure that is characteristic of obliquely deposited films.

2.4.2 Directional column growth: column tilt β

Another aspect of film growth critical to GLAD is that column growth is directional and biased towards the incident vapour source. In off-axis deposition, this directional growth produces a film morphology consisting of inclined columns tilted towards the incident vapour. The column tilt angle β, defined as the angle between the substrate normal and the column centre axis, points towards the incoming flux but is always less than the deposition angle α, with more oblique deposition geometries producing an increasingly inclined columnar microstructure (Figure 2.9). Thus, we see that, in addition to controlling the microstructural density transition described in the previous section, α also determines the column growth direction, allowing the fabrication of an aligned, tilted column morphology. Directional growth is exploited to enable column steering via substrate φ rotation, which is the topic of Section 2.5. Importantly, this column tilt produces a structural anisotropy that induces a corresponding anisotropy in many of the physical properties of GLAD films, a topic discussed at length in Chapter 4.

Several analytical equations of varying sophistication have been proposed in the literature providing different predictive relationships between α and β. The earliest and simplest of these is the empirical tangent rule proposed by Nieuwenhuizen and Haanstra [21]

$$\tan \beta = \frac{1}{2} \tan \alpha. \tag{2.1}$$

This equation fits well to multiple sets of data and thus provides a common starting point for predicting β [21, 22]. However, Equation 2.1 is not derived from a physical model and tends to overestimate β at large α (> 70°). Consequently, several alternative models have been reported, each attempting to provide a truer description of the growth mechanics and a more accurate β estimate. In a geometric analysis of the intercolumn shadowing geometry, Tait *et al.* derived the expression (known as the Tait or cosine rule) [23]:

$$\beta = \alpha - \arcsin\left(\frac{1 - \cos\alpha}{2}\right). \tag{2.2}$$

The predicted β values reproduced observed values in experimentally deposited MgF_2 and Fe films, as well as hard-particle simulations of ballistic film growth. In addition to providing a physical interpretation of the column tilt by tying β to the underlying local shadowing geometry, the agreement at high α was found to be superior to the tangent rule.

However, the derivation of Equation 2.2 is purely geometric and captures only the ballistic shadowing growth mechanism. This analysis, therefore, ignores growth kinetics, which play an important role in the morphology of individual columns. In practice, significant discrepancies are often found since the kinetics are influenced by numerous experimental factors (such as substrate temperature, deposition rate, crystallinity, etc.) [24]. The tilt angle can also be directly controlled via ion bombardment during deposition, as shown by Sorge *et al.* [25]. To provide greater flexibility in modelling different materials and experimental conditions, additional models of varying sophistication have been proposed. These models provide fitting parameters that may be used to capture material and process-related effects. Hodgkinson *et al.* added a fitting parameter E to Equation 2.1, yielding [26]

$$\tan\beta = E\tan\alpha. \tag{2.3}$$

In an experimental study of three different oxide materials, Ta_2O_5, TiO_2 and ZrO_2, the respective fitting coefficients were 0.322, 0.347 and 0.281, with films deposited at α from 40° to 70°. Lichter and Chen conducted a theoretical analysis of oblique deposition [27], employing a continuum modelling approach to relate β and α via

$$\beta = \frac{2}{3}\frac{\tan\alpha}{1 + \Phi\tan\alpha\sin\alpha}. \tag{2.4}$$

The coefficient Φ parameterizes the effects of surface diffusion and is given by

$$\Phi = \frac{4}{27}\frac{h_1 J}{D},$$

where h_1 is the shadowing feature height, J is the incident vapour flux deposition rate and D is the adatom surface diffusion constant. While the model can account for a variety of different deposition conditions, in the GLAD-relevant case of limited diffusion, Φ is small and Equation 2.4 approximates the tangent rule. More recently, Tanto *et al.* developed a model based on ballistic shadowing that accounts for the tendency of the columns to fan out as they grow (see Section 2.6.4 for more details) [28]. Their analysis of periodically arranged columns produced

$$
\beta = \begin{cases} \alpha - \arctan\left[\dfrac{\sin(\theta)-\sin(\theta-2\alpha)}{\cos(\theta-2\alpha)+\cos(\theta)+2}\right] & \text{for } \alpha \le \theta \\ \alpha - \dfrac{\theta}{2} & \text{for } \alpha \ge \theta \end{cases},
\tag{2.5}
$$

where θ is the fan-out angle of the columns, which varies with material and deposition conditions. For Ge films deposited onto periodically seeded substrates (see Chapter 3) at α from 0° to 85°, the authors measured $\theta = 50°$ and found good agreement with their model using this parameter. Zhu *et al.* extended this study to a broader material range, finding a relationship between θ and the material melting point [29]. As the melting point of the material rises, which correlates with decreased surface diffusion, ballistic shadowing effects increase, leading to larger fan angles.

For comparison, these five models relating β and α are plotted in Figure 2.10. Each model predicts that β increases monotonically with α and that $\beta < \alpha$, in agreement with long-standing experimental observations and the results of deposition simulations. In practice, the parameter-free models (Equations 2.1 and 2.2) are useful starting points for preliminary design. With ongoing testing and monitoring, it is useful to implement the more advanced models (Equations 2.3–2.5) and use the fitting parameters to construct calibration curves. Material dependencies and sensitivity to deposition parameters (substrate temperature, pressure, deposition rate, etc.) can thus be captured for more accurate process monitoring. Note that because of the inherent sensitivity to experimental conditions there is no 'correct' model, just perhaps one that will work well for a specific system/material/process.

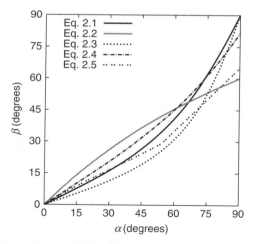

Figure 2.10 Column tilt angle β at different deposition angles α, as predicted by the different expressions in the text (Equations 2.1–2.5). Each model correctly predicts that β increases montonically with α and that $\beta < \alpha$. As plotted, Equations' 2.3, 2.4 and 2.5 fitting parameters are $E = 0.347$, $\Phi = 0.1$ and $\theta = 50°$ respectively.

2.5 ENGINEERING FILM MORPHOLOGY: COLUMN STEERING VIA φ ROTATION

During oblique deposition, nuclei that form during early film growth develop into tilted columnar structures oriented towards the source. GLAD capitalizes on this highly directional column growth by employing substrate motions to dynamically alter the growth direction during column formation. From the point of view of columns growing on the surface, rotating the substrate is equivalent to moving the location of the vapour source in the sky. The column growth direction is reoriented as the columns track the changing source location. This forms the basis of GLAD technology: manipulating the substrate angles to alter the apparent vapour direction, thus controlling the ballistic shadowing geometry and dynamically altering the column growth direction. With properly designed GLAD motions, the columns can effectively be *steered* along a desired growth path, leading to the formation of new and unusual columnar microstructures.

2.5.1 Controlling column architecture with φ: helical columns

Changing the azimuthal substrate angle φ produces a corresponding rotation of the line-of-sight shadowing geometry and the column growth direction. Using appropriate substrate movements, the columns are sculpted into different architectures as they track the apparent changes to the vapour source trajectory. Two parameters, the angular position φ and the rotation speed ω, provide basic control over substrate rotation and enable fabrication of a wide variety of microstructures. For microstructure control, it is convenient to define the rotation speed ω relative to the deposited film thickness t, typically measured by a calibrated rate-monitoring element in the deposition system (see Section 7.3 for discussion). An equivalent and convenient parameter is the pitch P, defined as the deposited film thickness per complete substrate revolution. The angle φ is typically zeroed at the beginning of deposition ($t = 0$), and all subsequent substrate positions are referenced to that orientation.

The impact of continuous substrate rotation succinctly illustrates the concept of column steering and its influence upon the film microstructure. When the substrate is slowly rotated, the growth direction smoothly rotates as well and the columns follow the migrating source position. The columns are thereby shaped into microscale helices, spiralling upwards from the substrate as they track the apparent source motion [31]. This unique microstructure is demonstrated in Figure 2.11. The structure of an individual helical column can be primarily characterized by several parameters, each controlled via α and the specific φ motion:

- The substrate rotation speed determines the structural pitch P of the helix and the helical radius of curvature. A faster rotation speed reduces the structural pitch and produces a more tightly wound helix (Section 2.5.2).
- The helix angle is equivalent to the column tilt angle β, which is material specific and can be controlled via α (Section 2.4.2).
- The column width w depends on both material and α, and is typically not uniform with thickness. Experimental calibration is required (Section 4.3.2).
- Rotating the substrate in the opposite direction reverses the chiral symmetry of the structure, producing a helix of opposite handedness.

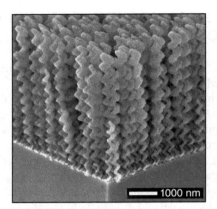

Figure 2.11 Continuously rotating the substrate about the surface normal effectively changes the incident vapour direction and the corresponding column growth direction. With this substrate motion, the columns are shaped into a helical geometry. The structural parameters of the helix, including vertical pitch, angle, radius of curvature and handedness, can be controlled through simple alterations of this basic substrate movement. Reproduced from [30]. With permission of A.C. van Popta.

The physical properties of these films are intimately connected to the specific helical morphology. For instance, microstructured helices exhibit mechanical characteristics similar to a macroscopic bed of springs (Section 4.8.3). The spring constant of the helical column can be tailored by adjusting structural parameters, allowing rational design of the film's elastic behaviour. In another example, the chirality of helical structures produces optical properties that are highly sensitive to circular polarization (Section 5.5). The structural parameters can be designed to control the optical response, enabling spectra to be tuned to specific wavelengths and for polarization selectivity to be optimized.

2.5.2 Controlling microstructure with rotation speed: vertical columns

The substrate rotation speed strongly influences the resulting film morphology, and different ω values can be employed to access particular microstructures [32]. Figure 2.12 shows the substrate motion algorithms corresponding to continuous φ rotation to realize the indicated P values. As P is decreased, the deposited film undergoes a dramatic microstructural evolution, as shown in Figure 2.13. These films, all consisting of TiO_2 deposited at $\alpha = 81°$, show a different microstructure depending on the ratio between P and the column width w ($\sim$30 nm). When $P \gg w$, the substrate rotation speed is very slow and the resulting microstructure is as presented in Figure 2.13a,e, where $P = 800$ nm. The helical morphology is substantially larger than the individual columnar structures and the film resembles a network of individual columnar fibres twisted into a helical configuration. As P is decreased yet remains greater than w, the overall helical structure remains, as shown in Figure 2.13b,f, where $P = 100$ nm. However, the faster φ rotation rate produces a more tightly wound helix with a reduced structural pitch and radius of curvature. The helical microstructure is gradually lost as the helical pitch approaches the column width, as shown in Figure 2.13c,g. In the $P \sim w$ regime, the distinct helical morphology transitions into a screw-like columnar microstructure,

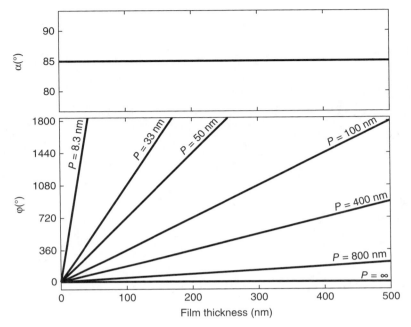

Figure 2.12   Substrate motion algorithms for continuous φ rotations at different speeds. Here, the rotation speed is controlled by setting the pitch P, as it has a direct impact on the resulting columnar microstructure (Figure 2.13). Large P (slow rotation) motion produces distinctly helical structures (Figure 2.13a,b), while smaller P (rapid rotation) can be used to fabricate vertically oriented columns (Figure 2.13c,d).

exhibiting fine ridge structures along the edges of the column as defined by the pitch. Once $P < w$, the helical structure degenerates completely and the film morphology fully transitions into one of individual, vertically oriented columns (Figure 2.13d,h). Under these conditions, the substrate rotation has become so rapid that the column steering effect is lost and the vapour flux effectively arrives from all φ positions simultaneously.

The microstructural transition between helical and vertical columns is highly dependent on w, a parameter sensitive to material properties and experimental conditions. Furthermore, columns deposited at high α tend to broaden during growth, following a power-law thickness scaling (Sections 2.6.2 and 4.3.2). Empirical calibration for a specific process is often necessary to accurately realize the intended microstructure. As a rough guide, w typically ranges between 10 and 50 nm; then, design parameters of $P > 100$ nm should produce helical columns and $P < 10$ nm should realize vertical columns. These starting parameters can then be optimized through systematic studies and taking into account column broadening effects.

2.5.3   Continuous versus discrete substrate rotation

Smooth, continuous substrate rotation provides access to films with helical and vertical column microstructures. However, there are also many GLAD substrate motion algorithms based on discrete, step-like substrate movements combined with linear growth intervals.

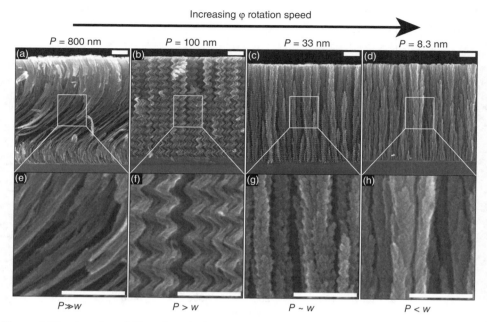

Figure 2.13 A series of films deposited under identical conditions (TiO_2, $\alpha = 81°$), except with different rotation speeds as quantified by the growth pitch P. As can be seen, the ratio of P to the column width w (~30 nm) directly controls the resulting columnar microstructure. In (a), the substrate rotation is slow and the microstructure consists of helically twisted, bundled fibres. (b) Increasing the rotation speed reduces both the helical pitch and the helical radius, as expected. As the growth pitch approaches the column width (c), the microstructure is no longer distinctly helical and degenerates towards a screw-like morphology. (d) With even faster rotation, the incident vapour effectively arrives at the column apex from all sides simultaneously and the microstructure completely degenerates into a vertically oriented column. (e)–(h) Higher magnification views of the columnar microstructure taken from the indicated sections of the film. All scale bars indicate 200 nm.

These discrete motions can be used to realize more complex column arrangements. Figure 2.14 presents a series of substrate motion algorithms, each designed with a fixed $\alpha = 85°$ and implementing a discrete φ rotation after every 100 nm growth interval. However, the rotation step is different in each case. In Figure 2.14b the substrate is rotated in $\Delta\varphi = 180°$ increments. With each rotation, the column growth direction is reversed to create a series of tilted columns with alternating orientations. This substrate motion creates a film with a zigzag columnar morphology, similar to that shown in Figure 2.15a. In Figure 2.14c the $\Delta\varphi$ rotation step is reduced to 120°. The resulting columns are oriented at 120° with respect to the previous column and the structure is repeated after three rotations, producing the three-sided helical microstructure depicted in Figure 2.15b. (Note that in these SEM images the $\Delta\varphi$ motions correspond to the substrate motions of Figure 2.14 but the column arm lengths are all different.) Figure 2.14d shows the φ motion algorithm used to create a square-spiral structure, where the substrate is rotated by $\Delta\varphi = 90°$ after each growth interval. The structure is thus repeated after every four rotations and the microstructure consists of four-sided helical

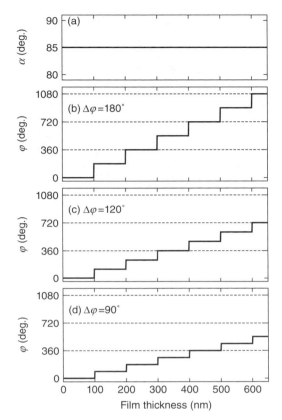

Figure 2.14   Sample motion algorithm implementing discrete φ rotations after 100 nm growth intervals. (a) The deposition angle is fixed at $\alpha = 85°$. The different φ rotation schemes produce different microstructures: (b) discrete $\Delta\varphi = 180°$ rotations produce the zigzag microstructure (Figure 2.15a); (c) $\Delta\varphi = 120°$ rotations produce a three-sided helical microstructure (Figure 2.15b); (d) $\Delta\varphi = 90°$ rotations produce a square-spiral microstructure (Figure 2.15c).

columns (Figure 2.15c). These motions can be readily generalized to the case of n-sided polygonal spiral architectures, which are fabricated by repeated $360°/n$ discrete rotations [34] and have been investigated for their unique optical properties (Section 5.5.3). Of these morphologies, the square-spiral column structure has been the focus of substantial research effort and structural optimization owing to its potential use in 3D photonic crystal devices [36, 37]. Such applications demand a high degree of structural uniformity and periodicity that can only be achieved using seeding techniques, which are explored in Chapter 3.

It is important to note that the substrate motions described in this section, while producing a range of GLAD architectures, comprise only a subset of all available substrate trajectories. Different φ motion concepts can be combined in a single deposition to realize an intricate, multicomponent morphology such as shown in Figure 2.16. This four-part microstructure consists of an RH helical layer, created by continuous, slow φ rotation; a zigzag column layer, created by a discrete 180° φ rotation; a vertical column section, realized with continuous,

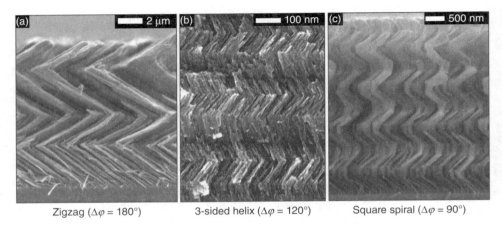

Zigzag ($\Delta\varphi = 180°$)   3-sided helix ($\Delta\varphi = 120°$)   Square spiral ($\Delta\varphi = 90°$)

Figure 2.15 Using discrete substrate rotations to create a series of linear growth segments with different orientations, it is possible to fabricate a diverse range of column structures. These SEM images show (a) zigzag columns, (b) three-sided helical columns and (c) square-spiral columns, each realized using a different φ motion algorithm as depicted in Figure 2.14. (a) Reprinted with permission from [33]. Copyright 2002, American Vacuum Society. (b) Reprinted with permission from [34]. Copyright 2005, AIP Publishing LLC. (c) Reproduced from [35] with permission from H.M.O. Jensen.

rapid φ rotation; and a final left-handed (LH) helical layer created by continuous, slow φ rotation in the opposite direction from the RH layer. By developing complex φ motions, it is possible to realize advanced engineering of the shadowing geometry and carefully control column growth. Ultimately, the range of possible motions is limited only by system geometry, motor hardware and ingenuity.

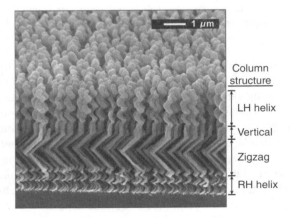

Column structure

LH helix

Vertical

Zigzag

RH helix

Figure 2.16 Complex, multicomponent microstructure consisting of a right-handed (RH) helix, a zigzag column, a vertical column section, and a left-handed (LH) helix. The entire structure is made in a single deposition run by combining a series of basic φ motions. Reproduced from [30] with permission from A.C. van Popta.

2.6 GROWTH CHARACTERISTICS OF GLANCING ANGLE DEPOSITION TECHNOLOGY FILMS

The combination of high α and low adatom mobility creates a distinctive shadow-dominated growth mode harnessed by GLAD to sculpt film growth and steer the column microstructure. Ballistic shadowing enables the interaction between growing columns since the shadow cast by one column can obscure a portion of a neighbouring column, thereby influencing how that column can grow. Ballistic shadowing thus creates a complex growth environment, established right from the earliest stages of nucleation and persisting throughout growth, continuing to affect the evolution of micrometre-sized columns. These shadowing dynamics have a profound impact on film evolution during GLAD, and many structural characteristics of GLAD films are critically related to the shadowing interaction. The following discusses several salient growth phenomena associated with column growth and microstructure in GLAD films. The qualitative discussion in this section is complimented by Chapter 4, which examines experimental efforts focused on the quantitative characterization of film structure and the growth process.

2.6.1 Evolutionary column growth

The stochastic nature of the initial nucleation and island growth stages produces not only random column locations, but also columns of varying dimensions. Further column growth is driven by the depositing vapour flux, and as the columns grow the shadows they cast grow commensurately longer. Occasionally, smaller columns fall into the shadows cast by their larger neighbours and are no longer able to grow, being cut off from the incident vapour flux. A shadowed column consequently becomes *extinct*, whereas the larger, shadowing column survives and continues to intercept vapour at the expense of its neighbour. The extinction process is schematically depicted in Figure 2.17, showing six columns (labelled 1–6) and their growth evolution (a–c). The columns initially exhibit similar shapes, but with a range of sizes, creating local variations in the shadowing environment. With continued growth (Figure 2.17b), column 1 falls under the shadow of column 2 and becomes extinct. In Figure 2.17c, column 3 becomes shadowed and dies off, and column 5 is at risk of extinction as well due to the growth of column 6. Extinct columns are quickly left behind by their surviving neighbours, which then compete with each other in turn. We thus see the operation of an extinction mechanism and the emergence of an evolutionary growth mode [38, 39]: the largest features tend to outcompete smaller ones for incident vapour and survive, akin to survival-of-the-fittest evolution.

The film microstructure evolves during growth in a continuous manner, shaped by the shadowing-induced competition between the columns. This competition-driven growth evolution is common to all the GLAD microstructures. Figure 2.18 shows cross-sectional SEM images of (a) slanted, (b) helical and (c) vertically oriented columns. As indicated, extinct columns can be readily identified in each geometry. Furthermore, extinct columns can be found of any height, indicating that competition operates continuously during growth.

The statistical distribution of column sizes produced by the nucleation stage is an important determinant of evolutionary growth dynamics. Larger columns, comprising the upper tail of the size distribution, cast larger shadows and tend to intercept greater amounts of

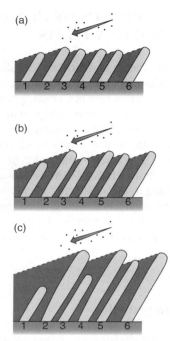

Figure 2.17 Competition between columns to capture incident vapour leads to the extinction of columns that fall in the shadows of their larger neighbours. (a) Initially, all columns receive flux and grow. (b) Column-to-column variations produce a nonuniform shadowing environment and some columns (1, 3) fall into the shadow of their neighbours (2, 4). (c) Shadowed columns are cut off from the source and become extinct, while the surviving columns continue compete via shadowing to further intercept vapour flux.

vapour flux. These columns thus have a greater probability of survival in comparison with smaller features, which are more likely to be obscured by a neighbour and become extinct. Therefore, over the course of film growth, the number of active columns decreases, as shown in Figure 2.19 [40]. The column statistics clearly change as growth progresses, as can be seen by comparing the series of top-down SEM images in Figure 2.19b–g. Initially, the film structure consists of a larger number of small columnar structures (Figure 2.19g). With the

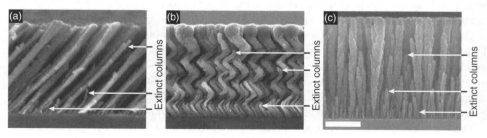

Figure 2.18 Examples of column extinction in different microstructures: (a) slanted posts; (b) helices; (c) vertical posts. The scale bar in each micrograph indicates 500 nm.

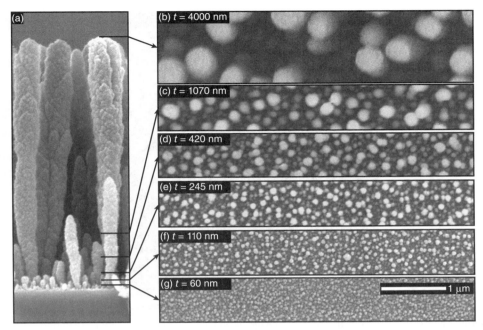

Figure 2.19 Microstructural evolution of a vertical-column GLAD film (SiO_2, $\alpha = 87°$, $P = 10$ nm) during growth. (a) Both the overall film microstructure and individual column profiles change continuously during growth. Looking at the film top-down after different growth intervals (b–g) reveals the evolution of the statistical distribution of columns. Initially (g) the film consists of many isolated, small columnar structures. Competitive forces drive many of these columns to die off as growth progresses, leading to steady reduction in the number of columns and an increase in the average column size. Reproduced with permission from [40]. Copyright S.G. Pandalai, Managing Editor.

onset of competitive growth effects, the column number rapidly decreases as deposition progresses. The average column size gradually increases as deposition progresses; and at later stages of film growth, fewer, larger columns are found (Figure 2.19b). These observations are supported by quantitative studies of intercolumn spacing and column number density (Section 4.3.3).

The competition-driven evolution of column growth creates a film morphology that, when viewed at different stages of growth, is remarkably scale invariant. Many researchers have experimentally investigated the statistical properties of this phenomenon using AFM techniques, finding that the film surface profile (i.e. the root-mean-square roughness of the column apexes) increases with thickness according to a complicated power-law scaling (Section 4.3.1). These morphology studies reveal the emergence of the columnar features and associated surface roughness, thus enabling quantitative investigations of microstructural evolution in GLAD films.

Significant research efforts have focused on controlling the early stages of film growth by depositing onto substrates with pre-patterned surface topographies. These engineered features act as column 'seeds' and can be employed to minimize competitive growth effects and improve the column-to-column uniformity of the resulting film. Chapter 3 examines the design and implementation of these seed layers in greater detail.

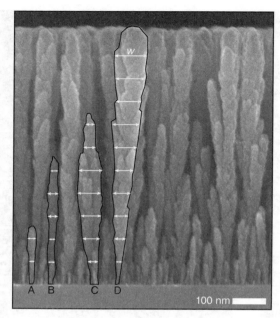

Figure 2.20   Column competition leads to nonequilibrium growth with a continuous evolution of column structural profiles. Here, a TiO$_2$ vertical column microstructure deposited at $\alpha = 85°$ is presented, showing the structural nonuniformity associated with column broadening. Four columns (A, B, C and D) have been outlined to highlight the change in the column width w as deposition progresses.

2.6.2 Column broadening

Because the competitive growth process and resulting column extinction leads to a continually evolving growth front, individual GLAD columns generally exhibit some degree of structural nonuniformity associated with the unstable local shadowing environment. While nonuniformity is observed in many GLAD structures, the most commonly investigated is column broadening in the vertical column morphology, where the column width w gradually increases with thickness. A representative vertical column microstructure is presented in Figure 2.20, showing a TiO$_2$ film deposited at $\alpha = 85°$. The column broadening is clearly visible, as are the column extinction events caused by competitive growth and the column-to-column structural variation associated with random film growth effects. Four columns (A, B, C and D) have been selected and highlighted for further discussion. Examining these selected column profiles produces several key observations regarding column evolution during GLAD growth. First, we note that, while each column broadens, the broadening rate differs from column to column. Column B broadens relatively slowly during deposition, whereas column C shows a rapid broadening before it becomes extinct. The broadening rate of column D appears to be intermediate between columns B and C. Therefore, it is necessary to average broadening measurements over many columns to ensure the reliability of qualitative/quantitative characterizations. Smaller columns that become extinct during initial film growth, such as column A, can require high-resolution studies for accurate structural characterization. Second, the extinction process can be either sudden, as shown by column A, or gradual, as shown by column C, reflecting how quickly the local shadowing environment can change. Third, it is difficult to predict at an early stage which columns will survive. For instance, while column C undergoes a rapid broadening at first and initially becomes a dominant local feature, it eventually falls into a shadow region and dies off.

The column width increases with film thickness t following a power-law dependence $w \propto t^p$, where p is the broadening exponent (Section 4.3.2). This expression has been used to describe the results of many experiments as well as the predictions of Monte Carlo simulations of ballistic film deposition and continuum growth models. On average, experimental investigations have shown that the broadening exponent p tends towards 0.5, in agreement with basic models of ballistic deposition. However, significant scatter in the reported data exists and the broadening rate is highly sensitive to the underlying growth kinetics, experimental conditions and measurement methods.

2.6.3 Column bifurcation

While the overall columnar morphology continuously changes during film deposition, the column apexes themselves also represent an evolving growth front and are equally subject to the effects of shadowing and diffusion. Incident vapour atoms condense on the exposed column apex, nucleating on the column and contributing to the roughening of the column surface. As any surface roughness acts as a source of ballistic shadowing, intra-column shadowing effects can become an important factor in the overall film microstructure. This roughness is also an important factor in the column broadening rate (Section 4.3.2).

The column apex roughness can become sufficiently large to create a shadowing-induced growth instability that splits the column into multiple sub-columns. Krause *et al.* have carefully studied this bifurcation process, using a combined focused ion-beam (FIB) and SEM technique to generate tomographic images of individual TiO_2 columns (Section 4.2.3) [41]. Figure 2.21 shows cross-sectional slices of a *single* column taken at different points along its height. The initial tomographic slice shows a single columnar unit, although with obvious surface features/roughness. Shadowing amplifies this surface roughness, which develops into larger surface mounds on the column apex separated by a cusp, as indicated. A divergent process is then initiated whereby the shadowing mechanism limits growth to the mounds and deepens the cusp between them. The mounds ultimately develop into distinct subcolumns, producing a complete bifurcation of the original column into multiple sub-columns.

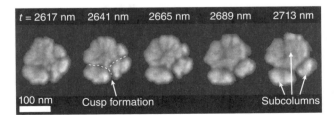

Figure 2.21 Tomographic slices through a single TiO_2 column showing the development of a shadowing instability in a growing column. Surface mounds separated by cusp regions initially emerge from the column surface roughness. Column bifurcation occurs as shadowing amplifies the mounds, which develop into sub-columns.

For the bifurcation instability to develop, sufficiently large surface mounds must be formed to overcome the smoothing effects of lateral adatom diffusion, which tend to fill in the cusps as they develop. Larger columns, which can support the formation of larger surface features, become more susceptible to bifurcation effects. Bifurcation is therefore a natural by-product of column broadening, and the occurrence of bifurcation events increases with film thickness. The bifurcation rate is also sensitive to the substrate temperature. At elevated temperatures, the surface diffusion length is increased, providing a mechanism to reduce surface roughness and prevent the formation of a shadowing instability. Consequently, controlling the substrate temperature could provide a method of preventing (or alternatively promoting) column bifurcation.

Zhou and Gall modelled the bifurcation process [42], proposing an expression for the bifurcation probability f_b that depends on the surface mound nucleation length l and the column width w following

$$f_b \propto \exp\left(-\frac{l}{w}\right). \tag{2.6}$$

A basic estimation of l is given by

$$\frac{D}{F} = \frac{l^6}{\ln(l^2)}, \tag{2.7}$$

where F is the deposition flux rate and D is the diffusivity (which has an Arrhenius-type temperature dependence). This expression agreed with measured bifurcation rates in Ta columns deposited at various temperatures (Figure 2.22). Note that Equation 2.7, being

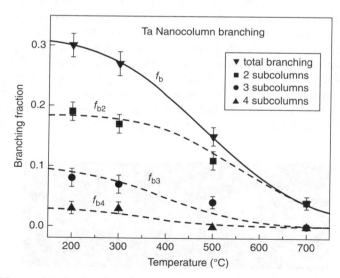

Figure 2.22 Measured (points) and predicted (lines) fraction f_{bn} of the total Ta column population exhibiting bifurcation into n sub-columns. Higher deposition temperatures lead to reduced column bifurcation rates as the increased surface mobility prevents cusp formation. Reprinted with permission from [42]. Copyright 2006, AIP Publishing LLC.

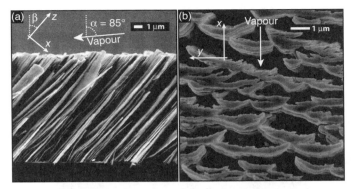

Figure 2.23   SEM images of slanted columnar TiO$_2$ structures taken (a) from the side and (b) looking down the column axis. The oblique vapour incidence leads to the formation of columns inclined in the xz-plane (first anisotropy). However, the lack of shadowing in the transverse direction (y-direction) leads to a pronounced elongation of the column cross-section as there is no shadowing mechanism to restrict growth perpendicular to the deposition plane. The column cross-section therefore fans out into an elliptical shape during column growth (second anisotropy). Reproduced from [16]. With permission from A.C. van Popta.

specific to 2D nucleation on flat surfaces, provides a qualitative approximation of the actual nucleation process. More complex estimates of l could be substituted when necessary to better characterize the growth process.

2.6.4 Anisotropic shadowing and column fanning

The ballistic shadowing mechanism, so critical to structure formation during GLAD growth, only acts along the incident flux direction. With respect to the film structure in Figure 2.23a, ballistic shadowing only restricts growth in the xz-plane (the deposition plane) and produces the characteristic inclined columnar structure. While the one-dimensional nature of the shadowing process yields highly directional column growth and enables column steering, there is no mechanism restricting transverse growth of the column cross-section (i.e. along the y-axis). This inherent shadowing asymmetry produces columns with an elliptical cross-section as shown by the columns imaged along the column axis ($-z$-direction) in Figure 2.23b. Along the x-direction, ballistic shadowing maintains the intercolumn separation and the column width is controlled. However, the column is able to grow along the y-direction, leading to a progressive fanning of the column cross-section.

Figure 2.24 illustrates the evolution of column cross-sections as a function of thickness, showing how the column fanning changes during film growth [40]. Although the cross-section is approximately symmetric in the x- and y-directions during the initial growth phase, the column features gradually fan out along the y-axis while shadowing prevents significant structural broadening along x. The cross-section thus becomes increasingly elliptical as deposition continues and the basic slanted columnar microstructure exhibits two separate structural anisotropies: the first being the column tilt with respect to substrate normal and the second being the development of an elongated column cross-section. Careful examination of the apex of tilted column films reveals that the cross-section is not quite elliptical but

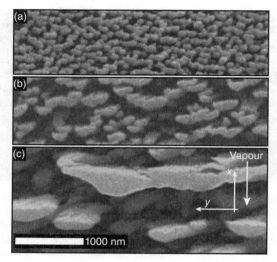

Figure 2.24 Looking down the column axis of slanted post films (SiO$_2$, $\alpha = 87°$) with thickness (a) 300 nm, (b) 1100 nm and (c) 3000 nm (each at the same magnification). The elongation of the column cross-section (aspect ratio) increases as deposition continues, driven by column fanning and the chaining together of adjacent columns. Reproduced with permission from [40]. Copyright 2002, Cambridge University Press.

is actually crescent shaped (Figures 2.23b and 2.24c). This additional asymmetry arises due to the shape of the shadow cast on the column by its neighbour at highly oblique angles. This leads to reduced growth on the underside of the column apex, promoting the formation of a crescent-shaped cross-section that is concave towards the deposition source. Along with fanning of the individual columns, the lateral growth allows neighbouring columns to touch and chain together, whereas shadowing prevents column merger along the x-direction. This produces a preferential bundling of the columnar microstructure in the y-direction. Together, the fanning and chaining mechanisms produce a highly elongated, fan-like columnar morphology. This pronounced structural anisotropy is reflected in the characteristics of the film, and many of the physical properties exhibit a related anisotropy. Advanced GLAD motion algorithms have been devised to provide 2D shadowing effects by controlled φ rotations. Motions such as phisweep can be used to suppress the fanning and broadening effects typically seen in basic GLAD (Section 2.7.3). Equally, motions such as serial bideposition can be used to maximize the shadowing and structural asymmetry, thus enhancing anisotropies in the physical properties (Section 5.3.3).

2.7 ADVANCED COLUMN STEERING ALGORITHMS

In the basic GLAD approach, simple substrate motions, tilting the substrate to glancing α and rotating the substrate in φ, are used to generate a number of distinct columnar geometries by manipulating the shadowing environment. It is possible, however, to go beyond these basics and develop substantially more sophisticated motion algorithms. These advanced methods are based on careful consideration of the shadowing geometry and the accompanying growth characteristics. Altogether, they provide greater control over film microstructure, thus increasing fabrication precision and providing access to alternative microstructures.

2.7.1 β variations in zigzag microstructures

When depositing onto a flat substrate surface the deposition angle α is simply equal to the substrate tilt relative to the vapour source. However, when depositing onto a previously grown columnar structure the pre-existing surface topography complicates the local deposition geometry. This leads to a situation where the effective deposition angle (α') can be different than α and can vary over the surface. Because the ballistic shadowing is geometric, it is sensitive to these local surface variations. As a result, the growth dynamics are altered which correspondingly impacts the columnar microstructure of the deposited film.

These effects arise in the zigzag column microstructure, as shown in Figure 2.25. This microstructure is produced by rapid 180° φ rotations after fixed growth intervals (Figure 2.14b), creating a series of tilted column sublayers with alternating orientation. Close examination of the column arms reveals that the column tilt angle β is not uniform but varies systematically from sublayer to sublayer. The second column in particular is inclined at a significantly greater angle than the first ($\beta_2 > \beta_1$), an observation that was conclusively verified by Harris *et al.* using image processing and autocorrelation analysis [33]. (This and other image analysis techniques are discussed in Section 4.2.2.) While the change is most dramatic between the first and second columnar layers, the subsequent columnar sections also exhibit β larger than the initial layer.

The β variation in the zigzag microstructure is related to the change in the local deposition geometry created when the substrate is rotated in φ. The initial column arm is deposited onto a flat substrate with a globally uniform α, yielding a corresponding β. After the first columnar arm of the microstructure is fabricated and the substrate is rotated 180°, the incident vapour encounters a surface with a different orientation as deposition is now performed on to the

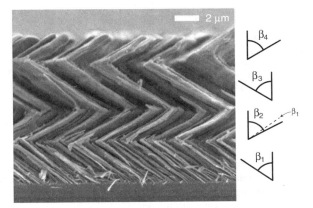

Figure 2.25   The zigzag microstructure, realized by depositing at a fixed oblique angle α with rapid 180° φ rotations after fixed growth intervals. Each growth interval deposits a columnar arm with a tilt angle β that varies systematically from arm to arm. Reprinted with permission from [33]. Copyright 2002, American Vacuum Society.

column sidewall near the apex. This altered deposition geometry is illustrated in Figure 2.26a). When depositing at α onto a column tilted at β, the effective local deposition angle is $\alpha' = \alpha + \beta - 90°$. Using the tangent rule (Equation 2.1), the deposited column will therefore grow with an effective tilt angle β' given by

$$\beta' = \arctan\left[\frac{\tan(\alpha')}{2}\right]. \tag{2.8}$$

Generalizing to the case of deposition onto the $(n-1)$th column arm and recasting this equation relative to the substrate normal leads to the recursive relationship

$$\beta_n = \arctan\left[\frac{\tan(\alpha + \beta_{n-1} - 90°)}{2}\right] - \beta_{n-1} + 90°. \tag{2.9}$$

Note that Equations 2.8 and 2.9 can be redeveloped for any of the previously discussed β models (Section 2.4.2) in a similar fashion. Here, the analysis follows Ref. [33].

Using Equation 2.9 we can recursively predict the column tilt angle of every arm in the zigzag microstructure. Figure 2.26b shows how β varies from between arms, depending on the deposition angle. In constructing these curves, the initial column angle was estimated via Equation 2.2 as the large α case is considered. After the initial layer, the local α' is $<70°$ and the tangent rule is used to predict subsequent β', as in Equation 2.9. The most significant β change occurs between the first and second column arms. Afterwards, β_n tends towards a fixed value with a settling time dependent upon the particular α. As can be seen,

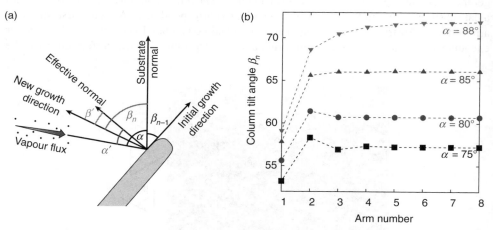

Figure 2.26 (a) A schematic depiction of the local growth geometry when depositing onto a pre-existing tilted column. The column apex surface defines an effective deposition angle α' based on the angle between the effective surface normal and the incident vapour flux. The new growth direction is tilted from the effective surface normal at an angle β', and subsequent columns are tilted at an angle β_n that is different from the previously deposited column β_{n-1}. (b) Because of the changing local deposition geometry, subsequently deposited column arms of a zigzag microstructure exhibit a systematic variation in column angle. The magnitude and behaviour of the β changes depend on the specific initial α.

the convergence is slow at very large α, with several sublayers required to reach equilibrium. At smaller angles, β_n displays a damped-oscillatory behaviour as it progresses towards a steady-state β value.

Based upon these observations and analysis, a corrected substrate algorithm can be designed in order to fabricate a zigzag microstructure with a fixed, uniform β. The concept here is to deposit the first column arm at the initial α and then deposit the subsequent arms at a modified angle α_m to compensate for the local change in deposition geometry. Mathematically, the corresponding requirement is $\beta_n = \beta_{n-1} = \beta$. When substituted into Equation 2.9, some simple algebra yields an expression for α_m:

$$\alpha_m = \arctan[2\tan(2\beta - 90°)] - \beta + 90°. \tag{2.10}$$

For example, depositing onto a pre-existing column layer with $\beta = 60°$ requires depositing all subsequent layers at $\alpha_m = 79°$ to maintain a constant β throughout the film. If the initial β is unknown, as is often the case in preliminary experimentation, it can be estimated using one of the relationships described in Section 2.4.2.

2.7.2 Spin–pause/two-phase substrate rotation: decoupling β and film density

As was previously discussed in Section 2.4, GLAD provides the ability to control the film density and column angle β over a wide range through the deposition angle α. However, it quickly becomes apparent that these two structural parameters are coupled through the growth mechanics and cannot be independently controlled through the substrate angle α. When manipulating the microstructure via α alone, a highly inclined columnar microstructure can only be realized in a low-density film. Conversely, a denser film always accompanies a more vertically oriented microstructure. To provide greater design flexibility and access alternative column geometries, it is desirable to uncouple these two structural parameters. A route to such decoupling is suggested by the vertical column microstructure (Section 2.5.2). In realizing this microstructure, rapid φ rotation of the substrate averages out the lateral growth direction of the columns, producing net column growth in the vertical direction and $\beta = 0$ for any α value. Therefore, the column orientation in this microstructure is entirely independent of α, which can be used then to separately control the film density.

Following this concept, Robbie et al. developed a generalized φ-rotation algorithm known as spin–pause that enables control over the column growth direction independently of the α-determined density [43]. The principle underlying this method is to rotate the substrate during deposition but with an angular velocity that changes over the course of a revolution. Whereas a constant, symmetric substrate rotation profile produces a time-averaged vertical column growth, applying an asymmetric angular velocity profile biases the net growth direction. A tilted microstructure can therefore be fabricated and, by controlling the rotation profile, the column tilt angle can be tuned anywhere between zero (vertical posts produced by symmetric rotation) and the natural inclination angle (tilted columns produced with no substrate rotation). In all cases, the deposition is performed at the same α and the density is constant between the resulting microstructures. The spin–pause approach was followed by the two-phase rotation technique, developed by Ye *et al.* [44], which is based on a similar asymmetric rotation profile as explained below.

In the spin–pause technique, each substrate revolution is divided into a spin and a pause interval. During the pause step, φ is held constant over a fixed growth interval P_1 (the major pitch). During the spin step, φ is rotated through $360°$ over a growth interval P_2 (the minor pitch). The major and minor pitches are the primary control parameters for the spin–pause algorithm. Varying the spin–pause duty cycle, corresponding to the ratio $P_1/(P_1 + P_2)$, sets the relative amounts of lateral and vertical growth per revolution. The net growth direction can thus be controlled, setting the β of the resulting microstructure. Figure 2.27 demonstrates the control provided by the spin–pause technique, showing columnar microstructures produced by SiO deposition at $\alpha = 86°$ with $P_1/(P_1 + P_2)$ equal to (a) 1, (b) 0.7 and (c) 0.5. The first result corresponds to basic GLAD, where the substrate is stationary and β is maximized. In

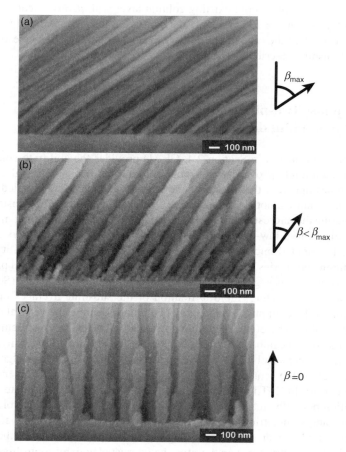

Figure 2.27   SEM images of SiO columar microstructures deposited at $\alpha = 86°$. Films were fabricated using the spin–pause technique with $P_1/(P_1 + P_2)$ equal to (a) 1, (b) 0.7 and (c) 0.5. The spin–pause duty cycle determines the column growth angle β without affecting the film density ρ_n, thus providing independent control over both parameters. Reprinted with permission from [43]. Copyright 1998, American Vacuum Society.

the second case, the spin growth interval pulls the net deposition angle towards the substrate normal, producing a film with equivalent ρ_n as in (a) but having a reduced β. In the third case, the 50% duty cycle yields net vertical growth and $\beta = 0$. Note that, to realize a well-defined column, the minor pitch P_2 should be less than the column width, as discussed in Section 2.5.2. Otherwise, the spin interval of growth will induce a helical component to the resulting microstructure. In Figure 2.27b and c, P_2 was set to 20 nm to observe this constraint. However, relaxing this restriction could yield interesting hybrid microstructures with helical and columnar components.

An example of the spin–pause motion algorithm is provided in Figure 2.28a and b, showing the α and φ trajectories as a function of deposited film thickness. The deposition angle remains fixed (at 85° in this example) throughout film growth. The pause and spin

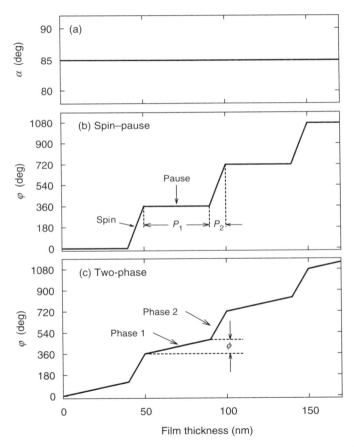

Figure 2.28 Substrate motions algorithms describing the spin–pause and two-phase substrate rotation techniques. (a) In both techniques, α remains fixed during deposition and is used set the film density. (b) During the spin–pause algorithm, the φ motion is divided into a pause interval of length P_1 where φ is stationary, and a spin interval of length P_2 where φ quickly rotates through 360°. (c) In the two-phase algorithm, the φ motion is divided into a slow rotation phase (phase 1) and a fast rotation phase (phase 2). As indicated, φ changes by ϕ during phase 1.

intervals are specified by P_1 and P_2, which are set to 40 nm and 10 nm respectively, yielding an 80% duty cycle. The substrate is initially fixed at $\varphi = 0°$ for the $P_1 = 40$ nm growth interval, making up the pause step. This is followed by the spin step, where the substrate is uniformly rotated to $\varphi = 360°$ over the $P_2 = 10$ nm growth section. This process is repeated every subsequent rotation.

In the two-phase rotation approach, the substrate revolution is partitioned in two intervals, phase 1 and phase 2, respectively covering angular ranges ϕ and $360° - \phi$ and having rotation speeds of ω_1 and ω_2. This approach produces the substrate motion algorithm shown in Figure 2.28a and c. As with the spin–pause example, α is fixed over the course of the deposition, while the φ rotation speed is changed between the phase 1 and phase 2 intervals, as indicated. Investigated values for these control parameters are ω_2/ω_1 from 3 to 20 (with $\omega_1 = 9.4$ mrad/s at a growth rate of 0.26 nm/s) and ϕ from 45° to 225° [44]. Changing the ratio ω_2/ω_1 and/or the angular range ϕ varies the growth direction duty cycle, allowing the column tilt angle to be controlled independent of α. In their report on two-phase rotation, the authors integrate the lateral and vertical growth components over the rotation cycle to develop a useful equation for the net deposition angle [44]:

$$\alpha_{net} = 2 \tan \alpha \sin \left(\frac{\phi}{2} \right) \frac{\frac{\omega_2}{\omega_1} - 1}{2\pi + \left(\frac{\omega_2}{\omega_1} - 1 \right) \phi}. \tag{2.11}$$

Note that α_{net} only affects β, and it is α that primarily determines ρ_n. Figure 2.29 plots this equation as the relative sector spin-speeds are varied for $\phi = 180°$, 90° and 20°. In all three situations, $\alpha = 85°$. In the special case of $\omega_2/\omega_1 = 1$ the substrate rotation is symmetric and the averaged growth direction is along the substrate normal. As ω_2/ω_1 increases, the rotation profile becomes increasingly biased towards the first sector of the rotation cycle, tilting the net growth direction away from the normal. Correspondingly, α_{net} increases but also remains

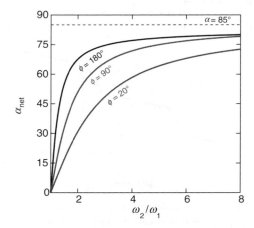

Figure 2.29 With two-phase substrate rotation, the net deposition angle α_{net} is controlled by both the spin speed ratio ω_2/ω_1 and the sector angular range ϕ, as shown by plots of Equation 2.11 of the text. By manipulating these control parameters, α_{net} can be varied from 0° up to the actual deposition angle α (85° in this case).

less than the true deposition angle α. Decreasing the angular range of the first section (ϕ) also decreases α_{net} as it alters the asymmetry of the rotation profile.

The similarity between the spin–pause and two-phase techniques is apparent when their respective motion algorithms are compared. The primary difference between the two methods is that the spin–pause approach implements a stationary interval whereas in the two-phase technique the substrate is always in motion. However, both techniques are successful in decoupling the column growth orientation β from the film density, providing greater freedom in realizing different microstructure designs.

2.7.3 Phisweep motion: competition-resilient structure growth

Ballistic shadowing is central to GLAD technology, enabling the directed self-assembly of a wide range of columnar morphologies. Shadowing seizes upon the structural variations that arise naturally between growing columns, thus initiating intercolumn competition and nonequilibrium growth. GLAD correspondingly exhibits growth characteristics that promote a nonuniform microstructure, such as column extinction and broadening. A general approach to reducing column competition and improving uniformity involves depositing onto engineered templates, a topic discussed in Chapter 3. Unfortunately, this approach introduces additional pre-deposition processing steps to GLAD, often at significant expense. However, it is also possible to use specific substrate motions to create a growth mode that is resilient to shadowing-induced instabilities. While this substrate motion algorithm, known as phisweep [45], has primarily been used to suppress column fanning in templated GLAD growth (see Section 3.4.2), it also leads to a remarkable suppression of competitive growth effects when used on flat substrates. A strikingly different columnar microstructure can be produced when the phisweep algorithm is correctly implemented.

During phisweep, the substrate is held at the desired α and then quickly rotated from side to side (between φ angles of $\pm\gamma$) after regular growth intervals q. The parameters γ and q are respectively called the sweep angle and sweep pitch. The phisweep motion algorithm is schematically depicted in Figure 2.30. From the initial starting position (Figure 2.30a; $\varphi = 0°$ and $t = 0$), the substrate is then rotated to $\varphi = \gamma$ (Figure 2.30b) and held there while an incremental growth interval of thickness q is deposited. The substrate is then rotated to $\varphi = -\gamma$ (Figure 2.30c) for another growth interval q. Returning the substrate to $\varphi = \gamma$ completes one period of the phisweep motion (Figure 2.30d). The algorithm can be repeated (Figure 2.30e) to progressively deposit the desired structure out of linear growth segments of thickness q (Figure 2.30f).

The segmented nature of the growth algorithm produces a remarkable effect when q is appropriately tuned. The early columnar features that form during preliminary GLAD growth are nominally on the order of tens of nanometres in diameter. For q smaller than these typical protocolumn widths, the vapour direction alternates so quickly that growth is haphazard and the protocolumns broaden and evolve as in traditional GLAD (Figure 2.31a). However, for q slightly larger than the width of these protocolumns, the growth interval allows a new protocolumn to develop at the apex of the previous column (Figure 2.31b). The sweep pitch must also be sufficiently small such that, before the next growth interval, the new columns do not grow long enough for shadowing instabilities to set in and initiate the competitive growth process. The sweep angle γ must also be large enough (typically greater than 30°)

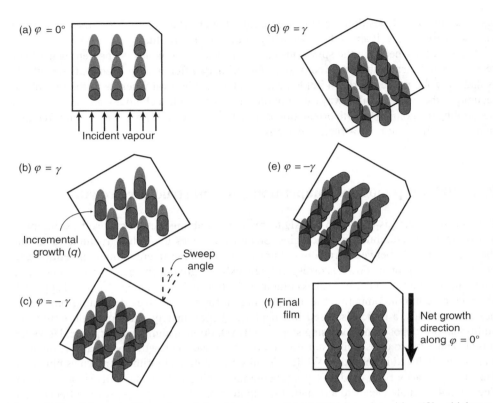

Figure 2.30 Schematic illustration of the phisweep motion. (a) Prior to deposition (film thickness $t < 0$), the substrate is homed to the $\varphi = 0°$ position. (b) At the start of deposition ($t = 0$), the substrate is quickly rotated to $\varphi = \gamma$. (c) After the first sweep pitch is deposited ($t = q$), the substrate is quickly rotated to $\varphi = -\gamma$ to deposit another incremental growth section, thus completing one period of the phisweep algorithm. The net film growth is along the central sweep axis since the two sweeps are symmetric. The algorithm is repeated (d,e) as many times as needed to (f) realize the desired microstructure.

that the new orientation is sufficiently orthogonal to produce a new shadowing environment and overcome the effect of one-dimensional shadowing (Section 2.6.4).

When the q and γ parameters are correctly tuned following these guidelines, a brand new column is nucleated at the start of each growth interval. The evolutionary growth process is effectively reset after every sweep, allowing a constant column width to be maintained and preventing column extinction events. Therefore, the phisweep algorithm realizes a growth mode that is resilient to competition and avoids the characteristic nonequilibrium growth generally observed in GLAD. The robustness of the phisweep growth mode is demonstrated by Figure 2.32, showing top-down SEM images of Si films grown using the phisweep technique at $\alpha = 84°$. The top row shows films deposited with $q = 45$ nm and $\gamma = 45°$ after 15 nm (Figure 2.32a) and 2645 nm (Figure 2.32b) of film growth. Because the competition is suppressed by phisweep, the structure remains remarkably consistent. The initial microstructure formed during preliminary growth, consisting of ~30 nm columnar features separated by a similarly sized void network, is incredibly well preserved even after significant growth has

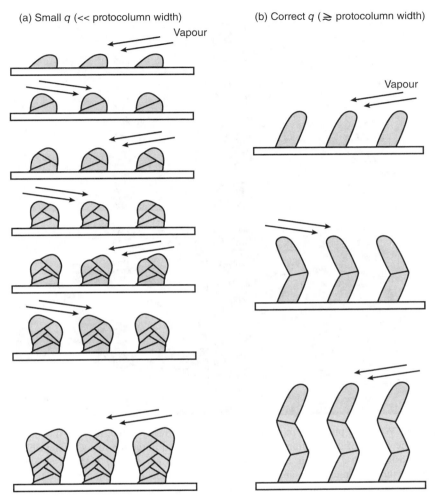

(a) Small q (<< protocolumn width)

(b) Correct q ($\gtrsim$ protocolumn width)

Vapour

Vapour

Figure 2.31 Schematic illustration of the phisweep growth mechanics. (a) When q is too small relative to the protocolumn width the incident vapour cannot fully form a new column and the structure gradually broadens. (b) However, when q is slightly larger than the protocolumn width a new column is formed at each growth interval. Column evolution is thus reset after every sweep interval and competitive growth characteristics are suppressed. Reproduced from [35]. With permission from H.M.O. Jensen.

occurred. However, this effect is only observed when the phisweep parameters are optimized, as shown by the second row of SEM images, where films were deposited with $q = 15$ nm and $\gamma = 45°$. In this second case, q is too small and after 270 nm of film growth (Figure 2.32c) intercolumn competition has already broadened the initial columnar features. Driven by competitive growth effects, the film morphology changes throughout growth and after 2730 nm of deposition (Figure 2.32d), the columnar microstructure is completely different.

Because the substrate rotation is symmetric, the phisweep algorithm yields a net-growth direction oriented along a central axis lying between $\pm\gamma$. Therefore, GLAD structures based

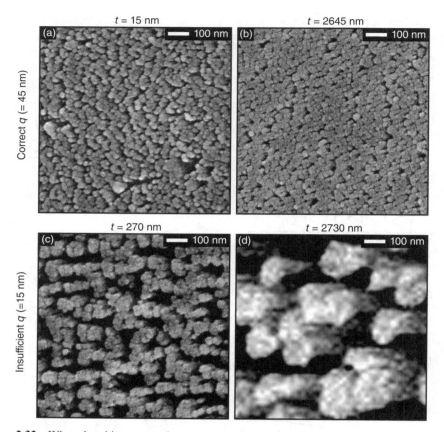

Figure 2.32   When the phisweep motion parameters are correctly tuned, the film microstructure is resilient to the effects of competition. Such a case is presented in the top row showing SEM images of films deposited with $q = 45$ nm and $\gamma = 45°$. After (a) 15 nm and (b) 2645 nm of film growth, the microstructure is virtually the same and unaffected by typical competition effects. If the motion parameters are incorrect, column competition is not suppressed and the film morphology will evolve as in conventional GLAD. This is shown in the bottom row of SEM images, where an insufficient sweep pitch ($q = 15$ nm) was used. Significant competition is apparent already after (c) 270 nm of growth, and extreme morphology changes are observed after (d) 2730 nm of growth. Aside from thickness and q, all four films are identical (Si phisweep square spirals deposited at $\alpha = 84°$). Reproduced from [35]. With permission from H.M.O. Jensen.

on linear growth sections, such as slanted columns, zigzag structures and polygonal helices, can be segmented into a phisweep equivalent. Implementations of the phisweep motion algorithm are depicted in Figure 2.33, showing the substrate motions required to fabricate (a) a slanted column and (b) a square-spiral helix. The dashed line shows the conventional GLAD algorithm used to produce these structures. The corresponding phisweep motion is created by breaking up the conventional motion into discrete intervals of length q with the substrate alternately rotated by in $\varphi \pm \gamma$.

While phisweep exhibits a remarkable suppression of column competition effects, the rapid alternation of the growth direction imprints a fine zigzag microstructure onto the

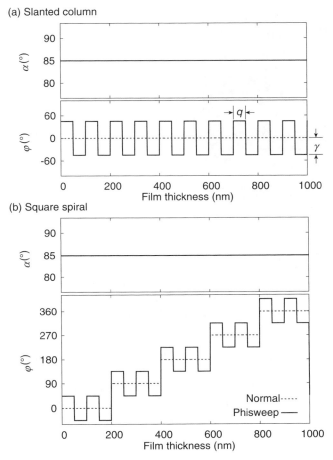

Figure 2.33 Substrate motion algorithms for phisweep (a) slanted columns and (b) square-spiral helices. The phisweep parameters in both algorithms are $q = 50$ nm and $\gamma = 45°$, and the substrate is held at a constant $\alpha = 85°$. The dashed line shows the corresponding basic GLAD algorithm.

nanoscale columns. Figure 2.34a shows an SEM image of the exposed cross-section of a phisweep film, revealing the fine structure that is superimposed on the tilted columnar morphology. The periodicity of the zigzag structuring (measured along the substrate normal) is on the order of $\sim 2q$, with the exact value depending on γ and the column tilt. The phisweep φ motion also changes the net deposition angle and affects the column tilt angle in a manner similar to the spin–pause and two-phase rotation algorithms discussed in Section 2.7.2. Because the two alternating growth sections are offset by 2γ, the column tilt angle will be different from that produced by the traditional GLAD algorithm. For instance, when $\gamma = 90°$ the net α and β will correspondingly be zero. Through geometric analysis, the phisweep column tilt angle β_{PS} is given by [46]

$$\beta_{PS} = \tan \beta_{TG} \cos \gamma, \tag{2.12}$$

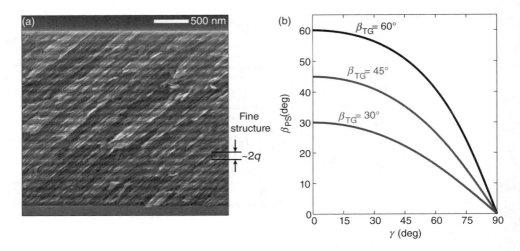

Figure 2.34   Secondary effects of the phisweep motion algorithm. (a) While the rapid substrate sweep creates competition-resilient column growth, an underlying zigzag nanostructure is also imposed with a period $\sim 2q$. (b) The alternating growth direction changes the resulting column tilt angle, producing columns with a tilt angle β_{PS} that is less than the column angle produced using traditional GLAD β_{TG}. The change depends on the sweep angle γ. (a) Reproduced from [47]. With permission from D.A. Gish.

where β_{TG} is the traditional GLAD column tilt angle. This expression is plotted in Figure 2.34b for different γ, showing that the resulting columnar features are increasingly vertical as the sweep angle increases. These secondary effects, the fine zigzag microstructure and the change in column orientation, are a product of the alternating growth direction fundamental to phisweep and must be considered when evaluating the phisweep algorithm.

2.8 ADDITIONAL CONTROL OVER FILM GROWTH AND STRUCTURE

2.8.1 High-temperature glancing angle deposition growth

GLAD is generally performed at low substrate temperatures to ensure adatom surface diffusion is limited and unable to smooth out the roughening caused by ballistic shadowing. These conditions produce films with a zone I columnar morphology in the context of conventional SZMs (Section 2.2.3) and are key requirements of basic GLAD structure formation. However, there has recently been a heightened interest in developing high-temperature GLAD (HT-GLAD) processes that couple the extreme shadowing effects present at high α with the high surface mobility provided by increased substrate temperature [48–56]. Several reports have shown that unique microstructures can emerge in HT-GLAD indicating that substrate temperature control provides an additional degree of freedom for morphology engineering. Furthermore, higher temperature deposition processes provide greater control over important atomic-scale material characteristics, such as crystallinity and electronic mobility, thus enabling optimization of these intrinsic material parameters for device applications. When developing HT-GLAD processes, it is critical to understand the morphological characteristics of films deposited under extreme shadowing and high-mobility conditions.

Mukherjee and Gall provided an important contribution towards this understanding by developing a general SZM for temperature-induced morphological changes in GLAD films [57] based on surveying microstructural data from several HT-GLAD investigations that examine many different materials. Their systematic investigation revealed that HT-GLAD films can be qualitatively classified according to five distinct morphological zones (Figure 2.35): high-aspect-ratio *rods*, highly broadened *columns*, *protrusions* that extend above the surrounding film, a dense *equiaxed grain* layer and long *whisker* structures. The specific morphology present in the GLAD film is primarily determined by the homologous substrate temperature ($\theta = T_s/T_m$) as shown by Figure 2.36a, which plots the microstructural information against θ and T_m for a wide range of materials [57]. The film morphology

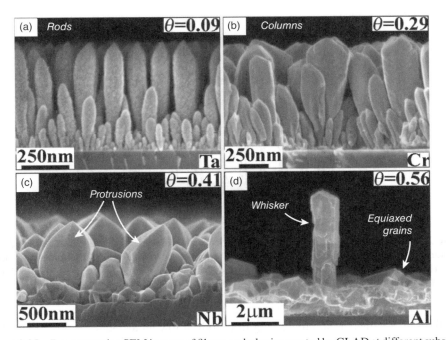

Figure 2.35 Representative SEM images of film morphologies created by GLAD at different substrate temperatures T_s, measured as $\theta = T_s/T_m$, where T_m is the material melting point. (a) The low-mobility conditions accompanying low-temperature deposition ($\theta = 0.09$, Ta material) lead to the formation of a *rod* morphology consistent with typical GLAD conditions. (b) Mobility increases with T_s and changes the nucleation dynamics during growth at a sufficiently high temperature ($\theta = 0.29$, Cr material). However, shadowing is still the dominant growth mechanism, and the resulting film consists of highly broadened *column* structures. (c) With further T_s increases ($\theta = 0.41$, Nb material), the increased surface diffusion rates drive rapid vertical and lateral growth of substrate features that outcompete neighbouring features in a runaway growth process, creating irregular *protrusion* structures on the surface. (d) At even higher T_s ($\theta = 0.56$, Al material), surface and bulk diffusion processes eventually overcome extreme shadowing effects and lead to the formation of an *equiaxed grain* film, with sparse high-aspect-ratio *whisker* structures. All films were deposited at $\alpha = 84°$ and with rapid φ rotation. Reprinted from [57] with permission from Elsevier.

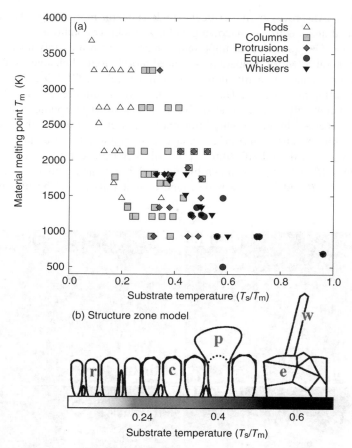

Figure 2.36 (a) Summary of reported morphological data observed in GLAD films, grouped into the indicated zones, as a function of the substrate temperature (normalized to the material melting point). (b) SZM constructed for HT-GLAD films based on the reported data. Reprinted from [57] with permission from Elsevier.

transitions as θ is increased, progressing from rods at low temperature ($\theta \lesssim 0.20$ to 0.26), to columns ($0.2 \lesssim \theta \lesssim 0.4$), to protrusions ($0.4 \lesssim \theta \lesssim 0.6$), to equiaxed grains and whiskers ($\theta \gtrsim 0.5$). Based on the clustering of the different morphologies and the consistent (although broad) transition temperatures between the morphology zones, Mukherjee and Gall developed an SZM for the HT-GLAD process shown in Figure 2.36b, providing a schematic depiction of the film morphology as a function of the substrate temperature.

The different morphologies and the strong dependence on substrate temperature are a product of the competition between extreme shadowing and atomic diffusion and the relative magnitude of these growth effects during film deposition. Figure 2.35 shows representative side-view SEM images of the different morphologies in the HT-GLAD SZM, showing (a) a Ta film deposited at $\theta = 0.09$, (b) a Cr film deposited at $\theta = 0.29$, (c) an Nb film deposited at $\theta = 0.41$, (d) and an Al film deposited at $\theta = 0.56$. Each film was deposited at $\alpha = 84°$, thus producing the extreme shadowing conditions, and with a rapid continuous φ rotation consistent with the vertical column motion algorithm (Section 2.5.2). As shown by these SEM images, modifying the substrate temperature during high α growth produces striking changes in the resulting film microstructure. Low-temperature growth conditions (Figure 2.35a; $\theta = 0.09$) produce the typical GLAD microstructure of vertically oriented columns, termed *rods* following the nomenclature of Mukherjee and Gall [57], and the film exhibits structural characteristics expected of a low-temperature GLAD process, such as extinction and structure broadening. Increasing the substrate temperature leads to a reduction in the structure broadening rate up until a threshold temperature ($\theta \sim 0.20$ to 0.26) whereupon the broadening rate increases dramatically, driven by an increase in surface roughening associated with a transition from a 2D to 3D nucleation mode (Section 4.3.2). GLAD films prepared under these conditions (Figure 2.35b; $\theta = 0.29$) exhibit highly non-uniform structures, termed *columns*, that constitute the second morphology class of the SZM. As the temperature is increased further, the extreme shadowing conditions continue to produce isolated structural features. However, new shadowing instabilities develop during film growth due to the increased surface diffusion rates associated with higher substrate temperatures. Larger structural features capture greater amounts of incident vapour flux and experience exacerbated vertical and lateral growth rates due to the high surface mobility. The enhanced growth rates further reinforce the runaway growth of these structures as they capture proportionally greater amounts of flux. The resulting film morphology is predominantly columnar (Figure 2.35c; $\theta = 0.41$), but contains many irregular structural features, termed *protrusions*, that are substantially wider and taller than the surrounding structures. The substrate temperature necessary to produce the protrusion morphology varies from $\theta \sim 0.3$ to 0.5, being primarily determined by the specific material system and the corresponding surface diffusion activation energy. The final two morphology classes are observed only at very high temperatures ($\theta > 0.5$), where the typical columnar film structure is lost as surface and bulk diffusion effects dominate over shadowing even as α approaches $90°$. In this temperature regime the substrate surface is covered by a rough and faceted layer (Figure 2.35d; $\theta = 0.56$), termed an *equiaxed grain* microstructure, which is qualitatively the same as the zone III morphology found in the conventional SZM. However, unique to the morphology of HT-GLAD films is the presence of unusual high-aspect-ratio structures, termed *whiskers*, that sparsely decorate the surface. While the origin of these whiskers, first observed by Suzuki *et al.* [54], is not clearly understood, both high surface mobility and extreme shadowing conditions are essential to their formation as the structures only form at $\theta > 0.5$ and large α. Suggested growth mechanisms include an increased whisker nucleation rate associated with surface roughness and oblique deposition angles, or the geometrically increased vapour flux that is captured by the whiskers at high α [54, 57].

2.8.2 Multimaterial structures: co-deposition processes

While the GLAD process is most often used to fabricate nanostructured films of a single material, it is possible to create nanostructures with varying composition either by sequential or simultaneous co-deposition of different materials. Because the wide material compatibility of the GLAD process (Section 1.2.2) enables so many possible material combinations, co-deposition approaches augment the nanostructuring capabilities of the GLAD technique with compositional engineering [59]. GLAD co-deposition therefore allows design of advanced functional nanostructures that simultaneously exploit different material properties. Many material combinations have been demonstrated, such as Co–Ag [60], Co–Cr [60,61], Co–Cu [62], Co–Ta [60], MgF$_2$–Ag [63], Pt–metal oxide [64], Si–Ag [65], Si–Cr [66], Si–Cu [67], Si–Ni [68], Si–Pt [65], Si–Ta [69] and TiO$_2$–Ni–Pt [70], thus clearly demonstrating the flexibility provided by GLAD co-deposition.

Implementing a co-deposition GLAD process requires a deposition system where the source can be changed, thus enabling sequential material deposition, or one with multiple sources, which enable both sequential and simultaneous co-deposition. Figure 2.37 presents a series of schematics and TEM images that illustrate the compositional and nanostructural engineering possibilities provided by GLAD co-deposition, using Si and Ta in this case [58]. In Figure 2.37a and d, the substrate is held at a large α and continuously φ-rotated to produce a vertical column structure. Then a series of Ta and Si sublayers are fabricated by sequentially depositing from Ta and Si vapour sources, with the sublayer thickness controlled by the deposition time. Multimaterial zigzag column structures can be fabricated as well (Figure 2.37b and e) by sequentially depositing Si and Ta vapour at high α from either side of

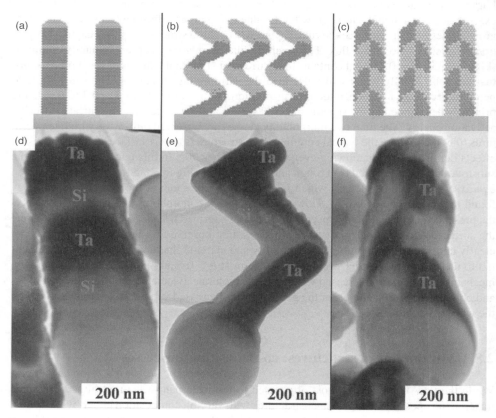

Figure 2.37 Schematics (a–c) and TEM images (d–f) of Ta–Si nanostructures fabricated using a co-deposition GLAD process. (a,d) Vertically oriented and (b,e) zigzag columns with alternating Ta and Si components are created by sequentially switching the source material during column growth. (c,f) A chequerboard column structure is created by simultaneously depositing Ta and Si from opposing sides. Reprinted from [58] with permission from Elsevier.

the substrate normal. Figure 2.37c and f presents a chequerboard patterned column created by co-depositing the Si and Ta vapour simultaneously at high α from opposite sides with intermittent $180°$ φ rotations. Each vertical column contains $100 \times 100 \times 100$ nm^3 Si and Ta subunits alternately stacked vertically and laterally, generating an unusual compositionally graded nanostructure.

REFERENCES

[1] Mattox, D.M. (2010) *Handbook of Physical Vapor Deposition (PVD) Processing*, 2nd edn, William Andrew, Oxford, UK.

[2] Ohring, M. (2002) *Materials Science of Thin Films*, 2nd edn, Academic Press, San Diego, CA.

[3] Chopra, K.L. (1969) *Thin Film Phenomena*, McGraw-Hill.

[4] Venables, J.A., Spiller, G.D.T., and Hanbücken, M. (1984) Nucleation and growth of thin films. *Reports on Progress in Physics*, **47**, 399–459.

[5] Zhang, Z. and Lagally, M.G. (1997) Atomistic processes in the early stages of thin-film growth. *Science*, **276** (5311), 377–383.

[6] Ratsch, C. and Venables, J.A. (2003) Nucleation theory and the early stages of thin film growth. *Journal of Vacuum Science and Technology A*, **21** (5), S96–S109.

[7] Petrov, I., Barna, P.B., Hultman, L., and Greene, J. (2003) Microstructural evolution during film growth. *Journal of Vacuum Science and Technology A*, **21**, S117–S128.

[8] Movchan, B.A. and Demchishin, A.V. (1969) Study of the structure and properties of thick vacuum condensates of nickel, titanium, tungsten, aluminum oxide and zirconium oxide. *Fizika Metallov i Metallovedenie*, **28**, 83.

[9] Thornton, J.A. (1986) The microstructure of sputter deposited coatings. *Journal of Vacuum Science and Technology A*, **4**, 3059–3065.

[10] Barna, P. and Adamik, M. (1998) Fundamental structure forming phenomena of polycrystalline films and the structure zone models. *Thin Solid Films*, **317**, 27–33.

[11] Messier, R., Giri, A.P., and Roy, R.A. (1984) Revised structure zone model for thin film physical structure. *Journal of Vacuum Science and Technology A*, **2**, 500–503.

[12] Steele, J.J., Taschuk, M.T., and Brett, M.J. (2008) Nanostructured metal oxide thin films for humidity sensors. *IEEE Sensors Journal*, **8** (8), 1422–1429.

[13] Jensen, M.O. and Brett, M.J. (2005) Periodically structured glancing angle deposition thin films. *IEEE Transactions on Nanotechnology*, **4** (2), 269–277.

[14] Karunasiri, R.P.U., Bruinsma, R., and Rudnick, J. (1989) Thin-film growth and the shadow instability. *Physical Review Letters*, **62**, 788–791.

[15] Summers, M.A. (2009) Periodic thin films by glancing angle deposition, PhD thesis, University of Alberta.

[16] Van Popta, A.C. (2009) Optical materials and devices fabricated by glancing angle deposition, PhD thesis, University of Alberta.

[17] Hrudey, P.C.P. (2006) Luminescent chiral thin films fabricted using glancing angle deposition, PhD thesis, University of Alberta.

[18] Robbie, K. and Brett, M.J. (1997) Sculptured thin films and glancing angle deposition: Growth mechanics and applications. *Journal of Vacuum Science and Technology A*, **15**, 1460–1465.

[19] Hawkeye, M.M. and Brett, M.J. (2006) Glancing angle deposition: fabrication, properties, and applications of micro- and nanostructured thin films. *Journal of Vacuum Science and Technology A*, **25**, 1317–1335.

[20] Amassian, A., Kaminska, K., Suzuki, M. *et al.* (2007) Onset of shadowing-dominated growth in glancing angle deposition. *Applied Physics Letters*, **91**, 173114.

[21] Nieuwenhuizen, J.M. and Haanstra, H.B. (1966) Microfractography of thin films. *Philips Technical Review*, **27**, 87–91.

[22] Dirks, A.G. and Leamy, H.J. (1977) Columnar microstructure in vapor-deposited thin films. *Thin Solid Films*, **47**, 219–233.

[23] Tait, R.N., Smy, T., and Brett, M.J. (1993) Modelling and characterization of columnar growth in evaporated films. *Thin Solid Films*, **226**, 196–201.

[24] Fujiwara, H., Hara, K., Kamiya, M. *et al.* (1988) Comment on the tangent rule. *Thin Solid Films*, **163**, 387–391.

[25] Sorge, J.B., Taschuk, M.T., Wakefield, N.G. *et al.* (2012) Metal oxide morphology in argon-assisted glancing angle deposition. *Journal of Vacuum Science and Technology A*, **30**, 021507.

[26] Hodgkinson, I., Wu, Q.H., and Hazel, J. (1998) Empirical equations for the principal refractive indices and column angle of obliquely deposited films of tantalum oxide, titanium oxide, and zirconium oxide. *Applied Optics*, **37**, 2653–2659.

[27] Lichter, S. and Chen, J. (1986) Model for columnar microstructure in thin solid films. *Physical Review Letters*, **56**, 1396–1399.

[28] Tanto, B., Ten Eyck, G., and Lu, T.M. (2010) A model for column angle evolution during oblique angle deposition. *Journal of Applied Physics*, **108**, 026107.

[29] Zhu, H., Cao, W., Larsen, G.K. *et al.* (2012) Tilting angle of nanocolumnar films fabricated by oblique angle deposition. *Journal of Vacuum Science and Technology B*, **30**, 030606.

[30] Van Popta, A.C., Sit, J.C., and Brett, M.J. (2004) Optical properties of porous helical thin films and the effects of post-deposition annealing, in *Organic Optoelectronics and Photonics* (eds P.L. Heremans, M. Muccini, and H. Hofstraat), vol. 5464 of *Proceedings of SPIE*, SPIE Press, Bellingham, WA, pp. 198–208.

[31] Robbie, K., Brett, M.J., and Lakhtakia, A. (1996) Chiral sculptured thin films. *Nature*, **384**, 616.

[32] Dick, B., Brett, M.J., and Smy, T. (2003) Controlled growth of periodic pillars by glancing angle deposition. *Journal of Vacuum Science and Technology B*, **21**, 23–28.

[33] Harris, K.D., Vick, D., Smy, T., and Brett, M.J. (2002) Column angle variations in porous chevron thin films. *Journal of Vacuum Science and Technology A*, **20**, 2062–2067.

[34] Van Popta, A.C., Brett, M.J., and Sit, J.C. (2005) Double-handed circular Bragg phenomena in polygonal helix thin films. *Journal of Applied Physics*, **98**, 083517.

[35] Jensen, H.M.O. (2005) Photonic crystal engineering in glancing angle deposition thin films, PhD thesis, University of Alberta.

[36] Toader, O. and John, S. (2001) Proposed square spiral microfabrication architecture for large three-dimensional photonic band gap crystals. *Science*, **292**, 1133–1135.

[37] Kennedy, S.R., Brett, M.J., Toader, O., and John, S. (2002) Fabrication of tetragonal square spiral photonic crystals. *Nano Letters*, **2** (1), 59–62.

[38] Messier, R. and Yehoda, J.E. (1985) Geometry of thin-film morphology. *Journal of Applied Physics*, **58**, 3739–3746.

[39] Bales, G.S., Bruinsma, R., Eklund, E.A. *et al.* (1990) Growth and erosion of thin solid films. *Science*, **249**, 264–268.

[40] Vick, D., Smy, T., and Brett, M.J. (2002) Growth behavior of evaporated porous thin films. *Journal of Materials Research*, **17**, 2904–2911.

[41] Krause, K.M., Vick, D.W., Malac, M., and Brett, M.J. (2010) Taking a little off the top: nanorod array morphology and growth studied by focused ion beam tomography. *Langmuir*, **26**, 17558–17567.

[42] Zhou, C.M. and Gall, D. (2006) Branched Ta nanocolumns grown by glancing angle deposition. *Applied Physics Letters*, **88**, 203117.

[43] Robbie, K., Sit, J.C., and Brett, M.J. (1998) Advanced techniques for glancing angle deposition. *Journal of Vacuum Science and Technology B*, **16** (3), 1115–1122.

[44] Ye, D.X., Zhao, Y.P., Yang, G.R. *et al.* (2002) Manipulating the column tilt angles of nanocolumnar films by glancing-angle deposition. *Nanotechnology*, **13**, 615–618.

[45] Jensen, M.O. and Brett, M.J. (2005) Porosity engineering in glancing angle deposition thin films. *Applied Physics A*, **80**, 763–768.

[46] Gish, D.A., Summers, M.A., and Brett, M.J. (2006) Morphology of periodic nanostructures for photonic crystals grown by glancing angle deposition. *Photonics and Nanostructures – Fundamentals and Applications*, **4**, 23–29.

[47] Gish, D.A. (2010) Morphology control and localized surface plasmon resonance in glancing angle deposition, Master's thesis, University of Alberta.

[48] Deniz, D. and Lad, R.J. (2011) Temperature threshold for nanorod structuring of metal and oxide films grown by glancing angle deposition. *Journal of Vacuum Science and Technology A*, **29**, 011020.

[49] Khare, C., Patzig, C., Gerlach, J.W. *et al.* (2010) Influence of substrate temperature on glancing angle deposited Ag nanorods. *Journal of Vacuum Science and Technology A*, **28**, 1002–1009.

[50] Khare, C., Gerlach, J.W., Weise, M. *et al.* (2011) Growth temperature altered morphology of Ge nanocolumns. *Physica Status Solidi (a)*, **208**, 851–856.

[51] Mukherjee, S. and Gall, D. (2009) Anomalous scaling during glancing angle deposition. *Applied Physics Letters*, **95**, 173106.

[52] Mukherjee, S. and Gall, D. (2010) Power law scaling during physical vapor deposition under extreme shadowing conditions. *Journal of Applied Physics*, **107**, 084301.

[53] Patzig, C. and Rauschenbach, B. (2008) Temperature effect on the glancing angle deposition of Si sculptured thin films. *Journal of Vacuum Science and Technology A*, **26**, 881–886.

[54] Suzuki, M., Nagai, K., Kinoshita, S. *et al.* (2006) Vapor phase growth of Al whiskers induced by glancing angle deposition at high temperature. *Applied Physics Letters*, **89**, 133103.

[55] Suzuki, M., Nagai, K., Kinoshita, S. *et al.* (2007) Morphological evolution of Al whiskers grown by high temperature glancing angle deposition. *Journal of Vacuum Science and Technology A*, **25**, 1098–1102.

[56] Suzuki, M., Kita, R., Hara, H. *et al.* (2010) Growth of metal nanowhiskers on patterned substrate by high temperature glancing angle deposition. *Journal of the Electrochemical Society*, **157**, K34–K38.

[57] Mukherjee, S. and Gall, D. (2013) Structure zone model for extreme shadowing conditions. *Thin Solid Films*, **527**, 158–163.

[58] Zhou, C.N., Li, H.F., and Gall, D. (2008) Multi-component nanostructure design by atomic shadowing. *Thin Solid Films*, **517**, 1214–1218.

[59] He, Y. and Zhao, Y. (2011) Advanced multi-component nanostructures designed by dynamic shadowing growth. *Nanoscale*, **3** (6), 2361–2375.

[60] van Kranenburg, H. and Lodder, C. (1994) Tailoring growth and local composition by oblique-incidence deposition: a review and new experimental data. *Materials Science and Engineering*, **R11**, 295–354.

[61] van Kranenburg, H., Lodder, J.C., Maeda, Y. *et al.* (1990) Microstructure of co-evaporated CoCr films with perpendicular anisotropy. *IEEE Transactions on Magnetics*, **26**, 1620–1622.

[62] Kar, A.K., Morrow, P., Tang, X.T. *et al.* (2007) Epitaxial multilayered Co/Cu ferromagnetic nanocolumns grown by oblique angle deposition. *Nanotechnology*, **18**, 295702.

[63] He, Y., Zhang, Z., Hoffmann, C., and Zhao, Y. (2008) Embedding Ag nanoparticles into MgF$_2$ nanorod arrays. *Advanced Functional Materials*, **18**, 1676–1684.

[64] Watanabe, Y., Hyodo, S.a., Motohiro, T. *et al.* (1995) Catalytic properties of thin films by simultaneous oblique sputter deposition of two materials from different directions. *Thin Solid Films*, **256**, 68–72.

[65] He, Y., Wu, J., and Zhao, Y. (2007) Designing catalytic nanomotors by dynamic shadowing growth. *Nano Letters*, **7**, 1369–1375.

[66] Kesapragada, S.V. and Gall, D. (2006) Two-component nanopillar arrays grown by glancing angle deposition. *Thin Solid Films*, **494**, 234–239.

[67] He, Y., Yang, B., Yang, K. *et al.* (2012) Designing Si-based nanowall arrays by dynamic shadowing growth to tailor the performance of Li-ion battery anodes. *Journal of Materials Chemistry*, **22**, 8294–8303.

[68] He, Y., Fu, J., Zhang, Y. *et al.* (2007) Multilayered Si/Ni nanosprings and their magnetic properties. *Small*, **3**, 153–160.

[69] Zhou, C. and Gall, D. (2008) Two-component nanorod arrays by glancing-angle deposition. *Small*, **4**, 1351–1354.

[70] Gibbs, J.G. and Zhao, Y. (2010) Self-organized multiconstituent catalytic nanomotors. *Small*, **6**, 1656–1662.

3 Creating High-Uniformity Nanostructure Arrays

3.1 INTRODUCTION

Many of the notable characteristics of GLAD growth, evolutionary growth, column broadening and competition, and the lack of planar order, originate from the randomness inherent in the nucleation phase of film growth. As material condenses from the vapour phase and nucleates on the substrate, nanoscopic features of varying size and random location are formed. Shadowing restricts growth to the existing nuclei and the nuclei develop into unevenly sized columns that then compete, again via shadowing, to intercept further incident vapour. However, certain applications demand a very high degree of uniformity and control over column order. To successfully create 3D photonic crystals [1], column arrangement must be periodic, the column-to-column structural variation should be minimal and the column shape must be maintained throughout the film thickness (Figure 3.1a). The boundaries between areas of highly uniform and disordered GLAD nanostructures can be used to define microfluidic channels [2], based on differing flow rates through the two microstructures (Figure 3.1b–d). Other applications, such as nanostructures for magnetic storage and high-surface-area electrodes for solar cells, may also benefit from improved growth control, column uniformity and structural ordering.

Improving structural uniformity requires not only carefully controlling column growth via shadowing, but also engineering the initial nucleation phase. Conceptually, the strategy for controlling nucleation is straightforward: whereas a flat substrate places no restrictions on nucleation sites allowing columns to grow anywhere, pre-exisiting features on the substrate act as initial nucleation sites and *seeds* for column growth. Pre-patterning a substrate with a microscale topography creates initial shadowing zones, thereby controlling where vapour can condense and establishing where columns will grow. Figure 3.2 illustrates this concept where a sub-micrometre topography is created on a flat substrate (Figure 3.2a), dictating the preliminary shadowing environment and controlling initial nucleation during GLAD (Figure 3.2b). Using appropriately designed, periodically arranged features, one can transition from the random columnar structure of typical GLAD films to arrays of high-uniformity, individual nanostructures (Figure 3.2c). Many of the structural characteristics discussed in Section 2.6, such as column broadening, extinction and bifurcation, are not seen in high-quality seeded growth as the shadowing instabilities associated with these effects can be suppressed or avoided.

Using these feature patterns, commonly called seed layers or seed patterns, provides a new level of control in GLAD technology. While the results of even basic seeding are striking (Figure 3.3), fully exploiting the potential of pre-seeded GLAD requires careful engineering

Glancing Angle Deposition of Thin Films: Engineering the Nanoscale, First Edition.
Matthew M. Hawkeye, Michael T. Taschuk and Michael J. Brett.
© 2014 John Wiley & Sons, Ltd. Published 2014 by John Wiley & Sons, Ltd.

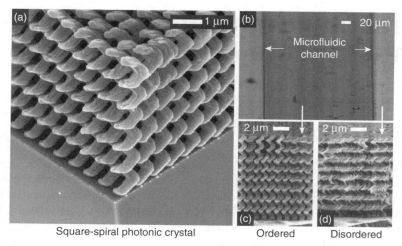

Figure 3.1   Selected device applications using GLAD-fabricated high-uniformity nanostructure arrays. (a) 3D photonic crystal devices are constructed from periodic arrays of uniform square-spiral columns. (b) Microfluidic devices can be created by selectively fabricating (c) ordered and (d) dis-ordered structural regions and exploiting the flow-rate differences between the two microstructures. (a) Reprinted from [3], with permission from Elsevier. (b–d) Reprinted with permission from [2]. Copyright 2005, American Chemical Society.

of the seed layer geometry. This chapter covers the entire patterning process, beginning with the optimization of seed pattern geometry in Section 3.2 to achieve uniform growth of both the individual nanostructure and the nanostructure array. Section 3.3 compares microfabrication techniques for pattern preparation, including optical and electron-beam lithography as well as new, unconventional approaches such as nanoembossing, colloid monolayer arrays and block co-polymer (BCP) templates. Important refinements to several of the substrate motion algorithms discussed in Chapter 2 are presented in Section 3.4 to take full advantage of the shadowing uniformity provided by seed arrays.

3.2 SEED LAYER DESIGN

A seed layer is used to tailor the nucleation process and engineer the shadowing environment from the onset of GLAD growth. With appropriate design, a single column grows from every seed and the columns adopt the structure of the underlying seed pattern. The impact of seeding on GLAD growth is clearly seen in Figure 3.3, which highlights the remarkable differences between seeded and unseeded GLAD over the entire patterned area. Column-to-column uniformity is greatly improved and competitive growth effects, such as broadening and extinction, are strongly suppressed. Significantly greater control over column growth is thus obtained with properly designed seed layers.

It is important to note that these are physical seeds: carefully engineered surface promi-nences that act as initial shadowing features to kick-start the ballistic processes. In seeded GLAD fabrication, the growth mechanism is predominantly physical and the vapour mate-rial directly condenses onto the seed surface without the need for chemical reaction. This

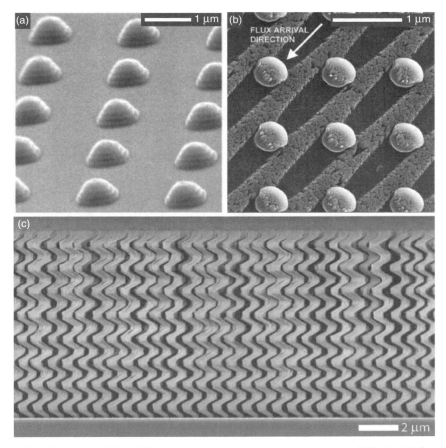

Figure 3.2 Fabricating high-uniformity arrays of nanostructures using seeded GLAD. (a) Creating a seed layer involves patterning sub-micrometre features (called seeds) onto the substrate surface before the GLAD fabrication step. (b) The seeds act as the initial shadowing features during growth. The effects of stochastic early growth are thus contained before they affect the film, and greater control over nucleation and the shadowing process is achieved. (c) Through precise design of the seed layer geometry and the subsequent GLAD motion algorithm, it is possible to realize high-uniformity arrays of uniformly structured nanocolumns. (a) and (b) © 2005 IEEE. Reprinted with permission from [4]. (c) Reprinted from [5], with permission from Elsevier.

growth process is different to chemical seeding approaches, commonly seen, for example, in nanowire and nanotube synthesis. In chemical seeding, a surface is decorated with nanoscale structures designed to catalyse a chemical reaction at the surface, causing nanowire growth. However, it is possible to combine the two seeding concepts in a GLAD process, where the the seed layer acts as both a physical shadowing feature and catalysis centre. Recent reports have shown success in this endeavour, notably in Ge nanowires catalytically grown from Au seeds [6, 7] and self-catalysed ITO nanowire growth [8, 9]. This new capability raises interesting prospects in nanowire engineering using GLAD, discussed in Section 4.6; we focus in this chapter on the physical aspects of seed design.

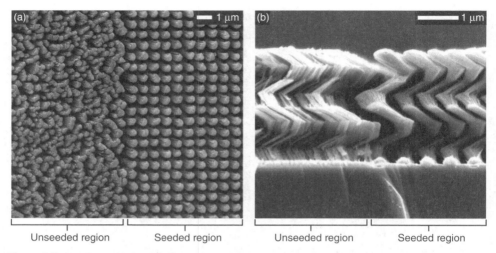

Figure 3.3   (a) Top and (b) side SEM views showing the boundary between growth onto seeded and unseeded substrate regions. Depositing onto the seed layer results in control over the column location as well as remarkably improved structural uniformity when compared with growth in the unseeded areas. © 2005 IEEE. Reprinted with permission from [4].

In practice, achieving high-uniformity nanostructure growth using seeds requires considering the available seed design parameters and understanding their influence over growth. As GLAD is primarily a low-temperature process, thermal effects are generally negligible and the chief concern in seed layer design is geometry. The correct topography must be established to properly control ballistic shadowing, and the size, shape and spacing of the seeds are of critical importance. These are then the first parameters examined when designing a seed layer to control growth, although secondary parameters, such as seed material, must often be considered to meet processing- or application-specific requirements (e.g. thermal stability, optical transparency, chemical properties).

3.2.1 Seed spacing and seed height

A preliminary objective of seeded GLAD growth is to establish a growth geometry where vapour flux deposits primarily on the seeds and deposition in the unseeded regions is limited as much as possible. This constrains the relationship between the seed spacing and seed height to produce a shadow length s long enough to reach the base of a neighbouring seed at the intended α. This condition provides a basic design parameter when constructing seed patterns [4].

Figure 3.4a illustrates the shadowing geometry imposed by an array of square seeds. For seeds of width w, height h and centre-to-centre spacing a, the shadowing constraint is

$$s = h \tan \alpha \geq a - w. \tag{3.1}$$

Other seed shapes (e.g. hemispheres) can be accommodated via geometric analysis in a similar fashion. While the seed shape generally has a small effect on the above constraint,

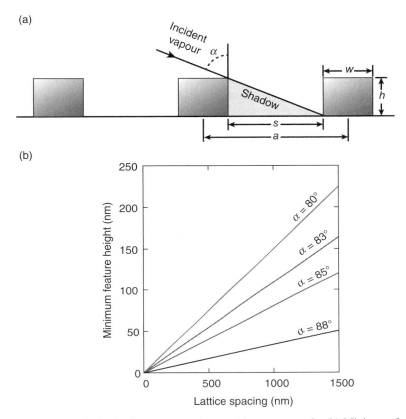

Figure 3.4   (a) The basic shadowing geometry imposed by square seeds. (b) Minimum feature size for full seed-to-seed shadow coverage at different α, as dictated by Equation 3.1.

it plays a large role in determining the 2D shadow coverage, as discussed in Section 3.2.6. Figure 3.4b plots the minimum feature sizes required to achieve nearest-neighbour shadowing at different lattice spacings for different α. Here, the geometric shadowing constraint is calculated for seeds with an aspect ratio $h/w = 1$. The shadow cast by the seed grows rapidly with increasing deposition angle and the necessary seed height drops off commensurately. At $\alpha = 80°$, a 150 nm tall seed is required to completely shadow a 1 μm lattice spacing. In contrast, the necessary seed height is only 34 nm at $\alpha = 88°$.

It is possible to increase the aspect ratio after seed fabrication by depositing a short columnar film at high α, thus amplifying the seed height and increasing the shadow coverage. For example, at $\alpha = 85°$, 50 nm tall seeds do not provide sufficient coverage for a 1000 nm lattice. However, because this seed height is acceptable at $\alpha = 88°$, an initial deposition at $\alpha = 88°$ can be used to amplify the seeds for subsequent use at lower α [10]. Note that the amplification step should only be used for small adjustments of the seed height, a restriction imposed because many of the seed design rules discussed below are α specific. Excessive deposition at an angle different than the design α could have a negative impact on seed geometry, rendering the seeds ill-suited for use at the intended α.

3.2.2 Seed lattice geometry

Because single columns grow from each seed in a properly designed layer, the seed lattice geometry controls the columnar order of the resulting GLAD film. The requirements for this order, such as lattice geometry and dimensional tolerances, are generally specific to the intended application. For example, square-spiral photonic crystals require a square lattice of columns, and any disorder in the arrangement (such as lattice spacing variation) quickly degrades the optical performance [11]. However, the lattice geometry does also influence aspects of shadowing and column growth, which we now identify and discuss.

The analysis of Section 3.2.1 depends on the orientation of the deposition plane relative to the seed lattice pattern. As can be seen in Figure 3.5, if the deposition plane aligns with the first nearest-neighbour of the lattice, a shadowing length based on the lattice constant a is sufficient. However, if the deposition plane aligns with the second or fourth nearest-neighbour seed, the shadow length must reach farther and the seeds must be correspondingly taller. (Note, we need not consider the third nearest-neighbours in these lattices as they align with and will be degenerately shadowed by the first nearest-neighbours.) This situation becomes more complex when dynamic φ rotation of the substrate occurs during growth as this will alter the orientation of the deposition plane and correspondingly change the shadowing requirement. A more general analysis of the dynamic shadowing environment is then often required (see Section 3.2.6).

The lattice structure also influences the column morphology as shown by Patzig *et al.* who investigated the effect of lattice pattern by depositing onto square, honeycomb-like and trigonal arrangements of seeds [12]. These three lattice structures are shown in Figure 3.6a–c. Using rapid φ rotation to deposit vertical columnar films onto the seed layers generates the resulting microstructures shown in Figure 3.6d–f. As can be seen, the columnar order is determined by the seed locations, with individual columns growing from each seed position as desired. However, the planar column cross-section is different in all three cases. The trigonal seed layer produced columns with a high degree of circular symmetry. The columns in the square lattice are also circular, although the symmetry is slightly reduced when compared with the trigonal case. However, the honeycomb seed layers produce remarkably

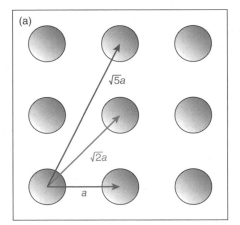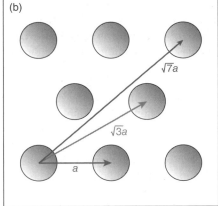

Figure 3.5   Nearest-neighbour distances in (a) square and (b) triangular lattices.

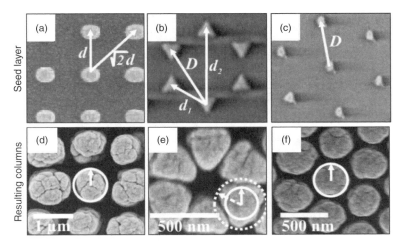

Figure 3.6 The geometry of the underlying lattice influences the structure of the resulting columns. This can be seen be depositing onto (a) square, (b) honeycomb-like and (c) trigonal seed patterns the seed lattices and observing the columnar morphology (d–f) after GLAD. Here, the d parameters are indicating nearest-neighbour distances and D is the diameter of the microspheres used in the seed fabrication process (see Section 3.3.2). Adapted with permission from [12]. Copyright © 2010. Wiley–VCH Verlag GmbH Co. KGaA, Weinheim.

different columns exhibiting a triangular cross-section. The authors attribute the difference to the underlying lattice symmetry, explaining the effect as follows. The trigonal lattice is highly symmetric with all the lattice points in the unit cell being equidistant (spacing indicated by D). This symmetry is responsible for the circularity of the column, as over a complete φ rotation the uniformity of the shadowing environment is maintained. In the square lattice, the inter-seed distance changes from d to $\sqrt{2}d$ over the rotation. This appears to be a sufficient level of symmetry as the resulting column retains a circular shape. In contrast, the inter-seed distance varies significantly in the honeycomb unit cell, ranging from $d_1 = D/\sqrt{3}$ to $d_2 = 2d_1$, as shown in Figure 3.6b. The shadowing effect from the neighbouring seeds is thus highly anisotropic and produces a corresponding asymmetry in the column cross-section.

A seed design guideline follows from this work as the column width is related to the reciprocal of the seed lattice. Under conditions of continuous and rapid φ rotations, the more equidistant the lattice points of the unit cell, the more rotationally symmetric the resulting columnar structure. Equally, low-symmetry seed-lattices can be used to induce structural anisotropies that may be useful in certain applications. We also note that this guideline also depends on the specific φ rotation employed. For example, with discrete φ rotations, such as used to produce polygonal helices (Section 2.5.3), the substrate motion symmetry need not match the lattice symmetry, which could lead to interesting growth effects.

3.2.3 Seed size

While, the seed height and width can be tuned to obtain sufficient shadowing coverage at a particular lattice constant, the seed width cannot be varied arbitrarily without considering

Figure 3.7 (a) Seed layer with excessively large seed widths that (b,c) allow multiple column growth. (d) Single columns grow from properly sized seeds. 500 nm scale bar in all images. (a) and (b) © 2005 IEEE. Reprinted with permission from [4]. (c) Reproduced from [13]. With permission from M.A. Summers. (d) Reprinted from [5], with permission from Elsevier.

additional growth issues. When trying to grow a single columnar feature from each seed it is important to match the seed width to the natural GLAD column diameter for the particular material [4, 14, 15]. The column diameter naturally broadens during GLAD growth, towards a quasi-equilibrium material- and process-specific diameter (see Sections 2.6.2 and 4.3.2). If the seed width is greater than this natural diameter, then multiple, isolated columns can grow from the individual seed, as shown in Figure 3.7a–c. Here, the 500 nm width seeds are significantly greater than the preferred width of the deposited columns, which is on the order of 100 nm. Consequently, multiple columns develop and grow from each seed, and the intended benefits of seeding – improved uniformity and control over column position – are lost. This loss is made particularly evident when the structures grown in Figure 3.7b and c are compared with Figure 3.7d, where the seed width closely matches the eventual column width. A single column is developed from each seed and the column ordering and uniformity are maintained throughout the film thickness.

At the onset of seed design, the preferred column width must be predetermined experimentally as it varies with material, α and substrate rotation speed. Process parameters, such as vacuum pressure and substrate temperature, can also influence the column width. A preliminary estimate can be made from growth on unseeded substrates, taking into account that columns grown on seeds tend to be wider than their unseeded counterparts (e.g. see Figure 3.3a). This starting estimate can then be iteratively refined into an accurate design value. Alternatively, a test sample can be prepared where the seed width is varied systematically over the substrate. Here, the seed width should be varied slowly such that the local shadowing environment remains uniform. From such a graded sample, the critical seed width can then be determined in a single deposition by inspecting the film for the onset of multiple column growth.

It is also possible to deposit smaller columns by purposefully scaling down the seed size and array geometry. However, a number of new challenges then arise. Practically, the fabrication tolerances become increasingly stringent in order to achieve the necessary shadowing uniformity, and particular attention must be paid to optimizing the deposition conditions and substrate motions to combat column broadening during growth (Section 4.3.2). Difficulties of a fundamental level also appear. As the seed size-scale approaches the diffusion length, thermal effects on adatom mobility will start to impact growth. Furthermore, with seeds on the order of tens of nanometres, atomistic noise (e.g. vapour flux shot noise, critical nuclei density) can introduce a relevant error term to the growth. Combating these fundamental issues requires very careful control over the deposition conditions to optimize growth and ensure repeatability.

3.2.4 Planar fill fraction

The GLAD film mean density ρ_{GLAD} is always less than the bulk material density ρ_{Bulk} owing to the significant porosity introduced by ballistic shadowing. This material-dependent reduced density is readily controlled through α, but it also depends on other variables, including substrate temperature, angular flux distribution and φ rotation. For seeded growth, the seed layer geometry must be designed to match the film mean-density [4]. If this criterion is not met, significant morphological changes can occur, such as column broadening and bifurcation, as the growing GLAD film tends towards its equilibrium density. The seed size and lattice geometry are therefore constrained to produce an appropriate planar fill fraction f_{seed} of the seed layer (defined as the fraction of the seed array unit cell covered by seed material) that matches the equilibrium film density.

This design requirement can be approached quantitatively by combining geometric analysis of the seed design with the previously developed expressions for GLAD film density. For example, in the case of circular seeds in a tetragonal lattice arrangement, the fill fraction is

$$f_{seed} = \frac{\pi w^2}{4a^2}.$$

(3.2)

To estimate the GLAD film density we can employ the basic density expression (Section 4.4.1)

$$\frac{\rho_{GLAD}}{\rho_{Bulk}} = \frac{2\cos\alpha}{1 + \cos\alpha}.$$

(3.3)

The design constraint is then achieved by combining Equations 3.2 and 3.3 to produce the requirement

$$\frac{\pi w^2}{4a^2} = \frac{2\cos\alpha}{1 + \cos\alpha}. \tag{3.4}$$

This requirement indicates that, for a typical deposition angle of $\alpha = 85°$ (where density is only 16% of bulk) and a lattice spacing of $a = 1000$ nm, the necessary seed width is $w = 452$ nm. Similar conditions can be readily constructed for different lattice geometries and seed shapes. We also note that expressions such as Equation 3.3 are useful for preliminary design and testing. For accurate fabrication, however, it is practical to calibrate the mean film density for the specific process conditions used to deposit the film. Such calibration methods are discussed in Section 4.4.

As discussed at the end of Section 3.2.3, one can also use a test seed layer with a graded lattice geometry where f_{seed} is slowly varied across the substrate. After the GLAD processing step, the sample can be inspected to identify the optimal lattice geometry for producing the highest quality structures. Note that multiple structural parameters can be investigated simultaneously with these calibration seed layers. For example, fabricating a seed layer with a systematic variation of w in one direction and f_{seed} in another will produce an effective phase map of the structural parameters after deposition. Samples such as these can be used to more rapidly optimize seed designs.

3.2.5 Seed shape

The important goals of seeding single-column growth, controlling column arrangement and preventing inter-seed growth are met by following the preceding design methods. When striving for optimal growth control, it is also important to examine the seed shape and consider its impact on seeded GLAD growth.

The design constraint in Section 3.2.1 is developed such that the shadow from one seed just reaches the base of the neighbour seed. While inter-seed deposition is thereby limited, deposition always occurs on the exposed sidewall, which can impact the morphology of the fabricated column. This introduces another geometric design parameter, the exposure height h_{sw}, quantifying the portion of the sidewall exposed to incident vapour [14, 16]. For square seeds as shown in Figure 3.8a, the exposure height is given by

$$h_{\text{sw}} = \frac{a - w}{\tan\alpha}. \tag{3.5}$$

This equation is specific to seeds with a square profile, but can readily be extended to other seed shapes via geometric analysis. It also depends on the lattice structure and the lattice location of the shadowing seed, as was discussed in Section 3.2.2.

It is necessary to consider sidewall deposition during seed design as, depending on the particular seed shape, the effective deposition angle α' (Figure 3.8a,c) can be different from α and even vary locally across the seed topography. This local variation of the deposition geometry produces corresponding structural differences between material deposited on top of the seed (the primary structure) and the sidewall (the sidewall structure), as indicated in Figure 3.8b. On the top of square seeds $\alpha' = \alpha$ over the entire surface, while $\alpha' = 90° - \alpha$

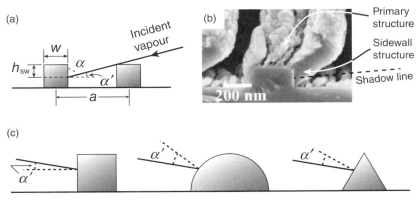

Figure 3.8 (a) A portion of the neighbouring seed sidewall is always exposed to the vapour flux, quantified by the exposure height h_{sw} as indicated. (b) The initial column growth consists of the primary deposition on top of the seed topography and the sidewall deposition. The primary and sidewall deposits will exhibit microstructural differences (column density, growth direction) caused by the variation in the local deposition geometry (defined here by α') over the seed surface. (c) The local deposition geometry is dependent upon the seed structure. (b) Reproduced with permission from [12]. Copyright © 2010. Wiley–VCH Verlag GmbH Co. KGaA, Weinheim.

over the entire seed sidewall. At large α, the sidewall structure is deposited at a near-normal deposition angle and a dense columnar microstructure consequently forms on the sidewall, in marked contrast with the higher intracolumn porosity of the primary structure. Furthermore, the primary and sidewall column growth orientation will also be different as determined by α' and the local surface normal. This sidewall deposit can thus contribute significantly to the initial broadening of the deposited column. As a guideline, the ratio of exposure height to seed width (h_{sw}/w) should be kept small to minimize the impact of sidewall deposition. Predicting the effects of α' can be more difficult for complex seed shapes, as illustrated by Figure 3.8c. For hemispherical seeds, α' varies continuously over all points of the seed surface. Triangular seed shapes are entirely sidewall with a consistent α' everywhere less than α.

3.2.6 Two-dimensional shadow coverage

Thus far, the analysis of the shadowing geometry has been strictly one-dimensional and we have ignored the complete 2D shadowing profile created by a given seed. However, examination of Figure 3.2b shows that while the seed height is sufficient to cast a shadow that reaches the neighbouring seed (satisfying Equation 3.1), adjacent regions of the substrate remain unshadowed and the 2D shadow coverage is incomplete. This inter-seed deposition represents vapour flux that is not controlled by the seed-induced shadowing environment. In the early stages of growth, this vapour condenses on the substrate as shown, whereas in later stages this vapour can deposit onto neighbouring column sidewalls or onto previously deposited column sections, thus contributing to undesired column growth. To optimize the seed design and minimize unwanted deposition, it is therefore important to consider the 2D shadow coverage provided by the seed, which depends on multiple factors. The structure of the individual seeds (height, width and shape) determines the shadowing profile at a particular α. However, it is also important to examine the alignment of the incident flux trajectories (as set by φ) with respect to the orientation of the seed lattice as this determines the shadowed fraction of the unit cell. We consider the individual seed shapes first.

The vertical cross-section of the seed determines the exact area shadowed. For example, a tapered, pyramidal seed provides subtly different shadowing than a cylindrical seed, particularly at the shadow apex. Summers performed a careful study examining seed shape and shadow coverage, developing a geometric model to parameterize the vertical cross-section and examining the portion of the seed lattice unit cell shadowed by seeds of different shape [13]. In this model, mimicking the shape of lithographically prepared seeds, the seed height h is given by

$$\frac{h}{h_0} = \frac{2}{\pi} \arccos\left[\left(\frac{x}{d_0}\right)^{1/k}\right], \tag{3.6}$$

where h_0 is the height at the seed centre, d_0 is the seed diameter at the substrate and k is a parameter controlling the seed curvature. The x parameter varies from 0 to 1 along the diameter of the seed (i.e. $x = 0$ and 1 at the seed edges and 1/2 at the seed centre). The 2D shadow profile $s(x)$ is then given by

$$s(x) = h(x)\tan\alpha, \tag{3.7}$$

which can be integrated to determine the 2D shadowed area A:

$$A = \int_0^1 s(x)\,dx. \tag{3.8}$$

This seed shape at different k is presented in Figure 3.9a–c, alongside the resulting shadow profile. Small values of k approximately reproduce a rectangular seed shape with a flat top. Increasing k rounds off the corners and creates a hemispherical cross-section for $k = 1/2$. For $k > 1/2$, a seed with a tapered centre point is produced. Increasing the squareness of the seed (via decreasing k) increases the 2D shadow coverage and reduces deposition in the unseeded substrate regions. The seed profile can be experimentally determined from images such as Figure 3.9d by measuring the shadow length at different points along the seed and then numerically fitting the data with Equation 3.6 to extract k and h_0. Equally, the total shadowed area can be measured and compared with the prediction of Equation 3.8.

In practice, the seed shape is difficult to tailor. Certain nanofabrication methods, such as electron-beam lithography or FIB milling, provide very high resolution, thereby offering a degree of design flexibility from the onset. For lithographically prepared seeds, photoresist reflow techniques based on thermal softening and feature rounding could be used to tune the seed cross-section. It is also possible to modify the seeds by adding an initial GLAD fabrication step to the process, as can be done to amplify seed height.

While maximizing the 2D shadowed area reduces uncontrolled deposition, further optimization is possible by examining how the shadow covers the unit cell of the seed lattice. This becomes particularly important when employing φ rotation during deposition as this

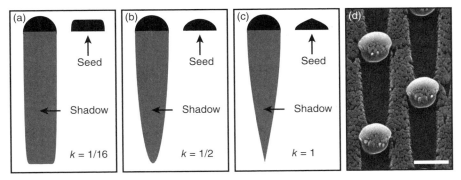

Figure 3.9 The 2D shadow coverage depends on the seed cross-section, which can be numerically described using structural models such as Equation 3.6. (a–c) The parameter k governs the curvature of the seed sidewalls. $k = 1/2$ describes a hemispherical seed; smaller k values produces more square-like seeds, and larger k values result in tapered seed structures. (d) Fabricated seed profiles can be experimentally determined by comparing SEM measurements of the initial shadowed regions with the numerical model predictions. The scale bar indicates 500 nm. (a–c) Reproduced from [13]. With permission from M.A. Summers. (d) © 2005 IEEE. Reprinted with permission from [4].

substrate motion alters the orientation of the deposition plane with respect to the seed-layer lattice vectors. The 2D shadowing geometry for a square-seed lattice is illustrated in Figure 3.10a, where the parameters are defined as before and θ is equal to the angle between the deposition plane (defining the incident flux direction) and the seed lattice vector. While the objective is to completely cover the unit cell in shadow, this cannot be attained by simply increasing w as this parameter is constrained to prevent multicolumn growth, as discussed previously. Instead, a combination of w, θ and h must be determined that maximizes the unit cell coverage. The minimum w for complete coverage occurs when the seed shadow reaches the fourth nearest-neighbour and tangentially grazes the first and second nearest-neighbouring seeds. From the geometry, the necessary parameters are determined to be [13]

$$\theta = \arctan(1/2) = 26.6°, \tag{3.9}$$
$$w = a \sin \theta = 0.447a, \tag{3.10}$$
$$h = \frac{\sqrt{5}a}{\tan \alpha}, \tag{3.11}$$

where we have assumed square seed profiles ($k = 0$). When these conditions are achieved, the unit cell is completely covered in shadows cast by nearby seeds, as shown in Figure 3.10b.

As demonstrated by Summers and Brett [11], analysing the 2D shadow coverage in this manner proved crucial in optimizing the shadow control, enabling high-precision growth with minimal column broadening and increased column ordering. This work is discussed further in Section 3.4.2. Note that the analysis can be extended, accounting for more complex substrate motions (in φ and α) and experimental seed shapes, in efforts to predict optimal seed shapes and arrangements.

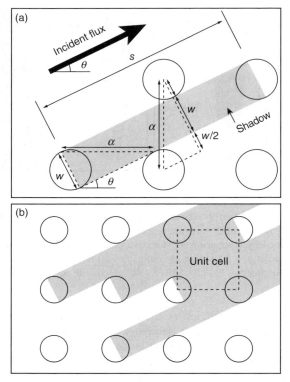

Figure 3.10   Maximizing the 2D shadow coverage of the seed lattice unit cell. Adapted from [13]. With permission from M.A. Summers.

3.2.7 Seed material

The preceding discussion of seed design focused on the seed layer geometry as these design parameters provide the key pathways for controlling ballistic shadowing, which dominates GLAD growth at low temperature. However, it is often important to consider seed material as proper selection may improve adhesion properties or be constrained by the particular application. For instance, in photonic crystal fabrication it is often desirable to have seeds that are transparent in the wavelength region of interest. Thermal stability can be a concern, whether for subsequent processing or in the intended application. As many seed-preparation techniques use organic-based materials (photoresist, epoxy), the seed pattern fidelity may present a concern when the substrate temperature approaches the seed material glass-transition temperature. The chemical properties of the seed material are also important, being a crucial aspect of vapour–liquid–solid GLAD techniques [6–9]. A general approach to modifying the seed material involves transferring the original seed pattern into the desired substrate material using common microfabrication procedures, such as wet/dry chemical etching. Another approach involves performing a short, high-α deposition of the desired material onto the existing seeds. This idea has already been discussed in terms of modifying seed height and shape, but given the material flexibility of the GLAD process it can also be used to selectively coat the seed with a different material.

3.2.8 Design parameter summary

This section has outlined a framework useful for designing seed layers in GLAD technology. The guidelines for seed layer geometry can be summarized as follows:

1. **Seed spacing and height:** The seed should be sufficiently tall so that the shadow at the desired α is sufficiently long to reach the neighbouring seed and prevent inter-seed deposition (Equation 3.1).
2. **Seed width:** The seed width should be matched to the natural column width of the deposited material to prevent multicolumn nucleation.
3. **Fill fraction:** The planar fill fraction of the seed lattice should match the equilibrium density of the deposited film to prevent column broadening and bifurcation (see Equation 3.4).
4. **Seed lattice:** The rotational symmetry of column cross-sections increases with the symmetry of the seed lattice. As a corollary, low-symmetry lattice structures can be used to fabricate alternative column shapes.
5. **Seed shape:** Sidewall deposition should be minimized by limiting the exposure factor (Equation 3.5).
6. **2D shadowing profile:** For high-precision growth, seed cross-section should be tailored to maximize shadow coverage of the seed-lattice unit cell (Equations 3.6–3.11).

Applying these guidelines will yield a seed geometry that controllably nucleates single-column growth and produces nanostructure arrays with significantly greater uniformity than traditional GLAD films.

3.3 SEED FABRICATION

Once the initial design has been obtained and the geometric parameters optimized to effectively control nucleation and shadowing, the challenge then becomes selecting a fabrication method with which to realize the seed layer. A wide range of micro- and nanofabrication methods have been applied to the preparation of seed layers, each having their particular advantages and disadvantages (Table 3.1). Before examining these techniques we outline some general requirements for seed fabrication.

The initial, basic factor to consider is the resolution of the fabrication technique, which must be compatible with the designed features. Achieving single-column growth from each seed typically demands sub-micrometre patterning resolution. Individual seed widths should be on the order of a few hundred nanometres and with inter-seed spacings of several hundred nanometres to a micrometre. Additionally, the features should be very uniform in shape, size and placement to maximize growth control. The tighter the fabrication tolerances, the more successful the seeding will be in controlling nucleation and mitigating column competition. The overall dimension ('die' size) of the patterned seed region produced by the fabrication method should be commensurate with the intended application. For practical handling and characterization, minimum patterned regions on the order of a square millimetre are generally convenient.

Table 3.1   Comparison of nanofabrication methods used to create GLAD seed layers.

Method	Resolution	Throughput	Design flexibility	References
Optical lithography	Low	High	High (requires new mask)	[1, 16–25]
Laser direct-write lithography	Low	Low	High	[2, 4, 26, 27]
Electron-beam lithography	Very high	Low	High	[5, 12, 16, 28–35]
Nanoimprint lithography	Very high	High	High (requires new master)	[36–38]
Colloidal monolayer	High	High	Low	[39–47]
Nanosphere lithography	High	High	Low	[12, 48–57]
Nanostructured alumina	High	High	Low	[58–60]
BCP template	High	High	Low	[61]

There are several additional factors that are often desirable in a fabrication method, depending on application and circumstance. During research and development, the ability to alter parameters and test new designs with minimal overhead is important. Implementing new designs should be rapid and relatively low cost, allowing initial results and prototypes to be produced as economically as possible. With an optimal design ready for manufacture, cost factors then dominate and the emphasis is on processing yield and throughput. Fabrication techniques that are scalable and can be developed into robust processes are favourable.

The fabrication techniques used for GLAD seed preparation can be broadly categorized as either conventional or unconventional approaches. Conventional here refers to techniques based on top-down microfabrication processes where tools are used to etch, mill or otherwise shape materials into the desired geometry. The unconventional approaches follow a bottom-up fabrication strategy where the desired structure is created from smaller constituents via directed self-assembly. While these technologies are significantly less mature than conventional fabrication methods, they offer potential low-cost, nanometre-scale resolution in the future.

3.3.1 Conventional techniques

Specific to GLAD technology, conventional techniques used in the past have included optical lithography, laser direct-write lithography (DWL), electron-beam lithography and nanoimprint lithography (NIL). Optical lithography is a very common microfabrication technique that involves transferring a pattern from a prefabricated master onto a substrate covered in a photosensitive resist layer. The patterned resist can then be used as a seed layer or the pattern can be transferred into the underlying material via etching or a subsequently deposited layer through lift-off. Optical lithography has been used by several groups to produce seed patterns for GLAD growth [1, 16–25].

For GLAD technology, optical lithography provides many benefits. Because the entire pattern is created at once, lithography is able to rapidly produce seed layers over wafer-sized areas. The ubiquity of the technique makes equipment and expertise easier to find, and well-established processes are readily available. These two advantages lend themselves to scalable and robust seed fabrication. However, the resolution of optical lithography is limited by diffraction, and fabricating features on the $\lambda/2$ size scale is challenging, requiring significant process optimization and expensive equipment. Additionally, changing the pattern design requires acquiring a new mask, which can be costly, especially for sub-micrometre features and complex designs.

In direct-write lithography (DWL), a focused laser- or electron-beam is rastered across the resist-coated substrate and features are written into the resist layer one at a time. This approach provides much greater latitude for design changes than normal optical lithography, as the technique is maskless and the patterned geometry can be adapted from run to run. This is an excellent advantage for research and development, allowing rapid design iteration and optimization. Although laser-based implementations are also diffraction-limited, electron-beam lithography can routinely pattern 10 nm scale features. Because of these advantages, both laser [2, 4, 26, 27] and electron-beam [5, 12, 16, 28–35] DWL have seen much use in GLAD research. The primary drawback of these approaches, however, is that the serial patterning process leads to lengthy fabrication times. This makes it difficult to produce the large-area patterns required for certain applications. Furthermore, these patterning systems are very expensive, both in terms of capital investment and operating costs. The combination of low throughput and high cost leads to poor production scalability and, barring potential high-value applications, the techniques are research focused.

A nanofabrication technique that provides high resolution while still processing large areas at high speed is NIL. In NIL, a master patterned with the desired features is prepared using a high-resolution technique such as electron-beam DWL. The master is then pressed into a resist material, mechanically deforming the resist and imprinting a negative reproduction of the master pattern. The resist reproduction can then be used as is, further processed via etch/lift-off, or used as a master itself and subsequently imprinted into a second, softer resist. NIL has been successfully applied to GLAD seed-layer fabrication in several reports [36–38]. The resolution of the NIL process is determined by the master feature dimensions. Because all the features are imprinted at once, NIL provides significantly greater throughput than DWL approaches do. Therefore, an expensive, high-resolution master may be fabricated and then used for subsequent low-cost, high-throughput processing. However, requiring a preformed master renders NIL inflexible to design changes and better suited to production than research.

3.3.2 Unconventional techniques

Driven by new discoveries and an economic desire to create nanostructures at low cost and high throughput, several alternative micro- and nanofabrication technologies have been adapted to prepare seeds for GLAD. Note that as this is a very active area of research, we expect the capabilities of existing techniques to improve rapidly and new approaches to be quickly applied as they are invented.

Several researchers have used self-assembled monolayers of submicrometre colloidal particles as seed layers [15, 39–42, 42–47]. In this approach, a monodisperse, colloid suspension is dried on a substrate and, as the liquid evaporates, surface tension pulls the settling particles together. With proper substrate preparation and under controlled drying conditions, the particles will spontaneously self-assemble into a hexagonally close-packed (HCP) monolayer, as shown in Figure 3.11a. The nanoparticles, made of silica or a polymer such as polystyrene,

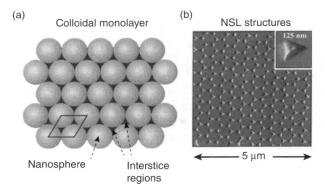

(a) Colloidal monolayer

(b) NSL structures

Nanosphere   Interstice regions

Figure 3.11 (a) Illustration of a colloidal monolayer, showing self-assembled HCP lattice of colloidal microspheres. These microspheres can act as individual seeds for subsequent GLAD fabrication steps. Nanosphere lithography (NSL) involves depositing through the colloid-lattice interstitial regions and onto the exposed substrate below. NSL thus generates periodic arrays of submicrometre features suitable for seeded GLAD growth. Reprinted with permission from [62]. Copyright © 2001, American Chemical Society.

are typically 100–500 nm in diameter, making them immediately suitable as seeds for GLAD growth. Furthermore, macroscopic (greater than square centimetre) particle arrays can be quickly formed at very low cost. Unfortunately, these techniques offer minimal design flexibility: they are limited to spherical seed shapes and there is limited control over the planar fill fraction. Moreover, the technique is limited to HCP lattice fabrication and unable to realize other lattice geometries or engineer geometric defects as required for many photonic crystal microfluidic devices. While this lack of flexibility is an obvious detractor, the low cost barrier involved makes it well suited to preliminary investigation and possibly for manufacturing.

Another seed preparation technique, known as nanosphere lithography (NSL) [62], uses colloidal monolayers like those discussed above as a deposition or etch mask for subsequent microfabrication steps. When viewed top-down, portions of the underlying substrate are left exposed through the interstices of the close-packed colloids, as indicated in Figure 3.11a. Depositing material (via PVD) through these interstitial areas creates nanoscale features arranged in a regular lattice pattern complementary to the HCP colloid lattice (see Figure 3.11b). This approach has been applied to GLAD technology by several researchers [12, 48, 50–52, 54–57]. Because NSL involves only one additional processing step, it possesses the same cost and throughput advantages as colloidal monolayer fabrication. NSL, however, offers superior design flexibility, as a wider range of materials can be used and GLAD-like substrate motions can be used to manipulate the size, shape and spacing of the deposited seeds [63].

Two less common self-assembly approaches for generating nanoscale seed structures are electrochemical oxidation of aluminium [58–60] and BCP templating [61]. In the former, HCP-arrays of hexagonal alumina nanotubes are spontaneously formed during oxidation, with diameters between 20 and 400 nm depending on process conditions. These structures can then be used as seed layers for subsequent GLAD growth. In BCP templating, two immiscible polymer blocks are chemically linked to form a longer chain (the BCP). The BCP then undergoes a self-organizing phase separation, forming micro-domains that minimize the interaction energy between the polymer blocks. Several unique geometries can thus be formed spontaneously with domain sizes from 10 and 200 nm which can be subsequently used as GLAD seed layers. Like the other self-assembly techniques, these two approaches provide significant cost advantages at the expense of design freedom.

3.4 ADVANCED CONTROL OF LOCAL SHADOWING ENVIRONMENT

While a properly designed seed layer tailors the nucleation phase, thus initiating uniform column growth, it is important to implement appropriate substrate algorithms that accurately control shadowing as growth continues. Otherwise, the preliminary order induced by the seed layer can be lost as columns grow. Because of the high initial column uniformity generated by seeded growth, detecting subtler structural defects is easier than in unseeded growth, where fine details are often masked by the inherent nonuniformity. Consequently, it is easier to diagnose the underlying mechanism creating the defect and develop an improved motion algorithm to prevent or mitigate the issue.

The significant efforts devoted to carefully tailoring the shadow environment have been primarily motivated by the fabrication of square-spiral photonic crystals, which place stringent uniformity requirements on the inter- and intra-columnar structure. As a result, the following motion algorithms are predominantly focused on the square-spiral geometry. However, the concepts behind these advanced motion algorithms provide much intuition about GLAD growth and are often generalizable to other structures.

3.4.1 Preventing bifurcation: slow-corner motion

In the square-spiral geometry, a series of connected columnar arms are fabricated with each arm oriented at right angles to the previous arm. The resulting structure spirals upwards like a microscale staircase. In the conventional GLAD approach to square-spiral fabrication, the substrate is held at the desired α to deposit an initial columnar arm of predefined length L. Once the arm is deposited, the substrate is quickly rotated 90° about its azimuth to deposit the next arm. This process is repeated to deposited the desired number of square-spiral periods, with four arms per period.

While this basic approach is successful over the first few turns, structural defects quickly appear and impact the growth leading to severe loss of uniformity and a degradation of the square-spiral performance. These are highlighted in Figure 3.12, showing an SEM image of a square-spiral deposited using the basic growth algorithm. As can be seen, structural defects emerge after only a single period of growth. Column bifurcation, where an individual column becomes unstable and splits into two distinct columns, is observed. The column fanning at the arm corners is pronounced, producing a highly elongated column cross-section. These problems result in an eventual loss of periodicity in the film structure, and all the effort in preparing the initial seed layer is ultimately squandered.

These growth defects originate at the corners of the square-spiral structure and are a subtle consequence of how the shadowing environment changes during the turn [21]. After a column arm is deposited and the substrate abruptly rotated 90°, the next column arm nucleates along

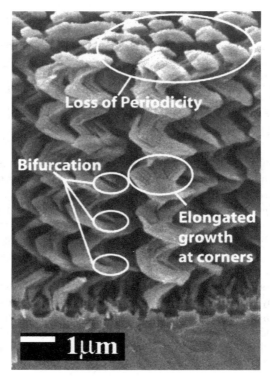

Figure 3.12 Build up of defects in a square-spiral created using basic GLAD substrate motion. Indicated on the SEM image are column bifurcation defects, where a single column splits into multiple columns, and a fan-like growth elongation at the square-spiral corners. The accumulation of structural defects with thickness also leads to a poor reproduction of the seed layer periodicity in later stages of film growth. Reprinted with permission from [21]. Copyright 2004, American Vacuum Society.

the extended, freshly exposed side of the previous arm rather than being confined to the column extremity. The bottom left panel of Figure 3.13a illustrates this process. Because the exposed area is larger than the column itself, two separate columns can nucleate and develop, causing a bifurcation of the arm and a loss of the intended structure. Additionally, the asymmetric cross-section of the exposed area (long and thin) creates a fanned columnar structure, one that is only amplified by the natural shadowing anisotropy inherent to GLAD process.

Preventing the formation of these defects requires a corrected substrate motion algorithm that tailors shadowing through the turn, leading to the development of the slow-corner technique [21]. Rather than a single, rapid turn, the substrate is slowly rotated through 90°, producing a smoother, continuous change in the local shadowing environment. This motion is illustrated in bottom right panel of Figure 3.13a. The columns are thus steered gently around the corner and the column shape is better preserved between arm segments.

The corresponding growth algorithm is shown in Figure 3.13b, which plots the substrate azimuthal angle φ as a function of deposited thickness. The black line represents the traditional GLAD approach with discontinuous substrate rotation, while the dashed line shows

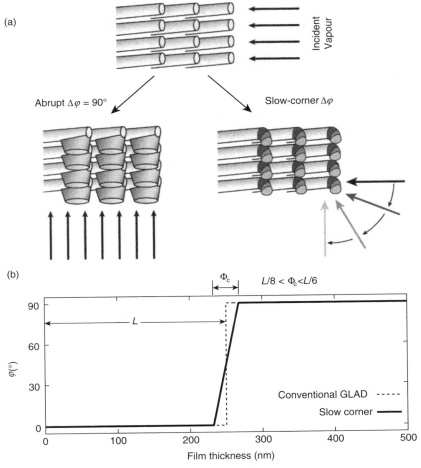

Figure 3.13 Depiction of the slow-corner technique. (a) When an abrupt 90° φ rotation is used at the corner, the incident vapour stream deposits onto the newly exposed length of the existing column, allowing a larger column to nucleate and grow. Because of its larger size, this column is unstable and has an increased bifurcation probability, allowing defects to quickly build up throughout growth. When the turn involves a slow φ rotation, the column shape is maintained through the corner and the bifurcation probability remains low and the defect formation rate is reduced. (b) Example substrate motion algorithm used to implement the slow-corner technique in a square-spiral corner. The dashed line shows the abrupt-turn approach where φ is rotated by 90° as quickly as possible after depositing an arm of length L (250 nm here). The slow-corner motion is achieved by rotating φ through 90° over a growth interval Φ_c, typically between $L/8$ and $L/6$. (a) Reprinted with permission from [21]. Copyright 2004, American Vacuum Society.

the slow-corner-modified motion. The slow-corner is implemented over a segment of vertical growth Φ_c, which becomes the control parameter determining the substrate rotation rate. Suggested Φ_c values for optimal growth are between $L/8$ and $L/6$. Smaller than this range, and the bifurcation problems return; larger, and the resulting structure devolves to a pseudo-circular helical column as opposed to the intended square-spiral.

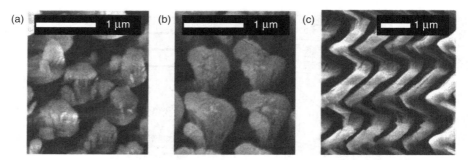

Figure 3.14   Compared with (a) traditional GLAD, slow-corner improves (b) the structure uniformity and (c) the fine structure of the column arms. Reprinted with permission from [21]. Copyright 2004, American Vacuum Society.

Implementing the slow-corner algorithm significantly improves the resulting square-spiral structure. The impact is shown in Figure 3.14, showing a square-spiral prepared with (a) abrupt turns and (b,c) slow-corner turns. Slow-cornering produces more uniform columnar structures that maintain the lattice symmetry of the underlying seed layer. The fine structure of the column arms is also maintained (see Figure 3.14c) and column bifurcation is strongly suppressed. Although to date the slow-corner technique has only been applied to square-spiral fabrication, its extension to other substrate motion algorithms is straightforward and may provide improved structures when implemented.

3.4.2 Preventing broadening: phisweep and substrate swing

The slow-corner algorithm corrects structural deficiencies introduced at the square-spiral corners. However, the issue of column fanning (Section 2.6.4) during column growth remains, which can lead to rapid structural degradation and loss of intended geometry. Column fanning is inherent to GLAD growth as the shadowing process is intrinsically anisotropic (Figure 3.15a): while shadowing controls growth in the deposition plane (parallel to the x-axis), growth in the orthogonal direction (y-axis) is unimpeded. The unrestricted lateral column growth causes the columns to broaden during growth and adopt fan-like cross-sections as a result (Figure 3.15a). This fanning can become very pronounced, as depicted in Figure 3.15b. In these SEM images, columns deposited onto a periodic seed layer (shown in inset) fan out during growth, broadening so severely that they eventually merge and chain together. While the square lattice symmetry remains apparent, the single-column structure is lost and any subsequent substrate rotation (e.g. to prepare a square spiral) would effectively precipitate nonuniform column growth along the rows. Note that, while column fanning is a negative effect in the pursuit of isolated nanostructure growth, extreme fanning and the resulting column chaining can be used to fabricate microribbon structures [31]. By allowing the deposited columns to fully merge into a cohesive unit, it is possible to form continuous sub-micrometre thickness ribbon structures with a length set by the dimensions of the patterned area. These microribbons can be made from a variety of materials and easily detached from the substrate after fabrication; for example, by mechanical stripping or ultrasonic agitation.

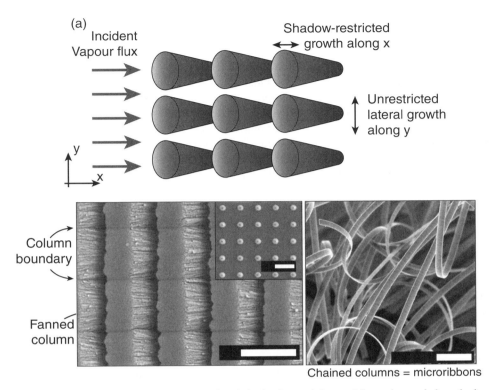

Figure 3.15 (a) Illustration of one-dimensional shadowing and the resulting anisotropic broadening. (b) SEM image of severely broadened slanted columns that lose the structure of (c) the underlying square lattice. Both scale bars indicate 1000 nm. (d) Severely broadened columns that chain together can be detached from the substrate to yield continuous, microribbon structures. (b) and (c) Reproduced from [20]. © IOP Publishing. Reproduced by permission of IOP Publishing. All rights reserved. (d) Reproduced from [31]. © IOP Publishing. Reproduced by permission of IOP Publishing. All rights reserved.

To solve the problem of column fanning, substrate motion algorithms have been devised to decouple the shadowing and growth directions. The first technique, known as phisweep [27] (discussed at length in Section 2.7.3), has proven very successful in controlling column broadening and maintaining column uniformity, as has the subsequently developed substrate-swing technique [24]. These two approaches, are similar in implementation and provide 2D shadow control to produce a more isotropic shadowing environment for the growing columns. During phisweep, the substrate is periodically rotated from side to side in φ between angles of $-\gamma$ and $+\gamma$ (the sweep angle). Each φ rotation occurs after prescribed amounts of incremental vertical film growth (defining the sweep pitch q). The equivalent substrate-swing motion algorithm is similar except that φ is varied continuously between $\pm\gamma$ over the growth interval, as opposed to the step-like rotations used in phisweep. Because the φ motions are symmetric in both phisweep and substrate swing, the net growth direction is oriented along $\varphi = 0°$, allowing a given columnar structure to be constructed from smaller growth segments. However, the main benefit of this motion algorithm is that the φ rotation also periodically alternates the shadowing direction. In reference to Figure 3.15, at each $\varphi = \pm\gamma$ position there is a nonzero y-component of the shadowing that acts to restrict the lateral column growth and prevent column fanning. The primary control parameters for these substrate movements are γ and q, which must be correctly chosen for optimal effect. The sweep pitch q should be of the same order as the intracolumn nanofibres (Section 2.7.3). This is a material- and process-specific quantity, dependent upon factors such as thermal mobility and nucleation characteristics. In general, it requires experimental determination, with typical values ranging from 15 to 50 nm. The sweep angle γ should be sufficiently large to induce shadowing in the direction orthogonal to the intended growth direction. Values between 20° and 60° have been shown to be effective.

Figure 3.16 presents implementations of the phisweep (solid line) and substrate-swing (dash–dot line) motion algorithms to produce tilted and square-spiral columns. The equivalent basic GLAD motion (dotted line) is shown as well for comparison. In all cases, the α angle remains fixed throughout the deposition (Figure 3.16a). To realize a slanted column structure (Figure 3.16b), the φ position is periodically rotated between $\pm\gamma$, where $\gamma = 45°$. As can be seen, the φ rotation is discrete in the phisweep case whereas it is continuous and sinusoidal in the substrate swing approach. In both instances, the motion is periodic and repeats after every $2q$ of growth (with $q = 50$ nm in this example). Because the φ rotations are symmetric and the q growth interval is sufficiently small, the net growth direction is equivalent to the basic GLAD case.

To develop the phisweep and substrate-swing motion algorithms for square-spiral column fabrication, the periodic φ motion is superimposed on the basic square-spiral motion, as shown in Figure 3.16c. Consequently, the phisweep/substrate-swing motion steps forward by 90° after each column arm is fabricated. The same concept can be further generalized to other GLAD motions as well.

The column broadening is greatly suppressed by implementing the phisweep or substrate swing growth algorithm. Figure 3.17 compares the growth of slanted TiO_2 column structures using both basic GLAD (a–c) and phisweep GLAD (d–f) onto seed arrays (square lattice) with 100, 200 and 300 nm lattice spacings [32, 64]. All depositions were performed at $\alpha = 84°$, with phisweep control parameters of $\gamma = 45°$ and $q = 45$ nm. The SEM perspective here is oblique, taken 45° from the substrate normal, thus providing a simultaneous view of the top and side of the arrayed nanostructures. As is clearly evident when using basic GLAD, the unrestricted lateral growth caused by one-dimensional shadowing leads to columns that fan out and chain together. In addition to the loss of desired column shape and poor uniformity, the column order does not follow the structure of the seed layer. In the $a = 300$ nm case (Figure 3.17c), the columns are arranged in lines and have merged together, similar to the structures shown in Figure 3.15b. The disorder increases as the lattice size is reduced, as seen in Figure 3.15a and b. Using phisweep, however, introduces 2D shadowing and suppresses the column fanning effect, greatly improving the column structure and uniformity. As can be seen in Figure 3.17, compared with the structure produced with basic GLAD, the phisweep-modified algorithm realizes columns with more symmetric cross-sections. Furthermore, the column order induced by the seed layer is better preserved by implementing the phisweep algorithm owing to the improved shadowing control provided. However, the overall fidelity decreases as the lattice size is reduced, suggesting that further process optimization is required to achieve sub-100 nm nanostructure arrays.

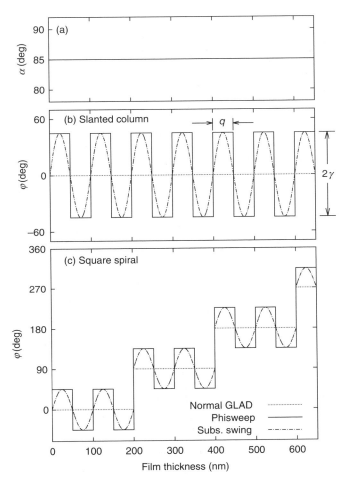

Figure 3.16 Slanted and square-spiral algorithms for basic GLAD substrate motions and phisweep and substrate-swing modified versions. (a) In each case α is fixed. (b) Slanted column case: the phisweep algorithm breaks down the normal GLAD motion into a series of smaller segments of length q, deposited at $\varphi \pm \gamma$. The substrate-swing approach replaces the discrete growth segments of phisweep with a continuous φ motion. Because the φ motion is symmetric, the net growth direction in the substrate plane is equivalent to that of the basic GLAD algorithm. (c) Square-spiral case: the phisweep and substrate-swing motions can be generalized to a more complex algorithm by superimposing the periodic φ motion onto the desired basic GLAD motion.

Optimizing the phisweep sweep angle γ

When implementing these complex substrate motions with an underlying seed lattice, there is an added subtlety that must be accounted for to optimize growth. As was discussed in Section 3.2.6, the shadow coverage depends not only on the seed size and shape, but also on the alignment of the incident vapour flux and the seed-layer lattice vectors. Summers

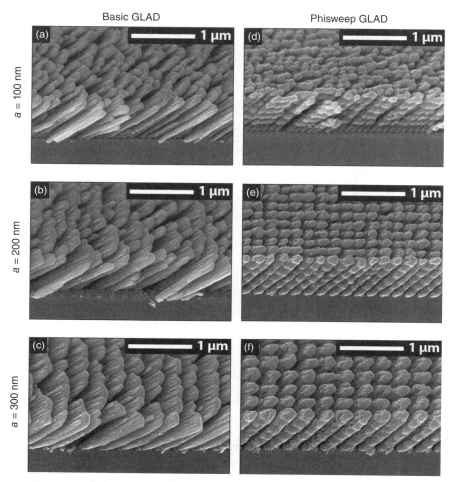

Figure 3.17   Result of (a–c) basic and (d–f) phisweep GLAD growth onto square-lattice seed layers with lattice periods of $a = 100$, 200 and 300 nm seed layers. Structures produced using the phisweep algorithm exhibit reduced column fanning, improved uniformity and increased ordering compared with structures produced using the basic GLAD algorithms. Reproduced from [64]. With permission from D.A. Gish.

and Brett investigated in detail how the selected γ parameter influences column uniformity when depositing onto a square lattice of seeds [11]. Figure 3.18 shows square-spiral columns deposited using phisweep with γ from (a) 0° to (e) 45°. Films were deposited onto identical lattice geometries, at $\alpha = 85°$ and with $q = 15$ nm.

In the $\gamma = 0°$ case (Figure 3.18a), which corresponds to conventional GLAD, the uncontrolled fanning leads to deleterious effects on the final film. The column uniformity is poor and significant broadening of the columnar structures is observed. For $\gamma = 45°$ (Figure 3.18e), the phisweep technique provides improved shadowing control and improved columnar structures are obtained in comparison with the previous case. However, column bifurcation and

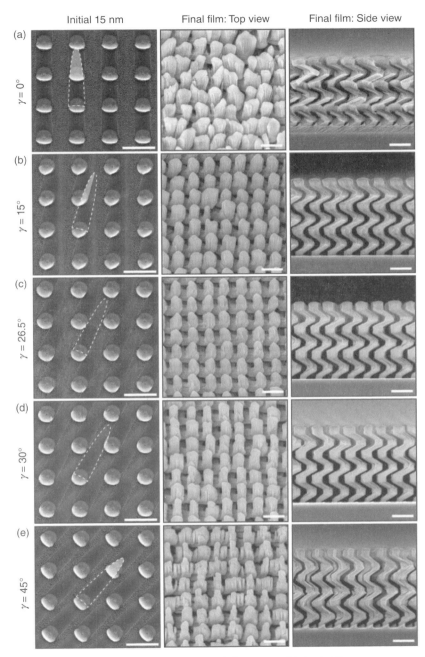

Figure 3.18 Effect of sweep angle γ on square-spiral structure fabricated from a square seed lattice. Structural uniformity in the square-spiral columns is achieved by maximizing the 2D shadow coverage, achieved for (c) $\gamma = 26.5°$. The scale bar indicates 1000 nm. Reproduced from [11]. © IOP Publishing. Reproduced by permission of IOP Publishing. All rights reserved.

nonuniformity are still observed, although to a lesser extent than at $\gamma = 0°$. Using intermediate γ values yields even greater structural improvements, as can be seen in Figure 3.18b–d, with maximum structural uniformity being achieved at $\gamma = 26.5°$. While this result is at first contradictory with the concept that a sweep angle of $\gamma = 45°$ produces the highest 2D shadowing uniformity, it is explained by the 2D shadowing analysis of Section 3.2.6. In that analysis, offsetting the incident vapour and the seed lattice vector by $26.5°$ maximizes the planar shadow coverage. This is clearly seen in the first column of Figure 3.18, which shows the initial deposition (to a thickness q) onto the seed layer. Of the set, the $\gamma = 0°$ case shows the most inter-seed deposition, followed by the $\gamma = 45°$ film. Inter-seed deposition represents vapour flux that is not restricted by shadowing and contributes to uncontrolled column growth. Rather than depositing onto the column apexes during growth as desired, this vapour instead condenses onto nearby column sidewalls, leading to column broadening and potential bifurcation effects. Moving to intermediate γ values improves the shadow coverage of the unit cell, with an optimum obtained at $\gamma = 26.5°$. For this seed configuration, this provides the best possible control over the local shadowing environment and yields precisely arrayed, high-uniformity nanostructures.

REFERENCES

[1] Kennedy, S.R., Brett, M.J., Toader, O., and John, S. (2002) Fabrication of tetragonal square spiral photonic crystals. *Nano Letters*, **2** (1), 59–62.

[2] Kiema, G.K., Jensen, M.O., and Brett, M.J. (2005) Glancing angle deposition thin film microstructures for microfluidic applications. *Chemistry of Materials*, **17**, 4046–4048.

[3] Taschuk, M.T., Hawkeye, M.M., and Brett, M.J. (2010) Glancing angle deposition, in *Handbook of Deposition Technologies for Films and Coatings* (ed. P.M. Martin), 3rd edn, Elsevier, pp. 621–678.

[4] Jensen, M. and Brett, M. (2005) Square spiral 3D photonic bandgap crystals at telecommunications frequencies. *Optics Express*, **13**, 3348–3354.

[5] Summers, M.A., Tabunshchyk, K., Kovalenko, A., and Brett, M.J. (2009) Fabrication of 2D–3D photonic crystal heterostructures by glancing angle deposition. *Photonics and Nanostructures – Fundamentals and Applications*, **7**, 76–84.

[6] Alagoz, A.S. and Karabacak, T. (2011) Fabrication of crystalline semiconductor nanowires by vapor-liquid-solid glancing angle deposition (VLS-GLAD) technique, in *Semiconductor Nanowires – From Fundamentals to Applications* (eds V. Schmidt, L. Lauhon, T. Fukui, G. Wang, and M. Björk), vol. 1350 of *MRS Proceedings*, Materials Research Society, Warrendale, PA, doi:10.1557/opl.2011.1005.

[7] Suzuki, M., Hamachi, K., Hara, H. *et al.* (2011) Vapor-liquid-solid growth of Ge nanowhiskers enhanced by high-temperature glancing angle deposition. *Applied Physics Letters*, **99**, 223107.

[8] Beaudry, A.L., Tucker, R.T., LaForge, J.M. *et al.* (2012) Indium tin oxide nanowhisker morphology control by vapour-liquid-solid glancing angle deposition. *Nanotechnology*, **23**, 105608.

[9] Tucker, R.T., Beaudry, A.L., LaForge, J.M. *et al.* (2012) A little ribbing: Flux starvation engineering for rippled indium tin oxide nanotree branches. *Applied Physics Letters*, **101**, 193101.

[10] Jensen, H.M.O. (2005) Photonic crystal engineering in glancing angle deposition thin films, PhD thesis, University of Alberta.

[11] Summers, M.A. and Brett, M.J. (2008) Optimization of periodic column growth in glancing angle deposition for photonic crystal fabrication. *Nanotechnology*, **19**, 415203.

[12] Patzig, C., Khare, C., Fuhrmann, B., and Rauschenbach, B. (2010) Periodically arranged Si nanostructures by glancing angle deposition on patterned substrates. *Physica Status Solidi (b)*, **247**, 1322–1334.

[13] Summers, M.A. (2009) Periodic thin films by glancing angle deposition, PhD thesis, University of Alberta.

[14] Ye, D.X., Ellison, C.L., Lim, B.K., and Lu, T.M. (2008) Shadowing growth of three-dimensional nanostructures on finite size seeds. *Journal of Applied Physics*, **103**, 103531.

[15] Patzig, C., Fuhrmann, B., Leipner, H.S., and Rauschenbach, B. (2009) Silicon nanocolumns on nanosphere lithography templated substrates: effects of sphere size and substrate temperature. *Journal of Nanoscience and Nanotechnology*, **9**, 1985–1991.

[16] Horn, M.W., Pickett, M.D., Messier, R., and Lakhtakia, A. (2004) Blending of nanoscale and microscale in uniform large-area sculptured thin-film architectures. *Nanotechnology*, **15**, 303–310.

[17] Kennedy, S.R. and Brett, M.J. (2003) Porous broadband antireflection coating by glancing angle deposition. *Applied Optics*, **42**, 4573–4579.

[18] Liu, D., Benstetter, G., Lodermeier, E., and Vancea, J. (2003) Influence of the incident angle of energetic carbon ions on the properties of tetrahedral amorphous carbon (ta-C) films. *Journal of Vacuum Science and Technology A*, **21**, 1665–1670.

[19] Singh, J.P., Liu, D.L., Ye, D.X. *et al.* (2004) Metal-coated Si springs: Nanoelectromechanical actuators. *Applied Physics Letters*, **84**, 3657–3659.

[20] Ye, D.X., Karabacak, T., Lim, B.K. *et al.* (2004) Growth of uniformly aligned nanorod arrays by oblique angle deposition with two-phase substrate rotation. *Nanotechnology*, **15**, 817–821.

[21] Kennedy, S.R. and Brett, M.J. (2004) Advanced techniques for the fabrication of square spiral photonic crystals by glancing angle deposition. *Journal of Vacuum Science and Technology B*, **22**, 1184–1190.

[22] Singh, J.P., Karabacak, T., Ye, D.X. *et al.* (2005) Physical properties of nanostructures grown by oblique angle deposition. *Journal of Vacuum Science and Technology B*, **23**, 2114–2121.

[23] Gaire, C., Ye, D.X., Tang, F. *et al.* (2005) Mechanical testing of isolated amorphous silicon slanted nanorods. *Journal of Nanoscience and Nanotechnology*, **5** (11), 1893–1897.

[24] Ye, D.X., Karabacak, T., Picu, R.C. *et al.* (2005) Uniform Si nanostructures grown by oblique angle deposition with substrate swing rotation. *Nanotechnology*, **16** (9), 1717–1723.

[25] Gaire, C., Ye, D.X., Lu, T.M. *et al.* (2008) Deformation of amorphous silicon nanostructures subjected to monotonic and cyclic loading. *Journal of Material Research*, **23**, 328–335.

[26] Jensen, M.O. and Brett, M.J. (2005) Embedded air and solid defects in periodically structured porous thin films. *Nanotechnology*, **16** (11), 2639–2646.

[27] Jensen, M.O. and Brett, M.J. (2005) Porosity engineering in glancing angle deposition thin films. *Applied Physics A*, **80**, 763–768.

[28] Malac, M., Egerton, R.F., Brett, M.J., and Dick, B. (1999) Fabrication of submicrometer regular arrays of pillars and helices. *Journal of Vacuum Science and Technology B*, **17**, 2671–2674.

[29] Dick, B., Brett, M.J., Smy, T.J. *et al.* (2000) Periodic magnetic microstructures by glancing angle deposition. *Journal of Vacuum Science and Technology A*, **18**, 1838–1844.

[30] Dick, B., Brett, M.J., and Smy, T. (2003) Controlled growth of periodic pillars by glancing angle deposition. *Journal of Vacuum Science and Technology B*, **21**, 23–28.

[31] Summers, M., Djufors, B., and Brett, M. (2005) Fabrication of silicon submicrometer ribbons by glancing angle deposition. *Journal of Micro/Nanolithography, MEMS, and MOEMS*, **4**, 033012.

[32] Gish, D.A., Summers, M.A., and Brett, M.J. (2006) Morphology of periodic nanostructures for photonic crystals grown by glancing angle deposition. *Photonics and Nanostructures – Fundamentals and Applications*, **4**, 23–29.

[33] Kesapragada, S.V., Sotherland, P.R., and Gall, D. (2008) Ta nanotubes grown by glancing angle deposition. *Journal of Vacuum Science and Technology B*, **26**, 678–681.

[34] Patzig, C., Zajadacz, J., Zimmer, K. *et al.* (2009) Patterning concept for sculptured nanostructures with arbitrary periods. *Applied Physics Letters*, **95**, 103107.

[35] Liu, Y.J., Chu, H.Y., and Zhao, Y.P. (2010) Silver nanorod array substrates fabricated by oblique angle deposition: morphological, optical, and SERS characterizations. *Journal of Physical Chemistry C*, **114**, 8176–8183.

[36] Dick, B., Sit, J.C., Brett, M.J. *et al.* (2001) Embossed polymeric relief structures as a template for the growth of periodic inorganic microstructures. *Nano Letters*, **1**, 71–73.

[37] Tanto, B., Ten Eyck, G., and Lu, T.M. (2010) A model for column angle evolution during oblique angle deposition. *Journal of Applied Physics*, **108**, 026107.

[38] Krabbe, J.D., Leontyev, V., Taschuk, M.T. *et al.* (2012) Square spiral photonic crystal with visible bandgap. *Journal of Applied Physics*, **111**, 064314.

[39] Wang, J., Huang, H., Kesapragada, S.V., and Gall, D. (2005) Growth of Y-shaped nanorods through physical vapor deposition. *Nano Letters*, **5** (12), 2505–2508.

[40] Zhou, C. and Gall, D. (2006) The structure of Ta nanopillars grown by glancing angle deposition. *Thin Solid Films*, **515**, 1223–1227.

[41] Zhou, C.M. and Gall, D. (2006) Branched Ta nanocolumns grown by glancing angle deposition. *Applied Physics Letters*, **88**, 203117.

[42] Kesapragada, S.V. and Gall, D. (2006) Two-component nanopillar arrays grown by glancing angle deposition. *Thin Solid Films*, **494**, 234–239.

[43] Zhou, C.M. and Gall, D. (2007) Competitive growth of Ta nanopillars during glancing angle deposition: effect of surface diffusion. *Journal of Vacuum Science & Technology A*, **25**, 312–318.

[44] Zhou, X., Virasawmy, S., Knoll, W. *et al.* (2007) Profile simulation and fabrication of gold nanostructures by separated nanospheres with oblique deposition and perpendicular etching. *Plasmonics*, **2**, 217–230.

[45] Zhou, C. and Gall, D. (2008) Two-component nanorod arrays by glancing-angle deposition. *Small*, **4**, 1351–1354.

[46] Dolatshahi-Pirouz, A., Jensen, T., Vorup-Jensen, T. *et al.* (2010) Synthesis of functional nanomaterials via colloidal mask templating and glancing angle deposition (GLAD). *Advanced Engineering Materials*, **12**, 899–905.

[47] Dolatshahi-Pirouz, A., Sutherland, D., Foss, M., and Besenbacher, F. (2011) Growth characteristics of inclined columns produced by glancing angle deposition (GLAD) and colloidal lithography. *Applied Surface Science*, **257**, 2226–2230.

[48] Kosiorek, A., Kandulski, W., Chudzinski, P. *et al.* (2004) Shadow nanosphere lithography: simulation and experiment. *Nano Letters*, **4**, 1359–1363.

[49] Fuhrmann, B., Leipner, H.S., Höche, H.R. *et al.* (2005) Ordered arrays of silicon nanowires produced by nanosphere lithography and molecular beam epitaxy. *Nano Letters*, **5**, 2524–2527.

[50] Zhou, C.M. and Gall, D. (2007) Growth competition during glancing angle deposition of nanorod honeycomb arrays. *Applied Physics Letters*, **90**, 093103.

[51] Zhou, C.M. and Gall, D. (2007) Surface patterning by nanosphere lithography for layer growth with ordered pores. *Thin Solid Films*, **516**, 433–437.

[52] Patzig, C., Rauschenbach, B., Fuhrmann, B., and Leipner, H.S. (2008) Growth of Si nanorods in honeycomb and hexagonal-closed-packed arrays using glancing angle deposition. *Journal of Applied Physics*, **103**, 024313.

[53] Patzig, C., Karabacak, T., Fuhrmann, B., and Rauschenbach, B. (2008) Glancing angle sputter deposited nanostructures on rotating substrates: experiments and simulations. *Journal of Applied Physics*, **104**, 094318.

[54] Zhou, C.M. and Gall, D. (2008) Development of two-level porosity during glancing angle deposition. *Journal of Applied Physics*, **103**, 014307.

[55] Khare, C., Fechner, R., Bauer, J. *et al.* (2011) Glancing angle deposition of Ge nanorod arrays on Si patterned substrates. *Journal of Vacuum Science and Technology A*, **29**, 041503.

[56] Khare, C., Fuhrmann, B., Leipner, H.S. *et al.* (2011) Optimized growth of Ge nanorod arrays on Si patterns. *Journal of Vacuum Science and Technology A*, **29**, 051501.

[57] Bauer, J., Weise, M., Rauschenbach, B. *et al.* (2012) Shape evolution in glancing angle deposition of arranged germanium nanocolumns. *Journal of Applied Physics*, **111** (10), 104309.

[58] Fujii, T., Aoki, Y., Fushimi, K. *et al.* (2010) Controlled morphology of aluminum alloy nanopillar films: from nanohorns to nanoplates. *Nanotechnology*, **21** (39), 395302.

[59] Kannarpady, G.K., Khedir, K.R., Ishihara, H. *et al.* (2011) Controlled growth of self-organized hexagonal arrays of metallic nanorods using template-assisted glancing angle deposition for superhydrophobic applications. *ACS Applied Materials and Interfaces*, **3** (7), 2332–2340.

[60] Tanvir, M.T., Fujii, T., Aoki, Y. *et al.* (2011) Dielectric properties of anodic films on sputter-deposited Ti-Si porous columnar films. *Applied Surface Science*, **257**, 8295–8300.

[61] Taschuk, M.T., Chai, J., Buriak, J.M., and Brett, M.J. (2009) Optical characterization of pseudo-ordered nanostructured thin films. *Physica Status Solidi (c)*, **6**, S127–S130.

[62] Haynes, C.L. and Van Duyne, R.P. (2001) Nanosphere lithography: a versatile nanofabrication tool for studies of size-dependent nanoparticle optics. *Journal of Physical Chemistry B*, **105** (24), 5599–5611.

[63] Haynes, C.L., McFarland, A.D., Smith, M.T. *et al.* (2002) Angle-resolved nanosphere lithography: manipulation of nanoparticle size, shape, and interparticle spacing. *Journal of Physical Chemistry B*, **106** (8), 1898–1902.

[64] Gish, D.A. (2010) Morphology control and localized surface plasmon resonance in glancing angle deposition, Master's thesis, University of Alberta.

4 Properties and Characterization Methods

4.1 INTRODUCTION

As film properties are highly sensitive to the underlying microstructure, it is therefore possible to carefully control many physical characteristics of the deposited film using GLAD technology. These properties can often be tuned over a very wide range, and novel physical effects not seen in conventional coatings can be realized as well. Many properties of GLAD films have been investigated, including the structural characteristics, film density, surface area, pore structure, film crystallinity, electrical properties and mechanical characteristics. (GLAD film optical properties have been heavily investigated and are examined in Chapter 5.) Significant experimental work has been conducted in these areas, driven by the desire to understand, control and exploit the properties of GLAD films. These efforts have revealed consistent experimental trends with respect to key processing variables, most notably α, film thickness and substrate temperature. Although these trends are highly reproducible, there are no 'handbook values' for the properties of GLAD films as the exact characteristics depend sensitively on the specific experimental conditions used during fabrication. Film characterization thus forms an integral part of GLAD research and development, necessary to quantify film performance, determine the effect of process variables on film properties, as well as optimize and monitor film fabrication. The experimental techniques employed are often well developed and/or highly specialized, and dedicated reference books exist for every method and variant thereof. Alongside the discussion of specific properties, this chapter summarizes and discusses the experimental approaches that are commonly and successfully employed in the characterization of GLAD films. While some of these approaches are fairly standard in materials science and nanotechnology, we identify the experimental subtleties that tend to arise in GLAD research due to the unique microstructures and characteristics of GLAD films.

4.2 STRUCTURAL ANALYSIS WITH ELECTRON MICROSCOPY

The SEM is a versatile tool for structural characterization, providing critical inspection capabilities for GLAD-fabricated nanostructures. Structural characterization via an SEM consists of three basic steps: sample preparation, image acquisition in the instrument and analysis of the captured image. To prepare a sample for the SEM it must often be fractured to expose the internal microstructure and then rigidly mounted in the desired orientation on

Glancing Angle Deposition of Thin Films: Engineering the Nanoscale, First Edition.
Matthew M. Hawkeye, Michael T. Taschuk and Michael J. Brett.
© 2014 John Wiley & Sons, Ltd. Published 2014 by John Wiley & Sons, Ltd.

a specimen holder. Image acquisition in an SEM involves scanning a focused electron beam over the sample and measuring the corresponding secondary electron emission rate, which is highly sensitive to the local surface topography. The SEM image quality is determined by many factors, including the instrument performance (e.g. electron optics, magnification, digitization), imaging parameters (e.g. accelerating voltage, electron beam current, working distance) and the sample type (e.g. material, pre-coated). Resolutions down to a single nanometre can be achieved in modern instruments. The imaging parameters can all affect subsequent measurements and should be systematically recorded and controlled to ensure standardization and repeatability of image acquisition. The image analysis step can involve qualitative examination of the film morphology and/or quantitative estimation of various structural parameters (e.g. thickness, column size, column spacing). This analysis can be either manual or computerized. Elemental information can also be obtained by collecting higher energy backscattered electrons, which are scattered at a rate proportional to atomic number. Many SEMs often provide additional elemental characterization tools, such as energy-dispersive X-ray spectroscopy for further elemental analysis. The following discusses SEM investigation of GLAD microstructures in greater depth; readers searching for greater detail are directed to the book by Goldstein *et al.*, which provides in-depth information on the theory and practice of scanning electron microscopy [1].

4.2.1 Practical aspects

Sample preparation

As-deposited samples are generally unsuitable for SEM inspection, and multiple preparation steps are often required, depending on the image and information ultimately desired. Therefore, deposited samples are commonly fractured to reduce the size of the sample and satisfy the physical clearance requirements of the SEM. To view the internal film microstructure, it is also important to fracture the sample and expose the columnar morphology at a point away from the substrate edge to avoid edge-related effects. The fracturing process is obviously destructive to the sample. A convenient approach is to fabricate additional witness samples on brittle, crystalline substrates, such as Si, for subsequent SEM analysis. These substrates fracture easily along cleavage planes, making it straightforward to prepare samples for the SEM. Note that the surface properties of the witness sample (such as roughness, surface chemistry and cleanliness) should be similar to those of the actual substrates. An initial experimental check should be performed to verify equivalent growth conditions and ensure reliable comparisons are possible. To cleave a crystalline substrate, make a small scratch (typically with a diamond scribe) on the substrate oriented along the desired cleavage plane. Gently bending the substrate applies a tensile stress to the scratch, creating a crack that propagates along the crystal plane. The sample can be cleaved multiple times in this manner to obtain the desired specimen size, which is then affixed to the SEM stub with a clamp or conductive adhesive. When depositing onto brittle, crystalline witness samples is not practical, other preparation methods can be used to expose the inner microstructure, such as substrate dicing or microtoming.

In all cases, it is important to minimize damage to the exposed microstructure that occurs during fracture and subsequent handling, to obtain a faithful characterization of the film morphology.

Sample mounting and scanning electron microscope views

In preparing the specimen, it is necessary to properly mount the sample to view the film microstructure from the intended perspective. It is important to carefully align the sample during mounting and inspection to ensure accurate characterization. Three basic viewing directions are common, summarized in Figure 4.1. The columnar morphology is quite naturally visualized by fracturing the sample and then imaging the freshly exposed edge from the

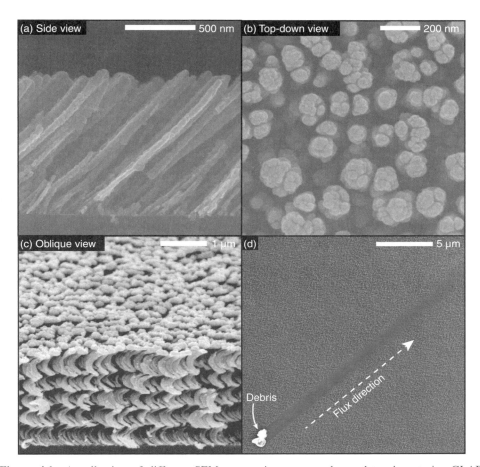

Figure 4.1 A collection of different SEM perspectives commonly used to characterize GLAD microstructures. (a) Side-view images, here of a tilted columnar film, reveal the columnar morphology and allow accurate measurement of various structural parameters such as film thickness, column tilt, and so on. (b) Top-down views, here of a vertical columnar film, show the film surface and column apexes and can be used to determine intercolumn spacing, column cross-section, and so on. (c) The oblique viewpoint, here of a helical columnar film, allows the column morphology and surface topography to be viewed simultaneously, thanks to the large depth-of-field (DOF) of the SEM. (d) Debris particles cast comet-like shadows during deposition ($\alpha = 88°$ in this case), useful for alignment of the SEM sample during inspection. (a) Reproduced from [2]. With permission from J.B. Sorge. (c) Reproduced from [3]. With permission from P.C.P. Hrudey.

side (Figure 4.1a). The side view allows straightforward inspection of the column structures and measurement of several structural parameters, such as film thickness, column width, aspect ratio and tilt angle. These measurements are most accurate when the film edge is oriented at right angles to the observer, thus eliminating geometric distortions. The ability to view the complete column as well as specific regions of interest at high magnification is very useful, providing information about the entire growth process, revealing extinct columns and allowing characterization of structural nonuniformities/defects. Note that background columns are readily visible due to the high DOF provided by the SEM, leading to a denser appearance. Top-down views (Figure 4.1b) are also useful, and several structural charac-teristics can be obtained via top-down inspection, including intercolumn spacing, column cross-sectional shape and porosity. An advantage of the top-down viewpoint is that it can be obtained without sample fracture, enabling nondestructive inspection of small samples. However, it only provides information about the tops of visible columns, including extinct columns, which often appear darker, as can be seen in Figure 4.1b). Tilted views of the sample are also useful, and tilting sample holders or angled SEM mounts (45° mounts are common) can be used for this purpose (Figure 4.1c). Thanks to the high DOF, the tilted view allows the surface topography to be inspected while simultaneously showing the columnar microstructure. Note that the non-normal viewing angle introduces a geometric distortion and complicates quantitative measurements. Excellent views can be obtained by examining points formed at cleavage corners of the fractured sample. Tilting mounts can be used to look down at the film along the column axis, a viewpoint that is particularly useful for characterizing the axial cross-section of inclined columnar structures (e.g. see Figure 2.24).

When mounting substrates for later SEM analysis in the deposition system, it is important to record the substrate orientation (including all crystal planes) relative to the GLAD depo-sition plane. This will ensure that, when preparing the sample for SEM, the microstructure is oriented as desired. For example, when preparing tilted columnar microstructures, it is convenient to align the deposition plane parallel to the crystal plane of a crystalline substrate. On transparent samples (e.g. glass, quartz), alignment arrows can be drawn on the bottom surface and aligned to the deposition plane. Aligning the substrate and cleaving the samples along alignment markers ensures that subsequent side views are oriented at right angles to the column tilt, thereby improving measurement accuracy. Debris particles on the substrate can be used to align the sample during SEM operation. Figure 4.1d shows the comet-like structure created by a micrometre-sized particle present on the substrate during GLAD. When depositing at high α (88° in this instance), small particles cast very long shadows that are easily detected and provide excellent alignment marks during SEM inspection of the film surface. On very clean substrates, controlled structures such as microspheres or lithographically patterned features can be placed on the substrate, typically in an innocuous place far from any sensitive, active region of the film. If the dimensions of the structure are accurately defined, this approach can also be used to characterize the deposition angle (via the shadow length) and vapour collimation (via the definition of the shadow boundaries).

Sample charging effects

When inspecting GLAD structures fabricated from electrically insulating materials or deposited onto nonconductive substrates (e.g. glass, polymer), charging effects can be prob-lematic. In this phenomenon, excess charges build up on the surface, creating local electrical potentials that distort the resulting image. The standard method to minimize charging is to sputter coat the sample with a thin (<10 nm) conducting layer, typically Au or Cr. Note that such coatings are granular and it may be difficult to achieve conformal coating on highly porous GLAD structures. The conductive coating may also mask very fine microstructural features. Decreasing the electron beam current and voltage and increasing the scan speed will also reduce the charging effect.

Uniformity and multiple inspection points

Deposition nonuniformity, related to system geometry (Section 7.4), produces variations in the film microstructure across the sample, and a methodical approach is thus required to ensure reliable, consistent data. For comparative purposes, witness samples should be mounted in the same place every run, and the same locations on the witness samples should be examined every time. It is particularly important to consider uniformity effects when comparing films produced by different substrate motion algorithms as the thickness/parameter uniformity patterns can be very different (e.g. see Figure 7.10). During SEM inspection, the sample should be examined at several different points to ensure that the images obtained are accurate representations of the whole sample.

4.2.2 Scanning electron microscope image analysis

SEM images are generally accompanied by a calibrated scale bar, the accuracy of which may be independently verified against a range of traceable standards. Direct, manual measurements can thus be performed to estimate relevant structural parameters, such as thickness, column width, intercolumn separation, helical pitch, and so on. During manual structure analysis, it is important to establish sound methodologies and consider several factors. The statistical reliability of the result should be established by making a sufficiently large number of measurements. Manual measurements always contain an element of observer subjectivity and the results can be strongly influenced by several factors, including the observer expertise, SEM resolution, magnification, image brightness and contrast. Mitigating this subjectivity requires consistent image output from the SEM and comparing measurement results from multiple, independent observers.

More recently, many researchers have pursued computerized numerical analysis of digital SEM images. This approach provides important benefits: the subjectivity inherent in manual analysis is reduced, the repeatability of the structural measurement is improved, the inherent automation provides greater throughput and a large number of measurements can be taken to increase the statistical rigour, and more sophisticated structural analysis can be performed. Where numerical parameters must be specified – for example, in setting pixel thresholding levels – sensitivity analysis can be easily performed to estimate the corresponding uncertainty. A range of image analysis techniques exists [4], and many software toolbox packages are available (e.g. Matlab, ImageJ, Mathematica, and Scanning Probe Image Processor SPIP). The following discusses several common image processing techniques used to analyse GLAD film microstructure.

Image preprocessing techniques

In numerical SEM analysis approaches, it is fairly common to apply one or more preprocessing steps to the SEM image before carrying out the intended analysis. A notable example is pixel thresholding, where the initial greyscale SEM image is converted into a binary, black-and-white version. The conversion is based on a simple algorithm: all pixels with brightness above a threshold value are mapped to white and all pixels below the threshold are mapped to black. Thresholding is useful in GLAD film analysis to separate the object of interest (typically the columns) from background features (e.g. the substrate or extinct columns). The key parameter is the threshold value, which can be set manually, guided by operator experience or based on features of the pixel histogram (distribution of pixel brightnesses), or using one of many automatic thresholding algorithms [5]. The automatic approach is generally preferable as it ensures repeatability and reduces subjectivity, and most image analysis programs provide several algorithms from which to choose.

In addition to thresholding, there are many specialized image processing algorithms available for denoising, edge enhancement, segmentation, and so on [4] that can be combined to provide advanced image preparation and analysis [6]. However, because each step can influence the result of subsequent numerical image analyses it is important to properly test and document the algorithms applied and the specific parameters employed to ensure the repeatability of the techniques (e.g. as done in Ref. [7]). When testing out new image analysis algorithms or applying them to new microstructures, the processed images should be compared against the original versions to ensure that unacceptable artefacts or distortions are not introduced during image processing.

Column tilt angle measurement

Investigating the relationship between the deposition angle α and the column tilt angle β has long been of interest in oblique deposition. Manual measurement of β is straightforward from side-view SEM images. However, the accuracy of the measurement is affected by rotational misalignment of the sample during SEM imaging as any deviation from a perfectly square viewpoint will lead to an underestimation of β. The measurement accuracy can be improved by taking two separate column tilt measurements β_1 and β_2 along nonparallel cleaved edges of the substrate. The corrected column tilt β_{corr} can then be calculated according to [8]

$$\tan \beta_{corr} = \frac{1}{\sin \psi} \sqrt{\tan^2 \beta_1 + \tan^2 \beta_2 - 2 \tan \beta_1 \tan \beta_2 \cos \psi}, \qquad (4.1)$$

where ψ is the angle between the substrate edges. Depositing onto (110) or (111) Si wafers can be useful in this regard, resulting in a ψ accurately defined by the orientation of the crystal planes.

Several researchers have also determined β using a rotational autocorrelation image analysis approach [9–11], a numerical technique that characterizes the entire image at once and reduces the measurement subjectivity. The greyscale SEM is first converted into a binary black-and-white image using a thresholding procedure (see Section 4.2.2). The processed image thus corresponds to an intensity map $L(x, y)$, where x and y are the spatial coordinates (ranging from zero to maximum values of X and Y respectively) and L is either one (film material) or zero (void). A column orientation function is then calculated according to [9]

$$F(\theta) = \frac{\cos^2 \theta}{L_a(1 - L_a)} \int_0^X \left\{ \int_0^{Y/\cos\theta} [L(x, y) - L_a] dy \right\}^2 dx, \qquad (4.2)$$

where L_a is the mean intensity of the binary image and θ is an angle measured from the substrate normal. $F(\theta)$ is maximized for $\theta = \beta$, thus providing an accurate β measurement

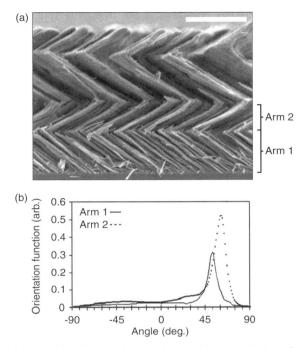

Figure 4.2 (a) SEM image of a zigzag column microstructure, consisting of a series of tilted column arms with alternating orientation. (b) The orientation function (Equation 4.2) peaks at an angle corresponding to β, enabling the orientation of the different column arms to be accurately determined. This numerical analysis confidently reveals the subtle β variation between arm 1 and arm 2, related to a change in their respective deposition geometries (Section 2.7.1). The scale bar indicates 5 μm. Reproduced with permission from [11]. © 2002, American Vacuum Society.

over the entire image. A larger image can be selectively cropped to perform segmented analysis on specific microstructural regions, and additional images of the same film can be quickly processed to further improve the analysis. Harris *et al.* used this computational approach to examine β in zigzag column microstructures [11]. Figure 4.2a shows an SEM image of zigzag columns fabricated by periodic 180° φ rotations to deposit a series of alternating column arms. The authors calculated Equation 4.2 after cropping the figure to isolate each arm segment, producing the results shown in Figure 4.2b for arm 1 and arm 2. As revealed by the peak position of the autocorrelation function, arm 2 ($\beta = 61°$) is tilted farther from the substrate normal than arm 1 ($\beta = 53°$). The β change is a consequence of the modified surface geometry when depositing onto a pre-existing column arm, as was discussed in Section 2.7.1. The result highlights the effectiveness of the autocorrelation approach in quantifying subtle microstructural variations in a robust way.

Fourier analysis techniques

Fourier analysis is a powerful tool for structural characterization, converting an image from the real-space domain into its complementary spatial frequency representation. Spatial patterns present in the image can be quantitatively analysed, allowing characterization of column arrangements and intercolumn spacing in GLAD films [12–20]. In Fourier analysis, the most commonly calculated quantity is the power spectral density (PSD), which determines the intensities of different spatial frequencies in the image content. Given an SEM image specified by a regularly spaced $N_x \times N_y$ array of brightness data values $L(x, y)$, the PSD is calculated as

$$I_{\mathrm{PSD}}(k_x, k_y) = \frac{1}{N_x N_y} \left| \sum_{x=1}^{N_x} \sum_{y=1}^{N_y} L(x, y) e^{-ixk_x \Delta x} e^{-iyk_y \Delta y} \right|^2, \qquad (4.3)$$

where k_x and k_y are spatial frequency components in the x- and y-direction respectively.

Figure 4.3 shows an example of applying Fourier analysis to quantitatively study GLAD film microstructure (from Refs [14] and [13]). Figure 4.3a shows a top-down SEM image of a vertical-column film (Si material, $\alpha = 87°$, 140 nm thick) with the inset (b) showing the calculated PSD of the image. The diffuse, ring-like structure of the PSD indicates that, while there is no long-range order in the film, there is a dominant spatial frequency at a value equal to the ring radius, which indicates short-range spatial order with a characteristic nearest-neighbour column spacing. The characteristic spacing can be estimated by calculating the radial average of the PSD as shown in Figure 4.3c for a series of films deposited at different α (curves offset for clarity). As α increases, the PSD peak shifts to shorter spatial frequencies, indicating an increase in the average column spacing. The FWHM of the PSD peak can also be measured, quantifying the variance of the intercolumn spacing. Note that while the lack of long-range order is typical of all GLAD films, a consequence of the random nucleation process during early film growth, the shape of the diffuse ring is sensitive to the specific symmetry of the microstructure. The characteristic column spacing can vary along different directions relative to the orientation of the column structure [17]. Other structural parameters can be measured from the PSD as well: the average column size can be estimated by determining the planar fill fraction [12], and the fractal dimension can be estimated by examining the high spatial frequency scaling of the PSD [15, 21].

When capturing SEM images for later Fourier processing, the SEM image magnification should be kept low to ensure that a sufficient number of columns are imaged to produce statistically reliable counts. However, the spatial resolution of the image (pixels per nanometre) must remain large enough to avoid aliasing artefacts when converted to the frequency domain. As with column tilt measurement described above, the SEM image can be cropped to selectively process specific microstructural regions. Previously discussed preprocessing techniques can also be applied to the image, with blurring algorithms used to reduce high spatial frequency pixel noise and thresholding procedures applied to better separate the column regions and background regions. However, such processing will affect the end results and should be properly documented to ensure repeatability.

As an added note, Fourier analysis is also commonly applied to surface height profile data obtained using atomic force microscopy to obtain similar information about structural periodicity in GLAD films [22–28].

Particle and pore measurement and analysis

Another useful image processing technique uses particle analysis methods to automatically locate, count and measure column features in the SEM image. Such particle characterization is laborious and subjective when performed manually, whereas computerized approaches allow

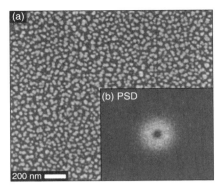

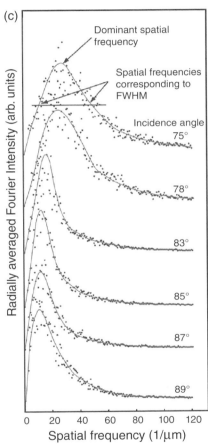

Figure 4.3 Using Fourier analysis techniques to quantify column distribution statistics. (a) Top-down SEM view of a 140 nm thick Si film deposited at $\alpha = 87°$ and (b) the corresponding 2D Fourier amplitude calculated for the SEM image. The diffuse, ring-like structure of the Fourier spectrum indicates the presence of a dominant spatial frequency in the original image, which corresponds to the average column spacing. (c) Radially averaged Fourier intensities calculated from SEM images of films deposited at different α. The dominant peak shifts to lower spatial frequencies as α increases, indicating that the average intercolumn spacing is increasing. The full width at half maximum (FWHM) of the peak also provides a useful estimation of the spacing variation as well. (a) and (b) Reproduced with permission from [14]. © 2005, AIP Publishing LLC. (c) Reproduced from [13]. © IOP Publishing. Reproduced by permission of IOP Publishing. All Rights Reserved.

rapid and repeatable quantification of many thousands of columns. Image preprocessing is a key factor in the particle analysis process, needed to segment the particle features from the background. Thresholding algorithms are useful in this regard, often with denoising, smoothing or edge-enhancing filters applied as well to improve the definition of the column edges [6, 29]. However, thresholding alone is often insufficient and may incorrectly group together closely spaced or bundled columns that are obviously distinct. Partially bifurcated columns can present a particularly difficult situation as it can be unclear when a single column transitions to multiple, distinct sub-columns. Additional processing may be required to algorithmically separate these features, using cut-at-local-minimum algorithms [14], aspect-ratio limiting [13, 14] or watershed separation algorithms [6, 29].

Figure 4.4 presents a collection of reported results using particle analysis techniques to examine GLAD-fabricated column microstructures. Figure 4.4a shows a segmented SEM image, where the automatically detected columns are highlighted using different grey-scale levels [14]. From such an image, statistical analysis of structural parameters can be performed, as shown in Figure 4.4) where the column diameter distribution has been calculated to reveal a mean diameter of 28 nm with a 9 nm standard deviation. Other parameters can also be analysed, such as porosity, column aspect ratio, orientation and separation [13, 29–31]. Specific microstructural motifs can also be selectively isolated and analysed with advanced object recognition methods [32]. Siewert *et al.* carried out a high-throughput analysis of column microstructures using a series of sophisticated image processing techniques [6]. Rather than examining typical top-down or side-view SEM images, the authors removed the columns from the substrate with a sonication treatment and dispersed them across a different wafer, producing a collection of isolated columns on the surface, as shown in Figure 4.4c. To extract the width profile of each column, a multistep workflow was devised. A denoising filter was first applied, before automatically identifying column structures using an advanced watershed-segmentation algorithm (results shown in Figure 4.4d). Each isolated column was then extracted (Figure 4.4e and f) and the column width profile measured (Figure 4.4g). With this procedure, the authors were able to analyse nearly 11 000 individual full-height and extinct columns to study growth scaling behaviours (Section 4.3.2).

4.2.3 Three-dimensional column imaging: tomographic sectioning

While SEM inspection is very useful in microstructural characterization and analysis, the internal film morphology of the GLAD film is hidden from view during regular imaging. Cleaving film sections does permit investigation of exposed cross-sections, but such imaging only provides a 2D slice through the 3D film morphology. In order to fully characterize the internal film morphology, Krause *et al.* recently developed a tomographic sectioning technique to obtain a 3D reconstruction of the complete GLAD film microstructure [33]. In this approach, the porous GLAD film (TiO$_2$, $\alpha = 85°$, vertical columns) was first filled with photoresist material via spin-coating and then mounted into a dual-beam instrument combining SEM functionality with FIB capabilities. Using the FIB, a thin layer of the film is removed over a specified area (Figure 4.5a), revealing a new planar cross-section of the microstructure for SEM imaging. The photoresist infiltration prevents columnar sections farther into the film from being imaged at the same time. Imaging a series of slices (e.g. Figure 2.21) through the film allows reconstruction of a highly detailed 3D tomographic image of the microstructure (Figure 4.5b). Numerical image analysis routines were applied

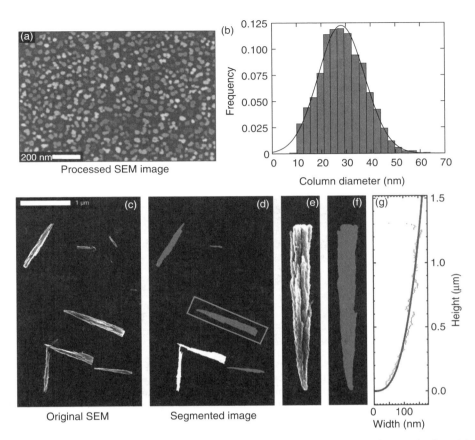

Figure 4.4 Examples of numerical particle analysis for column structure characterization. (a) A segemented SEM image, where specific particle detection algorithms have been employed to identify individual column features. (b) Various structural parameters can be automatically extracted from the segmented image and subjected to statistical analysis, such as this frequency distribution of the column size. (c–f) Image processing algorithms can be used to isolate individual columns dispersed on a substrate for subsequent characterization, used here to measure (g) the column width as a function of column height. Such automated approaches are well suited to high-throughput analysis of thousands of separate structures. (a) and (b) Reprinted with permission from [14]. © 2005, AIP Publishing LLC. (d)–(g) Reproduced with permission from [6]. Copyright 2012, Cambridge University Press.

to every slice to track individual column structural parameters with near-nanometre resolution in thickness. Figure 4.5c and d respectively show the column width and planar area as function of length for the four columns labelled in Figure 4.5b. Column 1 is an example of a column that survives to full height, whereas columns 2–4 all go extinct at different points of film growth. This behaviour is reflected in the column width and planar area thickness-scaling data. The diameter and area of column 1 increases steadily, a consequence of competition-induced broadening effects and unstable growth. Conversely, the other three columns diminish in size and eventually stop growing due to the shadowing-induced extinction processes. Tracking individual column growth with this 3D reconstruction technique also

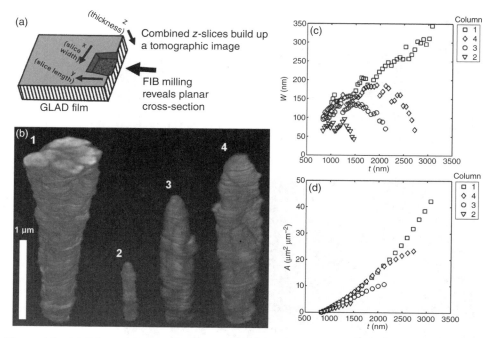

Figure 4.5 (a) Schematic of the FIB SEM tomography process. The FIB is used to progressively reveal a series of z-slices through the film, each of which is imaged via scanning electron microscopy. These separate SEM images are then combined to form (b) a 3D tomographic reconstruction of structures in the GLAD film, such as these four individual columns (1–4) shown beginning at $t = 840$ nm. Structures within a 3D image can then be analysed to provide high-resolution tracking of structural parameters such as (c) column width and (d) column area. The tomography provides a detailed view of the internal film microstructure that cannot be observed with conventional 2D scanning electron microscopy, enabling growth processes such as extinction and bifurcation to be studied in excellent detail. Reprinted with permission from [33]. © 2010, American Chemical Society.

revealed new insights into column bifurcation processes, as was discussed in Section 2.6.3. While the 3D tomographic imaging technique requires significantly more effort than normal SEM imaging, it provides unprecedented microstructural detail and understanding of GLAD growth processes.

4.2.4 Characterizing internal column structure with transmission electron microscope imaging

Although the SEM is capable of high resolution, it is a surface technique and unable to probe the internal column microstructure without resorting to the tomographic sectioning described in Section 4.2.3. However, the higher energy electrons used for imaging in transmission electron microscopy penetrate through the sample and provide superior characterization of the internal morphology. Figure 4.6 shows bright-field TEM images of four different

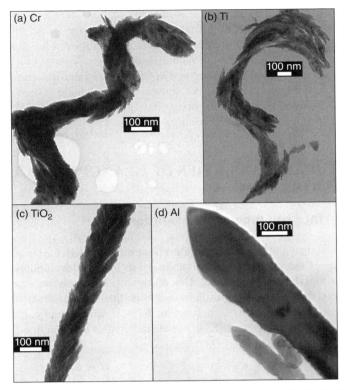

Figure 4.6 Bright-field TEM images of (a) a Cr helical column, (b) a Ti helical column, (c) a TiO$_2$ vertical column and (d) an Al vertical column. The high resolution provided by the TEM reveals internal columnar microstructure and highlights the different growth characteristics of the materials, related to their specific adatom mobilities and crystal structures. Reproduced from [34]. With permission from B.M. Djufors.

GLAD column structures [34]: (a) a Cr helical column, (b) a Ti helical column, (c) a TiO$_2$ vertical column and (d) an Al vertical column. The TEM images reveal significant microstructural differences between the columns that may not otherwise be revealed through basic SEM inspection. As explained through the evolutionary growth model (Section 2.6.1), the individual GLAD columns are themselves composed of smaller, bundled columnar fibres, a microstructure that can be seen in the TEM images. The separate fibre subunits can be clearly resolved throughout the column, giving the columns a feathered appearance under TEM inspection. This internal nanometre-scale fine structure is responsible for the microporosity found in GLAD films, which makes up a significant portion of the overall film surface area (Section 4.5). The exact fibre morphology depends on the larger column geometry. In the helical columns (Figure 4.6a and b), the orientation of the fibre units twists around a central axis following the chiral morphology of the larger column. In the case of vertical columns (Figure 4.6c), the fibres are inclined towards the column normal but also extend radially outwards, a consequence of the rapid φ rotation and vapour flux averaging. The TEM also reveals material-dependent microstructural differences, related to the specific growth kinetics of the deposited material. The fibre dimensions, bifurcation rates and crystal structures (obtained via the associated electron diffraction patterns) provide useful information as to the development of shadowing instabilities and surface mobilities in the growing columns. For example, the high surface mobility of Al during deposition promotes the coalescence of the fibres during column growth, resulting in a visibly denser microstructure (Figure 4.6d), and electron diffraction measurements (not shown) reveal a single-crystal morphology in the column. Transmission electron microscopy thus provides a useful tool for correlating morphology and crystallinity in GLAD columns (Section 4.5).

4.3 STRUCTURAL PROPERTIES OF GLANCING ANGLE DEPOSITION FILMS

4.3.1 Film surface roughness and evolution

Considerable theoretical and experimental effort has examined both the surface morphology and the dynamics of the surface growth in attempts to improve our understanding of and control over thin-film deposition process. This research is based on more general concepts of interface formation and nonequilibrium conditions that are universal to many growth processes [35]. For a surface with a height function $h(x, y, t)$, where x and y are the coordinates and t is the nominal film thickness, the surface characteristics are quantified using a pair of statistical parameters: the mean height $\overline{h}(t)$, defined as

$$\overline{h}(t) = \frac{1}{L^2} \sum_{x=0}^{L} \sum_{y=0}^{L} h(x, y, t),$$ (4.4)

and the root-mean-square roughness $\delta(t)$, given by

$$\delta(t) = \sqrt{\frac{1}{L^2} \sum_{x=0}^{L} \sum_{y=0}^{L} [h(x, y, t) - \overline{h}(t)]^2}.$$ (4.5)

In both these expressions, L defines the size of the region of interest. The δ parameter, also commonly called the interface width or the surface thickness, is of primary importance in surface morphology characterization and has been examined by several groups investigating GLAD processes, typically measured via AFM. While δ provides a standard measure of surface roughness, theoretical and experimental investigations into the scaling behaviour of δ have also revealed an underlying scale invariance and fractal geometry characteristic of ballistic deposition processes such as low-temperature film growth [36, 37]. The surface roughness of ballistically deposited films exhibits self-affine power-law scaling with t and L according to

$$\delta(L, t) \sim \begin{cases} t^b & \text{if } L \gg L_c \\ L^a & \text{if } L \ll L_c \end{cases}$$ (4.6)

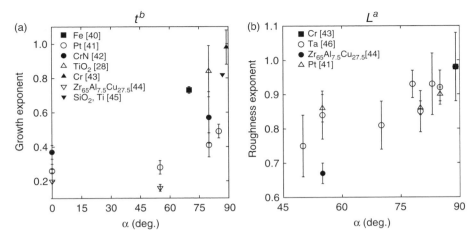

Figure 4.7 Experimentally measured values for the (a) growth exponent and (b) roughness exponent as a function of the deposition angle, which quantify the statistical evolution of the surface roughness during deposition. For both exponents, the general trend across the multiple data sets is to increase with α, indicating that the surface morphology of high-α films is significantly rougher and the roughening process is faster with respect to thickness, thus quantifying the emergence of columnar microstructure.

where L_c is the crossover point, scaling with t as $L_c \sim t^{1/z}$. The three scaling exponents[1] b, a and z are respectively called the growth exponent, the roughness exponent and the dynamic exponent, and are interrelated via $z = a/b$. Calibrating these growth exponents experimentally provides insight into the underlying growth mechanisms [38] and also enables fabrication of surfaces with controlled morphological characteristics. Furthermore, L_c is approximately four times the lateral correlation length of the system [39], providing an estimation of the column diameter and intercolumn separation.

Several experimental investigations [28, 40–46] have reported on the δ scaling behaviour of GLAD films, and Figure 4.7 summarizes the α dependencies of measured growth and roughness exponents. A strong α dependence is observed in both of the scaling exponents, and a sensitivity to material and experimental conditions can be discerned from scatter in the measured values. A number of useful, physical observations can be made regarding the α behaviour of each exponent. The data show that at low α ($<50°$) b is relatively small and insensitive to α, suggesting that surfaces prepared under these conditions are relatively smooth and the roughening rate is low. However, b quickly increases with α for films deposited at higher α, indicating that such films are significantly rougher and that the surface roughness increases with thickness at a much faster rate. The increase in b with α is a consequence of the dominant role of ballistic shadowing in generating surface roughness in the films. Under conditions of grazing incidence ($\alpha \rightarrow 90°$) b approaches unity, corresponding to a linear relationship between δ and t. The roughness exponent a characterizes the spatial scaling of local structural variations on scales smaller than the correlation length, which is on the order of the column diameter. The measured α dependence of a demonstrates that

---

[1] Our notation differs from that conventionally seen in the literature, where a is denoted by α and b by β. We have made the change to avoid confusion with the deposition angle and column tilt angle.

the small-scale surface morphology is sensitive to the deposition angle, with films deposited at higher α exhibiting increased nanoscale structural variations. Note that the experimental determination of a is more difficult as the scaling analysis requires several measurements in the $L \ll L_c$ regime. The lateral resolution of the AFM can thus be a limiting factor, particularly at low α, where the characteristic column dimension is small and the roughness measurement can be distorted by tip convolution effects.

The measured a and b trends are all consistent with qualitative SEM observations of GLAD film microstructures, which reveal the emerging columnar microstructure and associated roughness as α increases. However, the scaling measurements enable systematic investigations into various statistical properties of the surface morphology and are more useful for quantitative microstructural analysis and design.

4.3.2 Column broadening

It has long been recognized that GLAD-fabricated columns are characterized by a nonuniform structure and columns tend to broaden and fan out during the evolutionary growth process. Structural broadening is intrinsic to surface growth via ballistic deposition and shadowing, as opposed to being a by-product of secondary effects such as substrate heating or defect formation. Of the various GLAD structures, the broadening of vertically oriented columns has been the most well studied, revealing that as the columns grow the column width w increases according to a power law

$$w \propto t^p, \tag{4.7}$$

where p is a scaling exponent describing how quickly the column broadens and t is the film thickness (or column height, equivalently). Although qualitatively discussed in Section 2.6.2, where the general aspects and mechanisms of broadening were examined, many groups have systematically investigated the broadening phenomenon via quantitative determination of p [14, 33, 47–58]. The most commonly employed technique involves manual estimation of $w(t)$ from SEM or AFM images [47–49, 51, 53–58]. While the effectiveness of this technique has been demonstrated, a number of reproducibility concerns have been identified. Manual identification of column edges becomes subjective, and different observers can easily measure different w values from the same SEM image. This introduces systematic errors to the measurement that make absolute determination of p values challenging. It is also time consuming to characterize a sufficient number of columns for rigorous statistical analysis, adding uncertainty to the measured values. Several researchers have thus turned to computerized image analysis (Section 4.2.2) to improve the objectivity and throughput of the measurement [6, 14, 33, 49, 50, 52]. While automated processing enables more systematic analysis, it remains important to properly document each analysis step as the final p value is sensitive to the specific algorithm. For example, in analysis of top-down SEM images Siewert showed that small changes to the thresholding algorithm parameters can produce p value discrepancies of up to 40%, and adding a watershed algorithm into the preprocessing can change the measured p by 60% [7].

Figure 4.8 presents a survey of experimentally measured p values as reported by several different groups and spanning a diverse range of material classes (elements, compounds, metals, semiconductors, insulators, etc.). While certain groups have individually reported

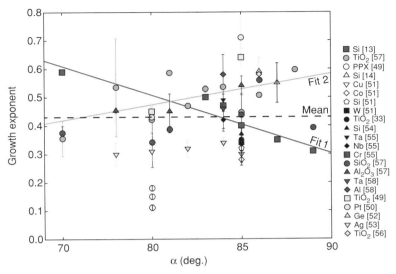

Figure 4.8 Summary of column broadening exponent p values reported in the literature. No consensus relationship between p and α has emerged; although specific studies report experimental trends, the results are often contradictory. For example, fit 1 and fit 2 are linear fits of the Si [13] and TiO_2 [57] data sets respectively, providing examples of negative and positive correlations between p and α. The lack of general agreement suggests that α has a weak effect on p and that additional experimental factors governing the growth kinetics (temperature, pressure, deposition rate, etc.) play a more significant role. Interestingly, the mean value across the data sets is $p = 0.43$ (dashed line), which is near the $p = 1/2$ value theoretically predicted by Karabacak *et al.* [51].

statistically significant relationships between p and α, the trends are not always in agreement. For example, the data of Buzea *et al.* (Si [13] with fit 1) showed a systematic decrease in p as α was increased [13], whereas the data of Taschuk *et al.* (TiO_2 [57] with fit 2) clearly reported the opposite tendency [57]. The inconsistency and significant scatter present in the surveyed data clearly indicate that a general consensus has not emerged regarding the effect of α on p and suggests that α is only a minor variable in this instance. This is an uncommon situation in GLAD research as the influence of α is generally predominant in a wide range of characteristics and experimental trends are often quite clear. The coexistence of different trends and scatter in the data can be attributed to the influence of additional, confounding factors (more on this below) and the measurement uncertainties associated with the common structural analysis techniques discussed above.

However, it is intriguing to note that the mean value across the assembled data sets is $p = 0.43$ (dashed line on Figure 4.8). Although the data variation is large, the experimental average is close to the $p = 1/2$ value predicted by Karabacak *et al.* [51], based on a rotational averaging of values calculated from the Kardar–Parisi–Zhang (KPZ) equation. The KPZ equation provides a continuum modelling of a growing interface (such as a thin film), and predicts that surface correlations (proportional to column width) will exhibit a power-law thickness scaling [35, 59]. In the diffusion-limited, stationary substrate case (fixed φ), the scaling exponents parallel and perpendicular to the deposition plane are 1/3 and 2/3, which yield 1/2 upon rotational averaging. Thus, the $p = 1/2$ regime can be ascribed to growth via ballistic deposition with negligible diffusion, a situation that accurately describes the low-temperature GLAD regime. Incorporating surface diffusion effects, via the Mullins–Herring (MH) model [35], and performing a similar rotational averaging of the scaling exponents predicts $p = 5/16 \approx 0.31$ [51]. As these growth models indicate the importance of surface diffusion in determining p, the broadening rate can be strongly linked to the thin-film growth kinetics, which are governed by material-specific and experimental factors. These factors are numerous, and a short list would include temperature, crystal structure and texture, surface energetics, deposition rate, vapour collimation, gas pressure, and so on. The often nonlinear and interacting nature of these factors further complicates the situation, and the end result is often a convolution of several variables that produces the experimental inconsistencies observed.

Temperature effects on column broadening

Because film growth kinetics are highly temperature dependent, as surface diffusion is a thermally activated process, the column broadening is sensitive to the substrate temperature T_s, as has been shown by a series of investigations [55,60,61]. While the KPZ and MH models suggest that p should decrease monotonically from 1/2 to 5/16 as T_s and, hence, surface diffusion increases, experimental data on a variety metals (Ta, Cr, Nb and Al) have revealed a complex relationship between p and T_s. These data are presented in Figure 4.9, plotting the measured p value against the T_s/T_m, where T_m is the metal melting point. By normalizing T_s in this manner, the temperature scaling behaviour of each metal becomes remarkably consistent. For low T_s/T_m (<0.25), the experimental trend is consistent with the KPZ- and MH-based predictions, with $p \sim 0.5$ at low T_s and then slowly decreasing as the rising T_s increases the surface diffusivity. However, by increasing the substrate temperature beyond a critical value

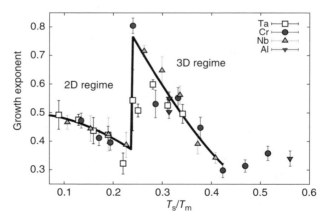

Figure 4.9 The growth exponent in a variety of metals exhibits a remarkably universal temperature scaling behaviour. Generally, the broadening exponent decreases with the substrate temperature T_s except for a rapid, anomalous increase associated with the transition between 2D and 3D nucleation on the column surface. This transition was found to occur for $T_s \sim 0.24T_m$ in several metals and is attributed to an increase in column roughness that enhances the resulting column broadening rate. The theoretical curve (black line) shows the prediction of Equation 4.8, using the fitting procedure discussed in the text. Reprinted with permission from [60]. © 2010, AIP Publishing LLC.

$(T_s/T_m = 0.24 \pm 0.02$ in Figure 4.9) an anomalous scaling behaviour is observed where p rapidly transitions to a large value ($p > 0.75$). After the critical transition, the monotonic behaviour is recovered, with p decreasing as T_s increases. In the high-temperature regime $(T_s/T_m > 0.4)$, p is relatively insensitive to T_s and ranges from 0.31 to 0.34, values consistent with the high surface diffusion regime. Analysis of this complex data provided significant insight into the broadening mechanisms. Using simple physical models, Muhkerjee and Gall argue that the surface roughness of the column apex enhances the radial growth rate of the column, leading to a progressive broadening of the structure [60]. As the temperature increases, surface diffusion smooths out the local surface roughness and decreases the column broadening rate, leading to a progressive decrease in the measured p values. The anomalous scaling behaviour is driven by an abrupt change in the nucleation process on the column surface, which transitions from a 2D to a 3D growth mode at the critical temperature necessary to overcome Ehrlich–Schwoebel energy barriers. The morphological transition associated with 3D nucleation dramatically increases the surface roughness amplitude and the corresponding broadening rate. The authors developed a numerical model motivated by these physical arguments that predicts

$$p = p_0 + G\langle s \rangle, \tag{4.8}$$

where p_0 is the scaling exponent at $T_s = 0$, G is a constant and $\langle s \rangle$ is the average separation between nuclei. Smaller $\langle s \rangle$ increases the effect of local shadowing, leading to rougher surfaces and increased column broadening. The temperature dependence of $\langle s \rangle$ can be estimated via mean field nucleation theory as [60–62]

$$\langle s \rangle = \begin{cases} \frac{1}{\sqrt{\eta}} \left(\frac{v}{4F} \right)^{1/6} \exp\left(-\frac{1}{6} \frac{E_m}{kT_s} \right) & \text{for 2D nucleation} \\[2mm] \frac{1}{\sqrt{\eta}} \left(\frac{v}{6F} \right)^{1/7} \exp\left(-\frac{1}{7} \frac{E_m}{kT_s} \right) & \text{for 3D nucleation} \end{cases}, \tag{4.9}$$

where η is a dimensionless constant, F is the incident vapour flux rate, v is the adatom hopping frequency, k is Boltzmann's constant and E_m is the surface diffusion activation energy. The authors fit Equation 4.8 to the experimental data, with $\langle st \rangle$ determined by Equation 4.9 and using a sigmoid function to capture the transition between the 2D and 3D nucleation regimes. In their analysis, $\langle s \rangle$ was normalized to the interatomic distance, making $\langle st \rangle$, D and G dimensionless. Fixed parameters in the model were a material-independent attempt frequency $v = 10^{12}$ Hz, a constant deposition flux $F = 2$ ML/s from the experiments and an activation energy of $E_m/kT_m = 2.46$, selected to represent experimental conditions. Fitting parameters were thus p_0 and G values for both the 2D and the 3D regimes and a transition point and width for the sigmoid function. The resulting theoretical curve is plotted as a solid line on Figure 4.9. The theory accurately predicts the T_s scaling in the two regimes as well as the critical transition in these elemental material systems, and suggests the underlying physical interpretations of the broadening mechanism are valid.

The link between broadening and microscopic nucleation supported by the temperature-scaling data indicates the general complexity of structural broadening as the nucleation dynamics are affected by a myriad of experimental conditions and are difficult to control. The data variation in Figure 4.8 can thus be attributed to many experimental factors (material, deposition rate, temperature, crystallinity), and the success of the model motivates further studies into complex material systems and other processing variables that may provide additional insights and understanding. In practice, precise control over column broadening demands excellent process repeatability and control.

φ rotation speed on column broadening

Of the various column morphologies, the broadening properties of the vertical column microstructure have been the most well studied to date. As discussed in Section 2.5.2, vertical columns are produced via continuous φ rotation at a rate determined by the growth pitch (vertical growth per revolution), which must be smaller than the column diameter w. The pitch is a key parameter in vertical column fabrication and, given that the pitch is directly controlled via the substrate motion algorithm, it is a particularly easy parameter to manipulate in a basic GLAD system. Siewert *et al.* investigated how the specific φ-motion algorithm impacts the structural broadening of TiO_2 vertical columns deposited at $\alpha = 81°$ [6]. Figure 4.10 presents the measured relationship between p and the growth pitch. As the φ rotation speed increases and the growth pitch is reduced, the column broadening rate is significantly increased. It seems reasonable to infer from the data that the broadening rate of vertical columns is minimized at the transition from helical-to-vertical morphologies, which occurs when the pitch is ~w. Designing a substrate motion algorithm to satisfy this condition would require the pitch be slowly increased during the deposition to match the inherent column broadening rate. Such an algorithm could be used to fabricate minimally broadening vertical columns, although the approach has not been verified.

Whereas the substrate temperature has a clear influence over surface diffusion and the nucleation process, it is less obvious how the growth kinetics are influenced by φ rotation. However, it has been shown that the growth pitch can strongly affect column evolution,

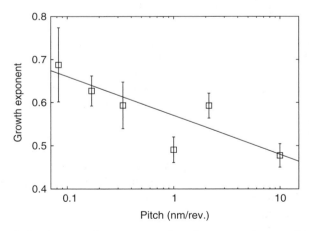

Figure 4.10   In vertical columnar microstructures, the substrate rotation speed (represented here as the growth pitch) influences the column broadening rate. Increasing the growth pitch by reducing the φ rotation speed decreases the column broadening. The exact mechanism here is not clearly understood, although growth simulations suggest that the growth kinetics are affected by a combination of geometric flux averaging and modified burial rate. Note that the growth pitch can only be increased up to the column diameter; further increase leads to formation of helical columns.

crystallinity and texture formation, indicating the rotation speed can influence adatom mobility during deposition [63]. While the exact mechanism is not fully understood, simulation results have suggested the rotation speed alters the burial rate, thereby reducing the effective surface mobility. The reduced diffusion increases the column roughness, thus leading to faster broadening as understood via the physical models discussed in the previous section. The increased radial averaging associated with faster rotation speed would also change the intercolumn shadowing environment, thereby affecting column competition and extinction rates. Further studies are required to develop this understanding.

Practical notes

The power-law structural broadening observed in GLAD columns is a natural consequence of ballistic shadowing growth, and experimental results indicate that the phenomenon is only weakly affected by α. On average, the experimental data are close to the theoretically predicted $p = 1/2$ value associated with diffusion-limited interface growth. However, the broadening is heavily influenced by kinetic growth factors, such as temperature, providing a strong dependence on experimental conditions. The column broadening rate can be reduced by increasing the deposition temperature (up to the 2D–3D nucleation temperature threshold) or by increasing the growth pitch (up to the column width). However, a general understanding regarding additional experimental factors is incomplete, and repeatability can be best obtained by implementing diligent and systematic process control. Process variations between groups can therefore lead to trends that are internally systematic but are difficult to reproduce elsewhere without additional information. Lacking prior calibration information on a specific material or process, the theoretical ($p = 1/2$) or experimental average ($p = 0.42$) broadening exponents could be used to provide an initial broadening estimate under conditions of low adatom mobility.

4.3.3 Intercolumn spacing and column density

Both the intercolumn spacing and column number density have been systematically investigated in a number of studies, employing either SEM or AFM analysis and often using one or more of the numerical image analysis tools discussed in Section 4.2.2. These reports, summarized in Table 4.1, have established several important relationships between these structural parameters and key processing variables: α, the film thickness t and the substrate temperature T_s. Each of these variables has a strong impact on the structural parameters, and the general trends observed across the various reports are remarkably consistent. The effect of each variable is discussed in turn.

As t increases, the average intercolumn separation increases and the column density falls off, a behaviour driven by the competitive growth process fundamental to GLAD. As film growth progresses, the competitive growth process drives column extinction as a portion of the column population becomes overshadowed by larger neighbouring columns. While the column extinction process is active continuously during growth, the extinction rate is significantly higher at small t owing to the greater number of columns and the larger column-to-column size variation. The higher extinction rate associated with early growth is therefore characterized by rapid changes in column separation and density. At higher t, the extinction rate decreases and these parameters become comparatively stable. The variation in the

Table 4.1 Experimental studies of intercolumn spacing and column number density with respect to different processing variables (deposition angle α, thickness t and substrate temperature T_s). SEM-3D refers to the SEM tomography technique described in Section 4.2.3. PPX: poly(p-xylylene).

Structural parameter	Ref.	Variable	Material	Method
Column spacing	[48]	α	Si	SEM
	[30]	α	TiO_2	SEM
	[28]	α, t	TiO_2	AFM
	[49]	t	PPX	AFM
	[33]	t	TiO_2	SEM-3D
	[14]	t	Si	SEM
	[23]	t	W	AFM
Column density	[48]	α	Si	SEM
	[33]	t	TiO_2	SEM-3D
	[49]	t	PPX	AFM
	[64]	t	Ag	SEM
	[65]	T_s	Si	SEM
	[66]	T_s	Ta	SEM

extinction process leads to both structural parameters exhibiting a nonlinear t dependence, which has been successfully modelled with exponential [23] and power-law [14, 33, 49] relationships. The structural variation can be quite dramatic. For example, a rough averaging over results reported by the studies in Table 4.1 shows the column density can change by more than an order of magnitude, varying from $>10^2$ μm^{-2} for $t < 100$ nm to ~ 10 μm^{-2} for $t > 500$ nm, in films deposited near $\alpha = 85°$. In terms of the column spacing, Karabacak *et al.* measured an average column separation of 70 nm at $t = 100$ nm, a value that increased to 160 nm at $t = 450$ nm [23].

Because α is central in determining the shadowing, it is immediately apparent that it will have a significant impact on film structure. The experimental reports have all found that, as α increases, the column separation increases and the column density decreases, an obvious consequence of the increased ballistic shadowing length. Changes of only a few degrees can produce significant structural differences, particularly in the extreme shadowing regime. For example, in 200 nm thick films Buzea *et al.* measured a column density of ~ 400 columns/μm^2 at $\alpha = 85°$, which decreased to ~ 100 columns/μm^2 at $\alpha = 89°$ [48].

As indicated by Table 4.1, only two studies have reported on the impact of T_s, with both finding that the column density falls off as the substrate temperature is increased. This trend was attributed to an increase in the average column size, driven by the increased thermally activated surface diffusion.

4.4 FILM DENSITY

Whereas traditional thin-film processing targets a dense deposit, GLAD exploits the open microstructures characteristic of oblique deposition to create novel, porous film morphologies with tunable physical properties. Compared with bulk values, the overall density of a GLAD film can be significantly reduced as a large portion of the film is occupied by void regions. Controlling void formation in the microstructure is a primary aspect of GLAD technology providing the ability to continuously tune the film density from values near bulk (at low α), all the way to a highly rarefied film consisting of isolated columnar nanostructures at glancing vapour incidence (at $\alpha > 85°$). Furthermore, the fabrication process is entirely bottom-up with the desired porosity level controlled by a single experimental parameter, the deposition angle α. Specifically, we define the normalized film density ρ_n as the fraction of the total film volume occupied by material (i.e. nanocolumns)

$$\rho_n = \frac{V_{material}}{V_{film}} = \frac{V_{material}}{V_{material} + V_{void}}. \tag{4.10}$$

The actual film density is then obtained by multiplying Equation 4.10 by the density of the deposited nanocolumns, which itself may be lower than the bulk material density.

4.4.1 Controlling density with α: theoretical models

The void fraction is primarily determined by the geometric shadowing length, itself controlled via the deposition angle α, providing a straightforward means of controlling the film density and several film properties. Theoretical predictions of film density can be made via geometric analysis of oblique deposition, considering the size of a substrate feature relative to the size of its shadow, a quantity that can be estimated using geometric analysis. For a hemispherical feature, representative of nanoscale islands or rounded column apexes, the normalized density at α is equal to [67]

$$\rho_n = \frac{2\cos\alpha}{1 + \cos\alpha}. \tag{4.11}$$

This expression captures the effect of geometric shadowing in defining void regions, although it neglects diffusion effects. More recently, Poxson et al. applied a similar geometric analysis but added a fitting parameter to account for variations of the shadowed area due to surface diffusion or material-dependent nanocolumn shape [68], resulting in

$$\rho_n = \frac{c}{c + \alpha\tan\alpha}. \tag{4.12}$$

The variable c in this equation is the fitting parameter, equal to

$$c = \frac{\pi V_c}{2A_{cs}h} \tag{4.13}$$

where V_c, A_{cs} and h are the volume, cross-sectional area and height of the shadowing nanocolumn. The authors fit the model to experimental data on SiO_2 and ITO films, deposited at α from 0° to 89°, finding best-fit values of $c = 3.17$ for SiO_2 and $c = 3.55$ for ITO.

Figure 4.11 plots Equations 4.11 and 4.12 for values of $c = 2, 3, 5$ and 7. As α increases in both models, greater fractions of the film are occupied by void and the normalized density falls off quickly. The rate of change is especially great at large α, and in the extreme shadowing regime ($\alpha > 85°$), where the film microstructure consists of isolated columns, the density is dramatically reduced and very sensitive to α. The prediction of Equation 4.12 is highly

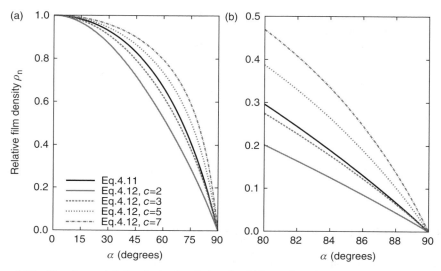

Figure 4.11   Density models for shadowing-induced void formation plotted over (a) the entire α range and at (b) the high α most relevant to GLAD. Equation 4.11 is based purely on geometric analysis, while Equation 4.12 contains a fitting parameter c, useful for capturing material dependence and process sensitivity. By adjusting α, the film density can be continuously tuned over a wide range, from relatively dense films produced at low α to void-dominated film microstructures at high α ($> 85°$).

dependent upon the exact value of the c parameter. Furthermore, when c is between 3 and 4, as was the case in the Ref. [68], the two models predict similar density values. This indicates that Equation 4.11 is a useful starting point for initial design estimates of normalized density, while Equation 4.12 becomes particularly useful for quantifying differences between materials, evaluating the impact of deposition parameters, and process calibration/monitoring. Recently, Krause *et al.* modified Equation 4.12 by adding a quadratic term to the denominator [69], resulting in

$$\rho_n = \frac{1}{1 + \frac{1}{c}\alpha \tan \alpha + \frac{1}{k}(\alpha \tan \alpha)^2}. \tag{4.14}$$

With the added fitting term k, the extended equation provided a superior fit of density data for TiO_2 helical microstructures. Although the physical basis for the additional term remains unclear, Equation 4.14 appears to capture additional experimental factors and may be useful for experimental calibration.

4.4.2 Experimental measurement and control of film density

The density measurement methods described below are minimally invasive and nondestructive to the sample under investigation. No physical contact of the film surface is required; this a particular advantage in high α films, which often exhibit poor abrasion resistance. Film density is also a relatively rapid and inexpensive parameter to characterize. The added benefit of reliable theoretical models for shadowing-induced porosity (Equations 4.11 and 4.12), makes density an attractive parameter to measure for simple characterization and monitoring.

Two approaches are commonly used to measure GLAD film density. The first is the standard gravimetric method, based on dividing the measured film mass by the film volume. The mass is usually determined by weighing the substrate before and after deposition. (Alternatively, the substrate can be weighed after deposition, the film removed via etch and then the bare substrate weighed. This approach has the disadvantage of being destructive to the film.) The film volume is estimated by multiplying the substrate area by the film thickness. The second method involves first performing an optical measurement of the sample, typically spectrophotometry and ellipsometry, and then applying effective medium approximations (EMAs) and optical simulations to estimate the film filling fraction (Section 5.1.2). The gravimetric method has the advantage of directly reporting absolute units, whereas the filling fraction is a dimensionless quantity equal to the fraction of the total film volume occupied by material. The filling fraction can be converted to absolute units through multiplication by the column material density, which is unknown. Bulk density values are often assumed, although the actual column density is generally less than bulk due to intracolumn microporosity. However, in cases where both gravimetric and optical measurements have been compared, the discrepancy between the results has been small [69].

To ensure accurate density measurements, several subtleties must be accounted for in the methodology. The accuracy of optical methods is directly related to how faithfully the optical model represents the film, a topic discussed in greater detail in Section 5.1.2. For gravimetric measurements, samples should be sufficiently large to provide a confidently measureable mass (determined by the sensitivity of the instrument used) and reduce the impact of defects, dust, chipping, and so on related to sample handling. The film volume must also be accurately estimated, requiring accurate film thickness information and the consideration of film uniformity, particularly in larger samples (Section 7.4). Standardized substrates of fixed size with good tolerances can be used to minimize sample-to-sample uncertainty. Edge effects may also contribute uncertainty to the measurement, depending on both the substrate size and how the substrates are fixed in the deposition system. The accuracy of both methods is also impacted by water adsorption into the film, which can be significant due to the high levels of porosity found in GLAD films. It is possible to quantify the adsorbed water mass and thus compensate for the additional load. Optical methods have been used to estimate the trapped water volume [28, 70]. The water load can constitute an appreciable portion of the total film mass, as demonstrated by results for GLAD-fabricated TiO_2 films (Figure 4.12) [70]. Note, however, that the actual water load will depend on several aspects of the deposited film, particularly film surface hydrophilicity and pore structure, as well as environmental factors such as relative humidity and temperature. When practical, the effect of water adsorption can be mitigated by baking-out trapped water (care must be taken to not affect film structure/crystallinity – see Section 6.2.1) and/or performing measurements in an inert or vacuum environment.

GLAD film density has been studied by many groups, and a range of materials has been investigated, including Al [71], CaF_2 [71], Cr [71, 72], Cu [71], Ge [73], ITO [68, 69, 74], MgF_2 [75], Mn [71], ATO [76], SiO [71], SiO_2 [68, 69, 75, 77], and TiO_2 [69]. These studies have primarily focused on how the density scales with α. As expected from the geometric nature of atomic shadowing, all the studies report a continuous decrease of the film density as α increases. The increased shadow lengths increase the inter-island separation during the

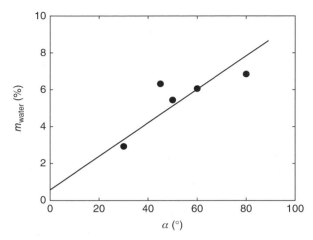

Figure 4.12   Contribution of adsorbed water to the total film mass for TiO_2 films deposited at different α, estimated using optical measurements. Moisture can form an appreciable portion of the total mass in films deposited at high α and should be accounted to ensure measurement accuracy. Reproduced from [70]. With permission from K.M. Krause.

initial nucleation phase, reducing the substrate coverage. This effect carries through to later film growth, and a larger fraction of the film microstructure will consist of void regions. The film density becomes especially sensitive to α in the extreme shadowing regime where an isolated column microstructure develops and small α changes strongly impact the average intercolumn spacing. Comparing density values reported by different groups for the same material is instructive, providing insight into the controllability of the GLAD process. Figure 4.13 plots measured densities for (a) SiO_2 and (b) ITO films deposited at a range of α by several groups. Equation 4.12 (first multiplied by the bulk material density) was fit to each data set via the c parameter. Each individual data set is well described by Equation 4.12, suggesting a good degree of consistency and control over the deposition process. The significant discrepancy between the data sets, however, indicates that the density is sensitive to the specific collection of processing conditions adopted by each group. These conditions, which are often incompletely specified in publications, include substrate temperature, deposition rates, system geometry and pressure, which together influence film density and hence properties. The collective influence of these conditions is parametrically folded into the c variable, potentially allowing Equation 4.12 to be used for process monitoring.

Amassian *et al.* characterized the film density as a function of deposited thickness [78]. The investigation compared the growth of Si films deposited at normal incidence and in the extreme shadowing regime ($\alpha = 87°$) using in-situ spectroscopic ellipsometry (SE), a commonly employed optical technique. The films were characterized over a wide range of thicknesses, with measurements taken from the initial nucleation phase through to the much later column evolution stage. The films were deposited at a slow rate (0.3 Å/s) to ensure that the initial growth phase was sufficiently well resolved. Figure 4.14 summarizes the thickness evolution of the film density, plotting the porosity ($1 - \rho_n$) on the vertical axis and the film thickness (determined via SE) in units of monolayers (ML) and nanometres, on the upper and lower horizontal axes plotting respectively. Because the SE apparatus was designed for

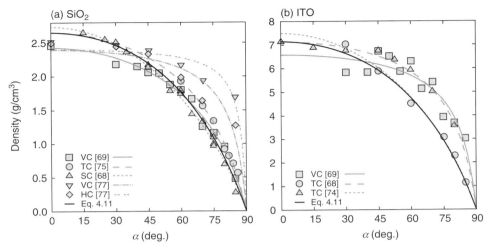

Figure 4.13   Measured film density as a function of deposition angle α reported by several groups for (a) SiO$_2$ and (b) ITO. Across these and other materials, the experimental data all exhibit the same trend: as α is increased, the amount of void in the film increases, causing a decrease in the overall film density. This trend is in general agreement with the prediction of Equation 4.11 (dashed line), based on parameter-free geometric modelling of shadowing and void formation. Better agreement with the individual data sets is obtained using Equation 4.12 (solid lines), which contains a fitting parameter that incorporates diffusion effects and allows modelling of material- and process-specific differences.

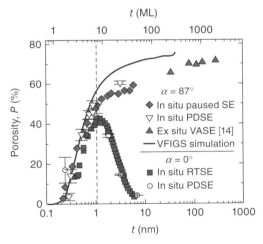

Figure 4.14   Film porosity evolution with thickness in films deposited at $\alpha = 0°$ and 87°, measured using several spectroscopic ellipsometry (SE) techniques (defined in text) and supported via 3D ballistic modelling (VFIGS). Ballistic shadowing prevents coalescence in the 87° film, leading to the divergent thickness scaling of the film porosity at $t \approx 1$ nm. For $t > 1$ nm, the film porosity slowly increases during column evolution and eventually stabilizes in very thick films ($t \gg 1000$ nm) [69]. Reprinted with permission from [78]. © 2007, AIP Publishing LLC.

use in the $\alpha = 0°$ configuration, the normal-incidence growth could be probed in real time (RT), producing the *in-situ RTSE* measurement. However, characterizing the $\alpha = 87°$ film required temporarily shuttering the vapour flux to halt film growth and then rotating the substrate to normal incidence for measurement acquisition, producing the *in-situ paused SE* data. In both the $\alpha = 0°$ and $87°$ cases, the RT and paused results agreed with measurements collected post-deposition (PD), showing the consistency of the various SE procedures and the noninvasiveness of the paused approach. Overall, the theoretical predictions of a 3D ballistic growth simulator (VFIGS, solid line on Figure 4.14) agree with the experimentally observed growth trend, supporting the validity of the measurements. Several important conclusions can be drawn from the experimental data. During the initial nucleation growth phase ($t < 1$ nm), the $\alpha = 0°$ and $87°$ films are practically indistinguishable and the porosity quickly increases with t. However, the growth evolution of the two films diverges dramatically at the $t \approx 1$ nm point. At normal incidence, the nucleated islands of film material coalesce as deposition continues, forming a dense deposit before $t = 10$ nm. The evolution is remarkably different in the extreme shadowing regime, as ballistic shadowing suppresses the coalescence process and drives the development of isolated columnar features. At $\alpha = 87°$, the porosity continuously increases with thickness well beyond the nucleation phase ($t > 1$ nm), although the rate of increase is significantly less than during preliminary growth. Similar work by Kaminska *et al.* using ex-situ variable-angle SE (VASE) methods continues the analysis well into the column evolution phase ($t > 100$ nm), where very high porosity levels ($>70\%$) are developed [14]. During this growth phase, the porosity was found to continuously increase at a slow rate, increasing from 63% at $t = 70$ nm to 83% for $t = 1000$ nm. Density measurements in very thick films ($t \gg 1000$ nm) indicate that the porosity asymptotically stabilizes [69], presumably related to reduced column competition and extinction.

As a final note, we would like to emphasize that these measurements examine the overall density of the total film. The density of an individual column has not been quantified, only examined qualitatively via electron microscopy techniques (Section 4.2.4). Such a measurement could provide fundamentally valuable information about nanoscale column structure and physical phenomena. This knowledge would benefit nanoscale devices based on single nanocolumn properties.

4.5 POROSIMETRY AND SURFACE AREA DETERMINATION

The ability to tailor film porosity using α is a noted advantage of GLAD technology. Increasing the substrate tilt angle α amplifies the shadowing effect, introducing significant porosity into the film microstructure. Furthermore, the columns themselves are porous entities, as revealed by the evolutionary growth model (Section 2.6.1) and SEM and TEM imaging (Sections 4.2.3 and 4.2.4). This hierarchical morphology leads to a complex pore structure spanning multiple size scales, rendering it difficult and time consuming to study via electron microscopy alone. However, developing a more detailed knowledge of the porous microstructures present in GLAD films is critical and would benefit the many technological applications (such as electrochemical devices, gas sensors and photovoltaics) that exploit the open microstructure and enhanced surface-to-volume ratios of GLAD films.

For this reason, several researchers have turned to porosimetry techniques to provide an improved description of porous structures in GLAD films. Porosimetry encompasses a set of analytical methods to quantify a range of important structural aspects, such as surface area, pore size and pore volume [79, 80], and has become a core tool for researchers seeking a detailed characterization of the internal structure of GLAD films. Multiple porosimetry techniques exist [81, 82], including mercury porosimetry, small-angle X-ray scattering, fluid-flow methods and thermoporometry. The most prevalent, however, are gas-adsorption-based techniques, and these are also the most commonly applied in GLAD studies.

In general, gas adsorption techniques rely on monitoring the controlled adsorption/desorption of gases onto/from the porous material under investigation. The gas adsorption/desorption characteristics of a material reveal detailed information about the surface structure. Two gas-sorption-based techniques have been applied to GLAD research. The first is based on analysis of measured adsorption isotherms that describe the amount of adsorbed gas relative to the saturation vapour pressure at a fixed temperature. The material surface area can then be estimated from the isotherm using the Brunauer–Emmett–Teller (BET) theory of multilayer gas adsorption [79, 80]. Adsorption isotherm analysis has been used to characterize the porosity in GLAD films composed of C [83], ITO [69], Mn [84], poly(p-xylylene) [85], SiO_2 [30, 69, 86], Ti [87], TiC [88] and TiO_2 [28, 30, 69]. Sorge et al. have also reported surface areas measured with this technique for a wide range of materials (Al_2O_3, CaF_2, ITO, MgF_2, Nb_2O_5, SiO, SiO_2, SnO_2, Ta_2O_5, TiO_2, Y_2O_3 and ZrO_2), although the study was limited to films deposited at $\alpha = 85°$ [89]. The other commonly used porosimetry technique in GLAD research is a standard surface science technique called temperature-programmed desorption (TPD). In TPD, a cooled sample is slowly heated in an ultra-high vacuum environment, thus freeing trapped gases from the surface. The gas desorption rate is monitored, typically by a mass spectrometer, allowing quantification of the total trapped gas volume and an estimation of the corresponding surface area. TPD has been used in several studies to estimate the maximum surface areas of tilted columnar Pd [90], TiO_2 [91], TiC [92], MgO [93], WO_3 [94] and amorphous solid water films [95–97].

4.5.1 Surface area enhancement in glancing angle deposition films

The surface area is often reported as the specific surface area (SSA), which is the total surface area of the porous material per unit mass. The surface area enhancement (i.e. the surface area per unit substrate area) is also a common figure of merit, especially in thin-film systems. Several systematic studies have examined the optimization of SSA in GLAD processes [69, 87, 88, 90–92, 94–97]. Because α is a key determinant of porosity in the film microstructure, investigating the α dependence has been an experimental focus of every study, and α has been identified as a primary control variable of surface area. These studies have found a consistent α dependence similar to that shown in Figure 4.15, although the absolute values are very sensitive to material and experimental conditions. Beginning at low α, the surface area increases with α as the longer shadowing lengths introduce more voids into the film morphology. This increase slows, however, and the surface area is maximized at a particular alpha α_{SA}. The surface area decreases for $\alpha > \alpha_{SA}$, although it remains significantly greater than the low α ($< 50°$) values. As shown in Figure 4.16, the SSA is maximized for $\alpha_{SA} = 70°$ when averaged over several experiments, a finding supported by 3D ballistic simulations of film growth [98]. Although there is variation in the data related to material and deposition conditions, the consistency suggests that the surface area maximum is related to the transition into the extreme shadowing regime discussed in Section 2.4. Entering this regime negatively impacts the SSA because the number of columns decreases and the

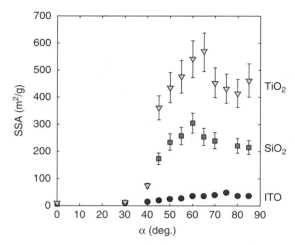

Figure 4.15   SSA measured for TiO_2, SiO_2 and ITO GLAD vertical columnar films deposited at different α. For $\alpha > 40°$, the film SSA is greatly enhanced compared with normally deposited films, often by multiple orders of magnitude. Although the peak SSA value depends on material and processing parameters (see text for discussion), the peak SSA is generally found at $\alpha = 70°$ (Figure 4.16). Reprinted with permission from [69]. © 2010, American Chemical Society.

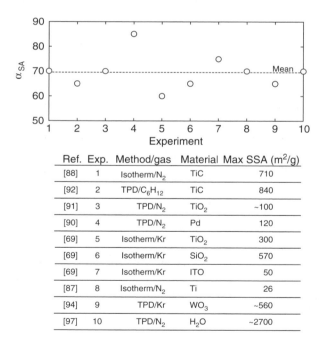

Ref.	Exp.	Method/gas	Material	Max SSA (m^2/g)
[88]	1	Isotherm/N_2	TiC	710
[92]	2	TPD/C_6H_{12}	TiC	840
[91]	3	TPD/N_2	TiO_2	~100
[90]	4	TPD/N_2	Pd	120
[69]	5	Isotherm/Kr	TiO_2	300
[69]	6	Isotherm/Kr	SiO_2	570
[69]	7	Isotherm/Kr	ITO	50
[87]	8	Isotherm/N_2	Ti	26
[94]	9	TPD/Kr	WO_3	~560
[97]	10	TPD/N_2	H_2O	~2700

Figure 4.16   A survey of systematic studies in the literature (presented in the table) indicates that the maximum SSA is highly sensitive to several experimental conditions, including material, substrate temperature and measurement technique. However, the plot shows that the surface area is consistently maximized for deposition angles near $\alpha_{SA} = 70°$ (dashed line). This α value thus provides a good starting point for further surface area optimization studies.

intercolumn spacing and column size increase, thus decreasing the surface-to-volume ratio of the microstructured film. While α_{SA} is fairly consistent, the peak SSA values for GLAD films vary widely, ranging from 26 m^2/g in Ti [87], to over 2700 m^2/g in films of amorphous solid water obliquely deposited at cryogenic temperatures [97]. Although these values are not as high as those seen in novel high-surface-area materials, such as porous metal–organic frameworks (SSA > 6000 m^2/g) [99], GLAD provides substantially greater microstructural control and material compatibility, thus providing greater design flexibility and functionality in device applications. Additionally, note that while the film SSA is generally maximized for $\alpha = 70°$, the film microstructure remains relatively dense at this α. The transition to an open microstructure observed at higher α can improve the mass-transport properties of the film and provide ancillary benefits on device performance; for example, improving the response of gas sensors based on porous GLAD coatings [100].

The relationship between surface area and other deposition parameters has been examined as well. Although the competitive processes characteristic of GLAD yield continuous structural changes during growth, as exemplified by the power-law thickness scaling of the surface roughness and column width (Section 4.3.2), the total surface area has been found to scale linearly with thickness with an α-dependent slope [69, 94]. A few studies have investigated the effect of deposition temperature on the surface area [91, 92, 94, 95]. In general, the surface area is enhanced by depositing at lower substrate temperatures; cooling the substrate from room temperature to cryogenic conditions (<100 K) has been shown to double the surface area of the resulting film (Figure 4.17) [91, 94]. This result is consistent with the understanding that reducing the substrate temperature decreases adatom mobility during deposition, thus enhancing the shadowing-induced surface roughening. However, Flaherty *et al.* noted an exception to this trend, finding that reactive evaporation processes may exhibit different behaviour due to the generation of reaction by-products on the surface that complicate the growth process [92]. Post-deposition annealing treatments have been found

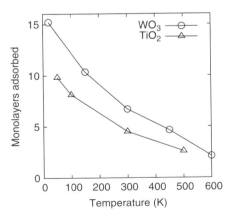

Figure 4.17   The surface area, here quantified as the N$_2$ adsorption capacity of the material measured in monolayers (where one monolayer is the amount of N$_2$ desorbed from a dense film), decreases with the deposition temperature as the increased adatom mobility reduces the surface roughness in the deposited films. The data sets correspond to WO$_3$ tilted columns deposited at $\alpha = 65°$ [94] and TiO$_2$ at $\alpha = 70°$ [91].

to decrease the total surface area by activating surface and bulk diffusion processes, with very high temperature anneals causing a severe loss of porosity (Section 6.2.1) [91,92,94].

Electrochemical measurements of active surface area

Section 4.5.1 focused on surface area measurements based on the capacity of material to physisorb various inert gases, which provides an accurate characterization of the total physical surface area. However, when evaluating structured materials for applications exploiting surface reactivity it is also important to consider the total *active* surface area that ultimately contributes to device performance. Electrochemical techniques, such as cyclic voltammetry, have been used in GLAD research to provide a complementary characterization of surface area. Applications examined include supercapacitor materials [84], high-surface-area electrochemical [83] and spectroelectrochemical [101] electrodes and oxygen-reduction electrocatalysis [102, 103]. Krause *et al.* systematically investigated the surface area of ITO GLAD films using both Kr adsorption isotherms and cyclic voltammetry [69]. The two measurements indicated that the electrochemically active surface area was reduced by a factor of 2–3 compared with the total surface area measured via Kr adsorption. The difference between the physical and active surface areas was α dependent, with a reduced discrepancy observed at higher α values. Although the source of the difference was unclear, the electrochemical activity of the ITO surface is highly sensitive to deposition conditions, surface condition, cleaning techniques and ageing. Moreover, the authors point out that ITO surfaces synthesized with other techniques show similar levels of surface inactivity. The complementary information provided by physical and active surface area characterization could be used to establish processing recipes to maximize the surface activity of the material. Post-deposition processes, such as annealing treatments, could be used to further tune the surface properties.

4.5.2 The pore structure of glancing angle deposition films

In addition to measuring the surface area, porosimetry can also provide a useful characterization of the pore structure. The pore size strongly influences the shape characteristics of the gas adsorption isotherm [104], and pore size information can be extracted from this data. The International Union of Pure and Applied Chemistry (IUPAC) provides a general classification of pore structures based on their width [104]. Pores with widths greater than 50 nm are called macropores, pores spanning the 2–50 nm range are termed mesopores and pores smaller than 2 nm are micropores. The conventional approach for pore size estimation from adsorption isotherms is the Barrett–Joyner–Halenda (BJH) method [105], which is based on the well-known Kelvin equation describing capillary condensation. The BJH technique can be used to determine the pore size distribution, which describes the statistical variation of pore sizes and is an important descriptor of a disordered, porous medium, such as one fabricated using GLAD.

The mesopore structure is particularly important with respect to vapour condensation into the film, and three studies have characterized this structure in SiO_2 [30], TiC [88] and TiO_2 [28, 30] GLAD films. All three reached similar conclusions regarding the influence of α on mesoporosity, finding that the average mesopore size increases with α (see Figure 4.18),

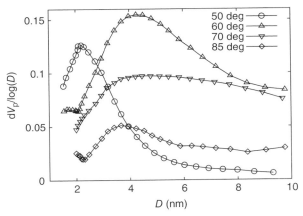

Figure 4.18   The pore size distribution measured for TiO_2 vertical column microstructures deposited at different deposition angles. The distribution characterizes the mesopore structure in the film and is presented as the portion of the total pore volume V_p made up of pores of size D. The mean mesopore size and the proportion of larger pores increase with α, a consequence of the increased shadowing and intracolumn roughness that accompanies high-α deposition. The results indicate that the pore structure can be controlled in the GLAD process using α [30].

as would be expected from an increasing shadow length at larger α. Furthermore, the pore size distribution becomes increasingly skewed towards larger pore sizes and the distribution broadens at large α. These findings indicate it is possible to tune the mesopore structure at the ensemble level by changing the deposition angle, which is useful for device design. For example, in a gas sensor application such a capability could be used to control capillary condensation and optimize sensitivity in specific pressure regimes. The micropore structure can also be further manipulated using post-deposition processes (discussed in Chapter 6), such as thermal annealing and surface coating. Note that the gas adsorption method and analysis is sensitive only for pores smaller than 10 nm. While GLAD morphologies exhibit structure at larger size scales, characterizing these pores requires alternative techniques, such as mercury porosimetry or electron microscopy analysis.

As a final consideration, it is important to recognize that extracting structural data from porosimetry measurements can often be troublesome, and ensuring accuracy is a difficult task [82, 106, 107]. The common analysis methods (BET, BJH) rely on basic models whose assumptions are often questionable or inaccurately represent complex pore structures. Fortunately, much can be done to improve the quality of porosimetry data, such as using better models and calibration data, implementing standardized measurement and analysis practices (e.g. ISO 9277). International standards organizations such as National Institute of Standards and Technology (USA), Bundesanstalt für Materialforschung und -prüfung (Germany), Institute for Reference Materials and Measurements (European Community) and LGC Standards (UK) provide certified reference materials for surface area and pore size analysis that can help improve the reliability of measured data. In all cases, it is important to understand the limitations of a given methodology to evaluate the reliability of a particular measurement.

4.6 CRYSTALLOGRAPHIC TEXTURE AND EVOLUTION

Most GLAD deposition algorithms implicitly assume that the material is interchangeable; small differences in column tilt angle, broadening and pitch threshold for the helical–vertical column transition aside, most GLAD materials will behave similarly under the same deposition algorithms. The exception occurs when the deposited material has kinetic effects for the deposition process being used; this is typically found in Zone II or Zone III growth, where adatom mobilities are high enough to achieve crystals in the growing thin film.

While such depositions are sufficient to produce crystalline films, additional opportunities are available in GLAD depositions. GLAD amplifies the small, stochastic variability between adjacent nanocolumns through the long shadow lengths that occur at high deposition angles. Crystal faces grow at different rates, and are expected to be randomly oriented in the initial phases of a GLAD thin film. Crystals with their fast-growth axes oriented closer to the substrate normal make efficient use of the vapour flux that they capture, producing a small advantage over neighbouring, misaligned crystallites. The amplification of small differences between growth surfaces allows crystallites that are just on the threshold of formation to experience a differential advantage during growth, giving the GLAD experimentalist access to different crystalline populations of a growing GLAD film, and can offer sufficient control to promote a desired population with good design of deposition algorithms.

Karabacak *et al.* have demonstrated selection between the α and β phases in W thin-film growth at normal and glancing angles [108]. In that work, they suggest that the growing W film is made up of two competing crystallite populations assigned to the α and β W phases: the first with high adatom mobility and faster growth in the substrate plane, and the second with low mobility and faster growth in the substrate normal direction. This model is shown schematically in Figure 4.19. The differences between the populations present an opportunity for GLAD to create an environment that favours a particular population. At normal incidence angles, crystallites with faster lateral growth (α phase) have a competitive advantage as they spread out in the substrate plane and capture more flux, accelerating their growth. However, with oblique-angle deposition, the second population with faster growth in the substrate normal direction (β phase) has a competitive advantage as they efficiently convert vapour flux to vertical growth, increasing their vapour capture cross-section. Competitive growth promoting texture formation has been studied for CaF_2 [109] and MgO [110].

There have been additional reports of similar effects at glancing deposition angles for rotating substrates [111, 112]; crystal texture has been shown to depend on α and ϕ rotation speed. The population of growing crystallites can be further broken down by breaking the symmetry in the ϕ dimension. Instead of using the vertical nanocolumn deposition algorithm, using a serial bideposition (SBD) algorithm or a threefold deposition [32] algorithm has been shown to induce a preferential azimuthal dependence, in addition to the out-of-plane orientation for the growing Fe crystallites. Through flux engineering, the growth environment experienced by the growing nanocolumns favours Fe crystals with the $\langle 111 \rangle$ growth axis aligned with the substrate normal and the (100) face oriented towards the vapour flux directions.

It is important to note that these texture effects do not necessarily require flux engineering to emerge. Strong texture effects have been reported by Okamoto *et al.* in a series of reports on oblique deposition of Fe [113, 114]. While these effects may be neglected for many materials

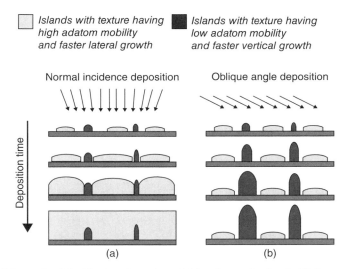

Figure 4.19 Schematic of the Karabacak *et al.* model for phase selection in W films under (a) normal incidence deposition and (b) GLAD conditions. The dark areas have lower mobility and faster vertical growth when compared with the light areas. This difference between nanorod populations provides an avenue for providing growth conditions that favour either population by changing the incidence angle. Reprinted with permission from [108]. © 2003, AIP Publishing LLC.

at typical GLAD deposition conditions, additional process development may be necessary once crystalline features such as angular column apexes or deviations from expected column tilt angles are observed.

However, for some material systems it is not necessary to take special care with the deposition conditions to achieve crystal texture. One such system is ITO, which can undergo self-catalysed vapour–liquid–solid (VLS) growth at elevated temperatures. VLS growth occurs when vapour flux accumulates within a molten droplet, achieving supersaturation. The vapour material crystallizes, producing nanocolumns with faceted structures. ITO is a self-catalysed system, which forms eutectic In-Sn droplets that act as the vapour flux aggregation sites, producing fascinating structures such as those shown in Figure 4.20. While the ITO system has strong preferential growth orientations and structures, it is possible to influence the growing structures by combining VLS with the engineered shadowing in GLAD to allow preferential branch orientation and height-dependent placement [31].

An interesting consequence of the VLS growth mode in a GLAD deposition is that the droplets which precipitate the nanotree trunks and branches can be shadowed, reducing the vapour flux accumulation rate of individual droplets. Using a vertical nanocolumn deposition algorithm, the droplets at the tips of branches are periodically shadowed by the central trunk. Since the branches continue to receive material from the droplet, the periodic reduction in material reaching the droplets reduces their diameter – reducing the diameter of the growing branch. This effect combined with the regular rotation in a vertical nanocolumn deposition produces ripples, seen in Figure 4.20b. This effect can be exploited to produce designed ripple patterns (Figure 4.20c) by engineering the vapour flux [115].

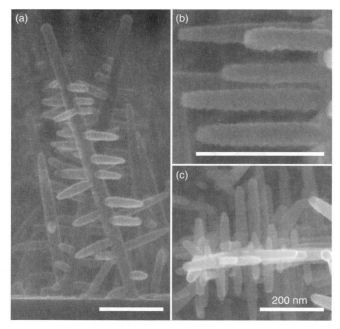

Figure 4.20 Images of (a) VLS–GLAD nanotrees showing structures possible in a material system with strong preferential crystalline orientation. ITO's cubic bixbyite structure produces nanotree structures with orthogonal geometries. Regular shadowing of the individual branches when using a vertical nanocolumn deposition algorithm produces (b) regular ripples in the branches as the VLS droplet is starved of flux. Exploiting this mechanism allows control over the (c) ripple pattern by modulating the vapour flux [115]. All scale bars indicate 200 nm. Reprinted with permission from [115]. © 2012, AIP Publishing LLC.

4.7 ELECTRICAL PROPERTIES

Controlling the electrical properties of a deposited layer, such as resistivity, dielectric constant, carrier mobility and lifetime, is an important topic in thin-film engineering. Notable technological applications relying on film electrical properties include high-conductivity metallization layers for electrical contacts and transparent conducting coatings for flat-panel displays. Thin-film electrical properties are highly sensitive to the structure of the deposited film, a fact that has prompted many researchers to use GLAD technology to manipulate film electrical properties. More recently, the material and structure flexibility afforded by GLAD suggests exciting prospects for developing nanoelectronic devices from the individual nanocolumns produced at high α.

4.7.1 Resistivity in microstructured glancing angle deposition films

Investigations have primarily focused on the film resistivity ρ and how it is affected by deposition at oblique incidence. Resistivity is generally measured using the four-point probe method, wherein two electrode pairs are used (one sourcing current and the other sensing voltage) to remove contact resistance effects and improve measurement accuracy. Note that while the four-point probe technique is well established in conventional film metrology, certain complexities arise when applying the technique to structured GLAD films, which we discuss below. Care should also be taken when physically contacting electrodes to the film surface, particularly on films deposited at higher α, which often display poor abrasion resistance. Investigating the properties of individual nanostructures demands different experimental methods, which are discussed towards the end of this section. An additional complexity arises from the anisotropic microstructure of many GLAD films, which creates a complementary anisotropy in the electrical resistivity and requires more careful analysis. Novel measurement geometries have also been developed to directly investigate the anisotropy (discussed below). Finally, because the four-point approach relies on lateral conduction paths through the film, it is an inappropriate method for characterizing films composed of highly isolated columns, as produced by deposition in the $\alpha > 80 - 85°$ range. Even if the column material is intrinsically conductive in such films, the column isolation will lead to negligible conductivity.

In GLAD research, the resistivity properties of a large number of materials have been investigated, including several metals (Cr [72,116,117], Fe [114,118], Ni [119,120], Ti [121]) and semiconductors (Ge [73], ITO [74,122,123], SnO_2:Sb (ATO) [76]). Figure 4.21 summarizes the measured resistivities of a range of materials deposited using GLAD, showing that α has a very strong effect on electrical conductivity. While reports on the same material by different groups vary, indicative of process-dependent factors, it is interesting that the α dependence is remarkably similar across each material and group. The resistivity is observed to increase rapidly with α. Dense films deposited at $\alpha = 0°$ exhibit electrical properties similar to bulk values, although with a process-induced variability. As α is increased, the film resistivity increases as well in a consistent manner across the different measurements. The rise is slow at first, but becomes increasingly rapid at higher α values. Regardless of the intrinsic material qualities (conductor versus semiconductor), the measured resistivity of a specific material can vary by several orders of magnitude over the full α range (0–80°). The general consistency of the observed scaling behaviour is due to the microstructural changes caused by ballistic shadowing at oblique incidence, which are (to a first approximation) the same from material to material. Conceptually, the rapid increase in film resistivity with α can be rationalized via percolation theory, where the resistivity is determined by the connectivity/continuity of the film material [124,125]. A GLAD film can be viewed as a highly disordered mixture of conducting (film material) and insulating (void region) phases. Depositing at larger α increases the void fraction of the film, reducing the number of 3D percolation paths through the film and reducing the conductivity. Progressing towards the extreme shadowing regime, the open film microstructure severely limits this electrical continuity. While the columns themselves remain intrinsically conductive, as demonstrated by specialized investigations (below), their isolation yields an overall film resistivity several orders of magnitude greater than typical dense films. Alternate electrical pathways, such as through the preliminary film nucleation layer, likely contribute significantly to the conduction at this stage and become the dominant conduction mechanisms in the extreme shadowing regime.

The importance of film continuity in determining the electrical conductivity is further reinforced by the thickness-scaling behaviour of the resistivity. Figure 4.22 shows measured resistivity values in Ge films of different thickness deposited at a range of α [73]. At all α, very thin films (<10 nm) are relatively poor conductors. This is generally understood to be because

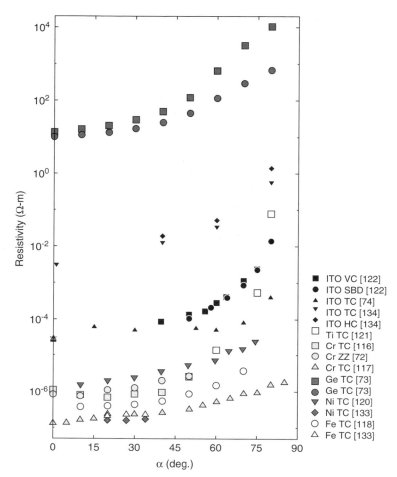

Figure 4.21 A survey of film resistivity ϱ in several materials as a function of deposition angle. The experimental α trend is generally consistent across the diverse materials: as α increases, the measured ϱ of the deposited film increases monotonically, typically by multiple orders of magnitude over the full α range. The ϱ change is particularly rapid at larger α. The general α dependence is a consequence of the changing film microstructures, which is (to a first approximation) material independent. VC: vertical columns; SBD: serial bideposition columns; TC: tilted columns; HC: helical columns; ZZ: zigzag columns.

the film microstructure is discontinuous in these early growth stages due to incomplete island agglomeration. At normal incidence, agglomeration is completed after minimal deposited thickness, quickly producing a low-resistivity film. As the deposition angle is increased, however, increased ballistic shadowing slows the agglomeration process. Connectivity of the film microstructure increases slowly, yielding a slower decrease in the film resistivity with thickness. The resistivity eventually stabilizes in thicker films (>100 nm), asymptotically approaching an α-dependent value determined by the film void fraction as discussed above. Because the resistivity exhibits nonlinear thickness scaling and slowly approaches saturation

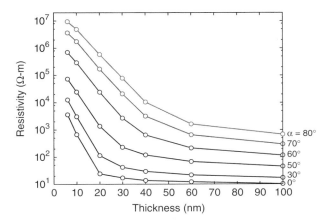

Figure 4.22 Resistivity scaling with film thickness depends strongly on the deposition angle α. At normal incidence ($\alpha = 0°$), the initial resistivity drop is rapid, being driven by island coalescence in early film growth. At higher α, the increased shadow length slows the coalescence process and leads to higher resistivities. Data are for Ge films, obtained from Ref. [73].

at higher α, expressing measurements in traditional thin-film units, such as Ω per square, is not generally recommended without accompanying thickness information.

4.7.2 Anisotropic resistivity

Thus far, the discussion has examined the dramatic rise in film resistivity accompanying the increasing void fraction of the film. However, there are additional GLAD-specific microstructural factors that impact film electrical properties and lead to unique effects. During tilted column growth, the columns progressively broaden in the directional orthogonal to the deposition plane, producing asymmetric, fan-like columnar structures (Section 2.6.4). Because the electrical properties are intimately connected to the columnar morphology, this structural anisotropy generates a corresponding anisotropy in the electrical behaviour. Columns preferentially bundle together in the fanning direction, increasing the connectivity and reducing the resistivity in this direction. Conversely, ballistic shadowing preferentially maintains the column separation in the deposition plane, thus increasing the resistivity in this direction. The microstructural asymmetry is illustrated by the SEM image presented in Figure 4.23a, examining an ITO tilted column film deposited at $\alpha = 85°$. In this SEM image, the film was viewed at an oblique angle to look down the column axis and reveal the structural anisotropy and asymmetric column bundling. Many researchers have investigated the phenomenon of anisotropic resistivity, quantified via the ratio

$$\xi = \frac{\rho_\perp - \rho_\parallel}{\rho_\perp + \rho_\parallel},$$

(4.15)

where $\rho_\perp$ and $\rho_\parallel$ are the film resistivity oriented perpendicular and parallel respectively to the deposition plane. This geometry and its relationship to the film microstructure are illustrated by Figure 4.23a.

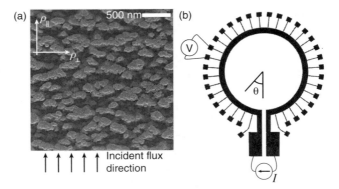

Figure 4.23 (a) SEM view of a tilted column ITO film imaged along the column axis, illustrating the microstructural asymmetry. The lack of shadowing perpendicular to the incident flux direction allows the columns to fan out and contact neighbouring columns. The microstructural asymmetry and increased connectivity in the fanning direction leads to anisotropic resistivity, where the resistivity along the perpendicular orientation ($\rho_\perp$) is less than the resistivity in the parallel direction ($\rho_\parallel$). (b) Circular electrode geometry devised by Kuwahara and Hirota to measure anisotropic resistivity. A series of four-point resistance measurements is taken at different angular positions θ and harmonic analysis of the resulting $R(\theta)$ data allows determination of the anisotropy coefficient [126]. (a) Reproduced from [2]. With permission from J.B. Sorge. (b) Reproduced with permission from [126]. © 1974 The Japan Society of Applied Physics.

As mentioned above, the commonly used four-point probe method becomes more complex when considering a film with anisotropic resistivity as it alters the current density distribution through the layer. Therefore, accurate interpretation of four-point probe data and extraction of $\rho_\perp$ and $\rho_\parallel$ require more complex analysis for specific sample and electrode configurations [127–132]. Repeated measurements with different electrode configurations allow estimation of Equation 4.15. In the most basic approach the resistance is measured twice, once in the parallel direction and once in the perpendicular direction [76, 114, 120]. Taking this method further, Kuwahara and Hirota devised a novel measurement geometry with multiple electrode locations in a ring-like arrangement (Figure 4.23b) [126]. Film material is confined to the ring section, either by post-deposition lithography processing [119, 126] or by shadow-mask during deposition [121]. Taking voltage measurements along different angular positions θ of the ring allows the resistance R to be expressed as

$$R(\theta) \propto (1 - A\cos\theta - \xi\cos 2\theta). \tag{4.16}$$

Taking the acquired resistance data and then using Fourier analysis or numerical fitting via Equation 4.16 yields the anisotropy parameter ξ. Importantly, such analysis returns the coefficient A, which captures the effects of thickness gradients over the electrode area that can introduce a separate, non-microstructural electrical anisotropy.

Anisotropic resistivity measurements have been reported for several materials, including Ni [119, 120, 133], Cu [119], Fe [114, 118, 133], Ti [121], and ITO [134]. These reports have all investigated the α dependence of ξ in tilted columnar microstructures, and the results are summarized in Figure 4.24. Across the multiple data sets and materials the experimental trends are generally consistent, finding that films deposited at higher α tend to exhibit a

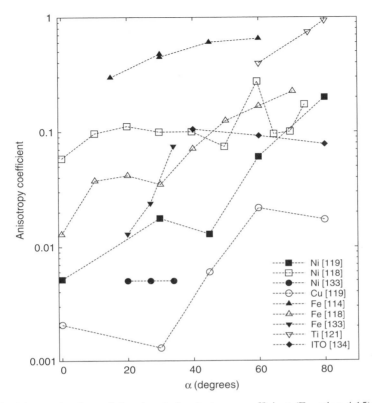

Figure 4.24 Measured values of the electrical anisotropy coefficient (Equation 4.15) in different materials as a function of α. The degree of electrical anisotropy present in the film tends to increase with α, due to the more pronounced microstructural asymmetry present in films deposited at higher α.

greater degree of electrical anisotropy and indicating that α can be used to control film anisotropy. In certain cases, ξ is very sensitive to α and can change dramatically over the full α range, with some of the data sets showing a greater than tenfold increase in ξ from $\alpha = 0°$ to 80° (Ni [119] and Cu [119]). Although the α dependence of the various data sets is fairly consistent, the exact ξ values exhibit significant variability. Material-specific differences are an obvious source of variation, although the measured values are inconsistent even when the same material is deposited, indicating that other experimental parameters contribute to the overall discrepancy. Because ξ is intimately linked to the microstructural anisotropy, factors impacting the growth kinetics, such as substrate temperature or crystallinity, will affect the resulting ξ. The film thickness should also be an important parameter in determining ξ given the nonlinear relationship between resistivity and thickness discussed above (Figure 4.22).

4.7.3 Modelling glancing angle deposition film resistivity

Overwhelmingly, the majority of studies to date are experimental and only a few attempts to model the electrical properties of obliquely deposited films have been reported. Vick and

Brett considered two simulation methods to predict the electrical conductivity of chevron-structured Ti [121]. The first was a relatively simple approach based on effective medium theories (EMTs). While EMTs are commonly used to describe the electrical properties of a variety of composite systems [124, 125], they were found to poorly describe GLAD film conductivity. The second approach was considerably more advanced, combining 3D film-growth simulations with a random walk model to statistically describe charge transport properties. The advanced approach proved more representative of the experimental data, although good agreement could only be obtained by adding a layer of native oxide on the Ti columns. Besnard *et al.* recently developed an analytical model for film conductivity in obliquely deposited films [117]. The theory extended previous models of electron scattering at column surfaces [135] to the case of tilted columnar morphologies, and accurately described the measured α-dependent resistivity of sputtered Cr films. However, anisotropic properties were not considered in their investigation.

4.7.4 Individual nanocolumn properties

Self-assembly of intricate electronic devices from nanoscale elements is a promising new concept, one being heavily pursued by the nanowire research community. With its broad material compatibility and scalability, GLAD technology presents an interesting approach to nanowire control and fabrication. Recent investigations have reported on nanoelectronics aspects of GLAD, and we expect further advancements in the near future.

Investigating the electrical characteristics of a single nanocolumn is a significant experimental challenge with many practical difficulties. Because the GLAD nanocolumns are oriented out of the plane, planar electrode geometries, commonly used in semiconductor nanowire and carbon nanotube characterization for example, are not immediately applicable without detaching the GLAD nanocolumns from the substrate. Other nanoscience tools, such as conductive atomic force microscopy and scanning tunnelling microscopy, are more convenient as they are based on top-down electrical contact to the structure. Singh *et al.* used a conductive AFM (cAFM) approach to pass electrical current through a Co-coated Si square-spiral (Figure 4.25) and actuated the nanospring architecture via the generated self-inductive electromotive force [136]. This interesting report focuses primarily on the mechanical aspects of the nanocolumns (see Section 4.8.3) rather than their electrical properties. However, we note that currents up to 20 mA were passed through the 10 nm thick Co coating with no reported damage.

Lalany *et al.* have recently reported a systematic investigation of column electrical properties using a cross-bridge electrode geometry [137]. In this approach, substrates are first patterned with a microscale bottom electrode layer. The desired GLAD structure is then deposited on the prepatterned substrate and then a post-deposition lithography process is used to define the top electrode. This configuration is displayed in Figure 4.26, showing (a) side and (b) top-down SEM views of the completed device. The top electrode is oriented at right angles to the bottom electrode, thereby defining device area A equal to the overlap between the two electrodes. The electrode configuration is compatible with four-wire measurements and passing current between the electrodes probes the vertical resistance of the columns, as opposed to their planar connectivity. An obvious concern with the approach, however, is the abnormal growth produced at the bottom-electrode border, which produces a modified structure along the electrode edge, as seen in Figure 4.26a. (This edge-growth

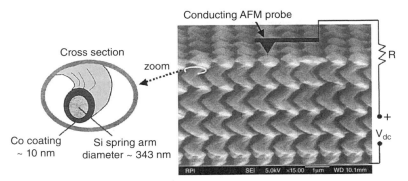

Figure 4.25 Schematic illustrating the use of a cAFM to probe the electrical and mechanical properties of individual columnar structures. In this instance [136], periodic Si square-spirals were coated with Co via chemical vapour deposition. Single columns were addressed with the cAFM tip and then inductively actuated by passing a current through the columns. Reprinted with permission from [136]. © 2004, AIP Publishing LLC.

phenomenon is discussed further in Section 6.6.1.) The experimental impact of the edge structure was investigated by systematically reducing A and measuring the device resistance R_{device} (Figure 4.26c). Two scaling regimes were observed, with a crossover at $A \sim 25\mu m^2$. In the first regime, R_{device} scales as A^{-1} and the electrical properties are determined by the device area. In the second regime, R_{device} scales as $A^{-0.5}$ and the electrical properties are edge limited. Devices in the first regime thus provide a more accurate characterization of the columns themselves. To estimate the column material resistivity, the authors treated the film as a parallel network of truncated cones (frustrums) with an areal density determined using an SEM. For ITO vertical columnar structures deposited at $\alpha = 83°$, the intrinsic column resistivity was found to be $(1.1 \pm 0.3) \times 10^{-2}\Omega$ m, which is comparable to the resistivity of dense, unannealed ITO films.

4.8 MECHANICAL PROPERTIES

While the mechanical properties of thin films are intimately related to the bulk material properties, it is widely recognized that the film microstructure is equally important in determining the overall mechanical behaviour [138–142]. It follows that the unique microstructuring capabilities provided by GLAD technology lead to the observation of interesting mechanical properties and also present opportunities to realize new devices, particularly in the isolated nanocolumn regime at high α. Many researchers have studied the mechanical properties of GLAD films, investigating more conventional thin-film quantities such as stress and hardness while also examining the fabrication of nanomechanical structures in functional devices.

4.8.1 α effects on film stress

In practice, thin-film materials always exhibit certain levels of tensile or compressive stress after deposition, even when no external load is applied. This residual film stress is an

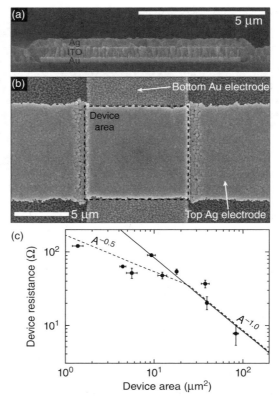

Figure 4.26   Investigating axial resistivity of GLAD columns using a cross-bridge electrode geometry wherein the microstructured columnar film is sandwiched between two stripe electrodes, each oriented at 90° to the other. (a) Cross-sectional and (b) top-down SEM images of the completed device. (c) Four-wire measurements of the device allow probing the axial resistance of the columns in the device area A. The resistance scaling with respect to A reveals two regimes: an A^{-1} regime where resistance is determined by the device area, and an $A^{-0.5}$ regime where resistance is dominated by edge effects. Analysing measurements from the former regime provides a better estimate of the intrinsic column resistivity. In a study of ITO vertical columns, an intrinsic resistivity $(1.1 \pm 0.3) \times 10^{-2}$ Ω m was reported, a value comparable to the resistivity of dense, unannealed ITO films. Reprinted with permission from [137]. © 2013, American Vacuum Society.

intrinsic by-product of the growth process and is highly sensitive to the material and deposition process variables, such as substrate temperature, gas pressures and deposition geometry. The film stress can be generated by a number of mechanisms, including [138] mismatched thermal expansion coefficients between the film and substrate, atom incorporation into the film (via burial of residual gases or chemical reactions at the substrate), film–substrate lattice mismatches, variations in the interatomic spacing, stress at crystal grain boundaries, phase transformations and void regions in the film microstructure. Although process parameter tuning can be used to influence each of these stress-generation mechanisms, the last mechanism is particularly important in GLAD, which provides control over microstructural porosity through algorithmic α and φ substrate motions.

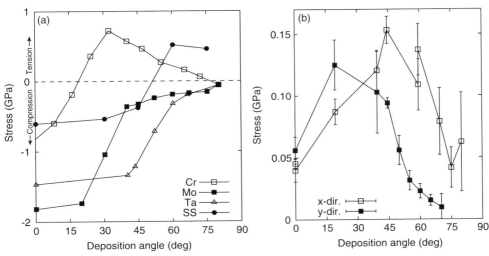

Figure 4.27   Measured residual stresses in obliquely deposited films. (a) A series of sputter-deposited films of different material [143]. Reported deposition conditions: 0.13 Pa Ar pressure, 0.4 nm/s deposition rate and 200 nm film thickness. (b) Tensile stress in thermally sublimated SiO films measured perpendicular to (x-direction) and along the deposition plane (y-direction) [144]. Reported deposition conditions: 1400 °C source temperature, deposition pressure 2×10^{-6} Torr, 100–600 nm film thicknesses.

Experimentally, stress is most often determined by measuring the substrate curvature induced by the film stress. This is a common thin-film characterization approach and is typically done optically or capacitively. Analytical expressions provide simple relationships between the total bending to the stress, although these ignore nonuniformity and anisotropy effects. These same techniques have been applied to examine film stress in microstructured GLAD films [143–149]. Because the film stress properties are highly process specific, the general aspects of the α scaling are discussed here based on reports by a number of researchers [143, 144, 148–151].

Figure 4.27 presents residual stress data measured for obliquely deposited films, showing the typical α dependence. The first plot (Figure 4.27a) presents measured film stress (normalized to film thickness) in sputter-deposited Cr, Mo, Ta and 304 stainless steel (SS) layers as a function of the deposition angle [143]. The films were deposited via DC-magnetron sputtering with a 0.13 Pa Ar working pressure. For near-normal incidence deposition, these processing conditions were found to produce films with compressive residual stress, with stress values between 0.5 and 2 GPa. The film stress becomes less compressive and even tensile as α is increased from 0°, with the exact α scaling being different across the materials. In all cases, however, the stress gradually diminishes towards zero as the deposition angle approaches grazing incidence ($\alpha \rightarrow 90°$). This stress reduction at large α is a consequence of the greatly increasing porosity, which uncouples the columnar microstructure and enables structural relaxation over short length scales. The overall film stress is thereby significantly reduced as α is increased.

As with many physical properties, the anisotropic microstructure of the obliquely deposited film can induce a corresponding anisotropy in the film stress. (These effects were not discussed in Ref. [143].) Figure 4.27b shows the measured residual tensile stress levels in SiO films deposited at various α and measured both perpendicular to (x-direction) and along (y-direction) the incident vapour direction [144]. Initially, the stress in both directions is observed to increase as the deposition angle rises. Film stress behaviour in this low-α regime is likely to be highly dependent on several experimental factors. As α is increased further, the stress measurements reveal a distinct anisotropy. Near $\alpha = 30°$, the stress in the y-direction begins to decrease, whereas the stress along the x-direction continues to rise before exhibiting a similar fall-off near $\alpha = 50°$. This anisotropic stress behaviour can be attributed to the anisotropic shadowing and column fanning effects (Section 2.6.4). Along the incident flux direction (y-direction), ballistic shadowing increases the column separation and allows for structural relaxation. Conversely, there is no shadowing along the orthogonal direction to restrict growth. The columns, therefore, fan out in the x-direction and have an increased tendency to chain together, thereby reducing the relaxation capability and maintaining stress levels in the film. However, the stress in both directions is minimized as α is increased further due to the continuous rise in microstructural porosity that accompanies high-α deposition; film stress is negligible in the extreme shadowing regime, where completely isolated columnar structures are formed. Note that, although the microporosity provides a stress-reduction mechanism in GLAD films, it has been shown that moisture condensation when the films are exposed to ambient conditions can generate significant compressive stresses and even lead to mechanical failure and delamination of the film [144]. However, such failure is likely a minor issue as it has only been reported in a few cases. Surface chemical modification (Section 6.2.5) and structural reinforcement techniques (Section 6.2.4) can be used to improve the mechanical stability of films in high-humidity and liquid environments.

The vanishing film stress in the extreme shadowing regime means GLAD films deposited at $\alpha > 80°$ can generally be fabricated without concern for stress-related mechanical failure. The microstructured films can also be used as stress-reduction layers in films deposited at normal incidence, as is discussed in Section 6.5.2.

4.8.2 Hardness properties

Hardness quantifies the resistance of the film to plastic deformations when subjected to mechanical loading and is an important parameter in evaluating the durability of a deposited film. GLAD film hardness properties have been evaluated using conventional micro- and nanoindentation instruments, where a standardized indenter tip is pressed into the film with micronewton-level forces. The film hardness is subsequently estimated through numerical analysis of either the size and shape of the plastic deformation left by the indenter, or by the measured force–displacement curves during the loading/unloading process. In GLAD research, investigations have all employed standard analysis methods in extracting hardness values from experimental data. The mechanical properties of the substrate should be accounted for when the penetration depth is an appreciable fraction of the total film thickness ($\sim$7–20%) [152]. Film surface roughness, which increases with α, or surface oxidation can also affect the measurements if the penetration depth is too small [153, 154].

Figure 4.28 summarizes experimentally determined GLAD film hardness values, showing how the hardness varies with deposition angle for several different materials. The film hardness is strongly influenced by α, with most of the data sets varying by more than an order of magnitude over the measured α ranges. The hardness also varies with material

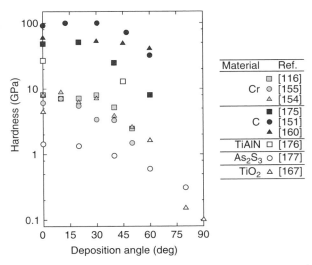

Figure 4.28 Measured film hardness values in obliquely deposited films reported by various groups.

according to the different intrinsic mechanical properties of the material, and the data variation within a given material set provides an indication of sensitivity to other processing conditions. However, the same general α dependence is observed across all the reports and materials, with the film hardness decreasing as α moves to larger angles. This general feature is a consequence of the increasing amount of porosity introduced to the film microstructure at higher α, which affects all materials in a similar fashion. The α scaling thus demonstrates that GLAD can be used to tailor the extrinsic microstructural factors determining the film mechanical properties. As can be seen, the majority of the data are for films deposited at non-glancing α ($<60°$), where the film is still relatively dense, albeit with a certain level of porosity. Only a few data points can found at high α ($>80°$) where the extreme shadowing lengths produce large void regions and distinct, isolated columns. In this regime, the hardness is greatly reduced compared with conventional films. While the column material may have a very large intrinsic resistance to plastic deformation, the micronewton indentation forces can induce large, complex stresses on the individual columns. The exact behaviour at high α can be quite intricate, as the transition to an open column microstructure also changes the elastic properties of the film (discussed below), leading to a drastic modification of the mechanical response compared with a film with a densely packed column microstructure. The specific columnar geometry (e.g. tilted, helical, zigzag) should strongly influence the mechanics in this regime.

4.8.3 Elastic behaviour of glancing angle deposition films

Effects of porosity in films deposited at mid-α

Because the columnar microstructure is densely bundled for moderate α values, any applied mechanical load is rapidly distributed to neighbouring columns, leading to an ensemble-averaged response. While the overall response is similar to that of a conventional thin film,

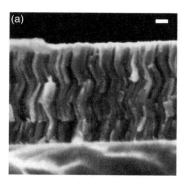
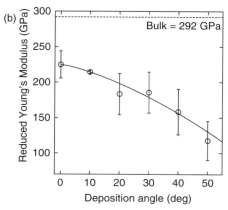

Figure 4.29   (a) A side-view SEM image of one of the characterized Cr films, showing the zigzag columnar morphology and the dense microstructure. The scale bar indicates 100 nm. (b) Reduced Young's modulus measured for Cr zigzag columnar films deposited at different α. The modulus for bulk Cr (292 GPa) is shown for reference, and the deposited films all exhibit moduli significantly less than bulk, an indication of porosity and surface roughness at all α. The solid line shows the prediction of Equation 4.17, using Equation 4.11 to model the film porosity and a best-fit value of $E'_r = 224 \pm 1$ GPa. Reprinted from [155], with permission from Elsevier.

the measured material parameters can be dramatically different owing to the presence of microstructural porosity. A series of investigations presented detailed characterization of the mechanical properties of slanted and zigzag column Cr films prepared via sputter deposition at α from 0 to 50° [72, 116, 154, 155]. Figure 4.29a shows a side-view SEM image of one such Cr film, where the zigzag columnar microstructure was fabricated at $\alpha = 30°$. The low-α geometry produces well-defined columnar features that, owing to their tightly packed configuration, will transfer mechanical loads to adjacent columns. However, the shadowing-induced microstructural porosity does reduce this coupling and allows increased structural relaxation when loaded. This effect can be seen in the measured α dependence of the reduced Young's modulus (defined as $E_r = E/(1 - v^2)$, where E is the Young's modulus and v is the Poisson ratio). As α increases, E_r was observed to monotonically decrease, reaching approximately half the initial value when $\alpha = 50°$. This represents a remarkable enhancement of the film compliance and demonstrates that α can be used to tailor the elastic properties of the resulting film. At higher α, the transition to an open column microstructure produces an even greater enhancement of film elasticity, as discussed in the following section.

To further quantify the effect of porosity on the elastic modulus in films deposited at moderate α, Lintymer et al. used a mixing rule given by [155]

$$E_r = E'_r(1 - f^{2/3})^{1.21}. \tag{4.17}$$

In this expression, E_r is the reduced Young's modulus of the porous film, E'_r is the reduced Young's modulus at $\alpha = 0°$ (which is typically less than the bulk material value) and f is the void fraction of the microstructure, calculated using the normalized density via $f = 1 - \rho_n$. This last parameter can be determined from experimental measurements (Section 4.4) or using a theoretical model (Section 4.4.1). Equation 4.17 provides a useful starting point in predicting E'_r as a function of α, as shown in Figure 4.29. To obtain this curve, Equation 4.11 was used to determine ρ_n and E'_r was numerically fit to the measured modulus data. As is evident from the plot, Equation 4.17 accurately describes the α dependence of the data, demonstrating that the mixing rule provides a useful method of predicting the mechanical properties of GLAD films. The only parameter that must be initially specified is the $\alpha = 0°$ modulus, found in this instance to be 224 ± 1 GPa. This value is significantly less than the expected bulk Cr value, a common discrepancy in thin films and attributable to several factors, including decreased material density, surface roughness and surface oxidation. In this specific report, films deposited at $\alpha = 0°$ were found to have $\sim 15\%$ porosity levels, and surface effects contribute an estimated 12% error [155]. When making preliminary design predictions of GLAD film properties, it is thus recommended that an initial calibration film be fabricated to determine the $\alpha = 0°$ properties. Failing that, approximate values on the order of 70–90% of bulk could be used, but only if a large margin for error is included.

In a follow-up report, the authors use mechanical analysis of the zigzag microstructure to develop a complex model of the films from bulk material properties [154]. While the model appropriately reproduces various experimental trends, discrepancies between measured and calculated values suggest improvements to the model are required for accurate a priori predictions.

Measuring the mechanical response of individual microstructures

As α increases and the GLAD film transitions to an open microstructure, the large intercolumn spacing prevents mechanical coupling between columns and the isolated columns behave as individual nanomechanical elements. The elastic properties of a microcolumn depend on both the intrinsic material parameters (Young's modulus and Poisson ratio) and the extrinsic structural parameters defining the geometry. The material parameters are only somewhat controllable, being sensitive to general processing conditions such as substrate temperature and working gas pressure. However, the structural parameters can be directly manipulated over a wide range by adjusting the α and φ motions used in film fabrication. The unique microstructures found in the high-α regime combined with the ability to carefully tune their properties have prompted many investigations into their mechanical response. The combinations of materials, column microstructures and α conditions that have been investigated in the literature are summarized in Table 4.2.

Figure 4.30a and b presents the basic intendation measurement approach applied to GLAD films. The indenter tip with a radius of curvature R is pressed into the microstructured film, measuring the force F and displacement d during the loading and unloading cycle. Depending on the intercolumn separation and R, the indenter probe may load either a single column or several columns simultaneously. In the latter case, the mechanical response can generally be considered as a parallel assemblage of individual springs. To evenly distribute the force across the contacted columns, several researchers have added a planar encapsulation layer (Section 6.5) atop the separated helical microstructures [156, 164, 166]. This dense overcoat minimizes the lateral bending of helices towards the indenter edges and prevents the tip from slipping between the columns. Flattened and cylindrical indentation probes can also be used to minimize the edge effects and yield more uniform vertical loading. As R is reduced, it becomes possible to apply mechanical loads to single columns, an approach that generally requires AFM instrumentation [136, 160–162]. With an AFM, the surface topography can

Table 4.2 Experimental investigations into the mechanical response of isolated column microstructures. PS: periodic structures (see Chapter 3); Alq_3: tris(8-hydroxyquinoline) aluminium.

Material	Ref.	Microstructure	α (°)	Notes
Alq_3	[156]	Helical	85	Encapsulated film; AFM and electrostatic actuation
Cr	[157]	Helical	85	
	[158]	Helical	85	Macroscopic loading
Ni	[157]	Tilted	85	Periodic structures
	[159]	Helical	$\beta = 43$	Inverted structure (Section 6.8)
Si	[160]	Helical	85	PS and random structures
	[136]	Helical	85	PS; Co coated; AFM and inductive actuation
	[161]	Tilted	85	PS
	[162]	Sq. spiral	85	PS; fatigue analysis
	[163]	Helical	85	
SiO	[164]	Helical	84	Encapsulated film
	[157]	Helical, tilted	84	
SiO_2	[165]	Tilted	85	Nanoparticle reinforcement
Ta_2O_5	[166]	Helical	84	Encapsulated film; shear response measured
Ti	[157]	Helical	85	
TiO_2	[167]	Tilted	~90	

also be imaged at high resolution, allowing an individual column to be selected and loaded in a controlled and repeatable manner.

The open microstructure produces a greatly reduced mechanical stiffness when compared with a dense film. For example, Seto et al. measured a reversible 260 nm deformation in an array of SiO helices (Figure 4.30c) under an applied load of 112 μN, which was an ~10^3 greater deformation than observed in a dense $\alpha = 0°$ film (Figure 4.30d) [157]. Furthermore, the elastic properties can be tailored by controlling the columnar microstructure with α and φ. For a given microcolumn geometry, the elastic properties can generally be analysed with equations developed for macroscopic structures. For example, helical GLAD columns are often modelled using an expression for the force constant k of macroscopic springs [156, 157, 160, 163, 164, 166]:

$$k = \frac{Gw^4}{64R^3N}\left[1 - \frac{3w^2}{64R^2} + \frac{3+v}{2(1+v)}\tan^2(90° - \beta)\right]^{-1}. \qquad (4.18)$$

The intrinsic parameters here are the shear modulus G, equal to $G = E/[2(1 + v)]$, and the Poisson ratio v. The remaining parameters are structural: w is the column diameter, R is the helical coil radius, N is the number of helical turns and β is the column tilt angle (measured from the substrate normal). These structural parameters can all be controlled by adjusting the α and φ motion algorithm used to fabricate the helical structure (Section 2.5.1). For example, α can be tuned to obtain structures with specific β values, and the φ rotation speed can be changed to control R. Similarly, the elastic properties of tilted column microstructures have been modelled using [157, 161]

$$k = \frac{3EI}{L^3\cos^2(90° - \beta)}, \qquad (4.19)$$

where L is the column length and I is the axial moment of inertia of the column cross-section. These macroscopic expressions have typically provided predictions accurate to within an

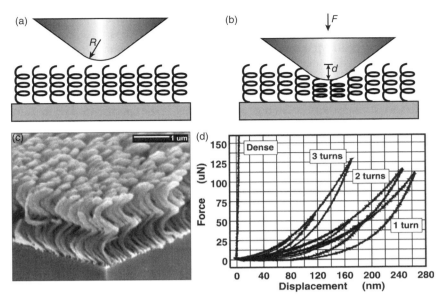

Figure 4.30 Schematic description of using nanoindentation to probe the mechanical properties of GLAD microstructures. (a) An indenter, typically with a spherical tip (radius of curvature R), is (b) controllably pressed into the film microstructure to a displacement d while recording the applied force F. This approach can be used to load the GLAD columns with forces in the micro- to nanonewton range and investigate the mechanical response. (c) The similarity between helical columns and a bed of springs is more than visual, and the elastic behaviour of microscale helices mimics their macroscopic counterparts. (d) The helical columns are dramatically less stiff than a dense film, as shown in this measured F–d curve. Furthermore, the elastic properties can be controlled by modifying the microstructure (e.g. number of turns, column rise angle), allowing fabrication of tailored nanomechanical elements using GLAD. (a) and (b) Reproduced from [168]. With permission from M. Seto. (c) and (d) Reproduced from [157]. © IOP Publishing. Reproduced by permission of IOP Publishing. All rights reserved.

order of magnitude using structural parameters obtained from SEM images. The intrinsic parameters are often assumed to be equal to bulk values or are estimated by characterizing an $\alpha = 0°$ film deposited under similar conditions. The above macroscopic equations can therefore provide preliminary estimates of mechanical properties, useful during the design of nanomechanical column structures. Improving the design accuracy requires developing more representative structural models (taking into account effects such as column broadening) and performing advanced mechanical simulations, typically based on finite-element analysis (FEA). This approach is more demanding but provides superior agreement and analysis capabilities, as demonstrated by several groups [136, 157, 159, 160, 162, 169, 170].

4.8.4 Additional mechanical properties

Nanocolumn mechanical failure

Although limited work has been done examining the mechanical failure characteristics of microstructured GLAD films, a number of interesting results have been reported. Gaire *et al.*

investigated the deformation and failure characteristics of Si square-spiral columnar structures subjected to monotonic and cyclic loading [162]. Using AFM tests and FEA, stresses on the order of 440 ± 80 MPa (~ 1.6 µN) applied to a single column led to failure under vertical loading. Single-column fatigue testing showed that lifetime (measured in number of loading cycles) increased monotonically as the force was reduced, and the columns survived for $\sim 10^4$ cycles at ~ 0.8 µN loads (50% the failure force). The failure characteristics can be expected to vary strongly with column geometry owing to the specific stress concentrations that develop in each case. However, the AFM methodology developed by Gaire *et al.* could be readily adapted to further investigations.

Whereas most reports examine column properties under vertical loading, Hirakata *et al.* investigated shear loading of GLAD structures to measure the strength of the column–substrate interface [170]. A novel microstructure was developed for this study. A typical GLAD film was fabricated, consisting of tilted Ti columns deposited at $\alpha = 85°$ and then encapsulated with a dense upper layer (Section 6.5). Micrometre-sized brick-like structures were then defined in the sample by removing the surrounding film with a focused ion-beam. Mounting the sample on its side in an AFM instrument permitted lateral loads to be applied directly to the dense layer, generating shear stresses in the structured column layer. Under increasing lateral loads, the films were found to eventually fracture at the column–substrate interface. Interestingly, the critical fracture force was found to depend on the orientation of the applied load with respect to the column tilt. A larger critical load ($\sim 2\times$) was required when the load was applied along the tilt direction (forward direction) than when applied against the tilt direction (reverse direction). Using FEA simulations of the microstructure, the asymmetric interface strength was attributed to locally increased stress singularities at the column–substrate interface under the reversed loading configurations.

Anisotropic friction in structured glancing angle deposition films

The tribological properties of a surface are known to be sensitive to micro- and nanoscale structures, which has generated much interest in nanoengineered surfaces. The ability to flexibly manipulate surface structure using GLAD has prompted several reports examining the frictional properties of GLAD films [171–173]. In general, these reports have found that the frictional properties depend on whether the relative motion is with or against the orientation of the GLAD columns. Some intriguing applications have been identified based on these findings and related works. Stempflé *et al.* suggest the ability to tune the ratio between the dynamic to static friction coefficients could be used to realize microgripper devices [173], whereas Malvadkar *et al.* exploited the anisotropic surface friction and wettability to control water-droplet movements [174].

REFERENCES

[1] Goldstein, J., Newbury, D.E., Joy, D.C. *et al.* (2003) *Scanning Electron Microscopy and X-ray Microanalysis*, 3rd edn, Springer.

[2] Sorge, J.B. (2012) Argon-assisted glancing angle deposition, PhD thesis, University of Alberta.

[3] Hrudey, P.C.P. (2006) Luminescent chiral thin films fabricted using glancing angle deposition, PhD thesis, University of Alberta.

[4] Russ, J.C. (2011) *The Image Processing Handbook*, CRC Press, San Diego, CA.

[5] Sezgin, M. and Sankur, B. (2004) Survey over image thresholding techniques and quantitative performance evaluation. *Journal of Electronic Imaging*, **13**, 146–168.

[6] Siewert, J.M.A., LaForge, J.M., Taschuk, M.T., and Brett, M.J. (2012) Disassembling glancing angle deposited films for high-throughput, single-post growth scaling measurements. *Microscopy and Microanalysis*, **18**, 1135–1142.

[7] Siewert, J.M.A. (2012) Disassembling glancing angle deposited films for high throughput growth scaling analysis, Master's thesis, University of Alberta.

[8] Wakefield, N.G., Sorge, J.B., Taschuk, M.T. *et al.* (2011) Control of the principal refractive indices in biaxial metal oxide films. *Journal of the Optical Society of America A*, **28**, 1830–1840.

[9] Sikkens, M., Hodgkinson, I.J., Horowitz, F. *et al.* (1986) Computer simulation of thin film growth: Applying the results to optical coatings. *Optical Engineering*, **25**, 142–147.

[10] Brett, M.J. (1988) Structural transitions in ballistic aggregation simulation of thin film growth. *Journal of Vacuum Science and Technology A*, **6**, 1749–1751.

[11] Harris, K.D., Vick, D., Smy, T., and Brett, M.J. (2002) Column angle variations in porous chevron thin films. *Journal of Vacuum Science and Technology A*, **20**, 2062–2067.

[12] Fan, J.G., Dyer, D., Zhang, G., and Zhao, Y.P. (2004) Nanocarpet effect: pattern formation during the wetting of vertically aligned nanorod arrays. *Nano Letters*, **4**, 2133–2138.

[13] Buzea, C., Beydaghyan, G., Elliot, C., and Robbie, K. (2005) Control of power law scaling in the growth of silicon nanocolumn pseudo-regular arrays deposited by glancing angle deposition. *Nanotechnology*, **16**, 1986–1992.

[14] Kaminska, K., Amassian, A., Martinu, L., and Robbie, K. (2005) Growth of vacuum evaporated ultraporous silicon studied with spectroscopic ellipsometry and scanning electron microscopy. *Journal of Applied Physics*, **97**, 013511.

[15] Fan, J.G. and Zhao, Y.P. (2006) Characterization of watermarks formed in nano-carpet effect. *Langmuir*, **22**, 3662–3671.

[16] Zhao, Y.P. and Fan, J.G. (2006) Clusters of bundled nanorods in nanocarpet effect. *Applied Physics Letters*, **88**, 103123.

[17] Fan, J.G., Fu, J.X., Collins, A., and Zhao, Y.P. (2008) The effect of the shape of nanorod arrays on the nanocarpet effect. *Nanotechnology*, **19**, 045713.

[18] Kwan, J.K. and Sit, J.C. (2010) The use of ion-milling to control clustering of nanostructured, columnar thin films. *Nanotechnology*, **21**, 295301.

[19] Patzig, C., Miessler, A., Karabacak, T., and Rauschenbach, B. (2010) Arbitrarily shaped Si nanostructures by glancing angle ion beam sputter deposition. *Physica Status Solidi (b)*, **247**, 1310–1321.

[20] Shah, P.J., Wu, Z., and Sarangan, A.M. (2013) Effects of CO_2 critical point drying on nanostructured SiO_2 thin films after liquid exposure. *Thin Solid Films*, **527**, 344–348.

[21] Amar, J.G., Family, F., and Lam, P.M. (1994) Dynamic scaling of the island-size distribution and percolation in a model of submonolayer molecular-beam epitaxy. *Physical Review B*, **50**, 8781–8797.

[22] Knepper, R.A. and Messier, R.F. (2000) Morphology and mechanical properties of oblique angle columnar thin films, in *Complex Mediums* (eds A. Lakhtakia, W.S. Weiglhofer, and R.F. Messier), vol. 4097 of *Proceedings of SPIE*, SPIE Press, Bellingham, WA, pp. 291–298.

[23] Karabacak, T., Wang, G.C., and Lu, T.M. (2003) Quasi-periodic nanostructures grown by oblique angle deposition. *Journal of Applied Physics*, **94**, 7723–7728.

[24] Karabacak, T., Wang, G.C., and Lu, T.M. (2004) Physical self-assembly and the nucleation of three-dimensional nanostructures by oblique angle deposition. *Journal of Vacuum Science and Technology A*, **22**, 1778–1784.

[25] Tang, F., Karabacak, T., Li, L. *et al.* (2007) Power-law scaling during shadowing growth of nanocolumns by oblique angle deposition. *Journal of Vacuum Science & Technology*, **25**, 160–166.

[26] Martín-Palma, R.J., Ryan, J.V., and Pantano, C.G. (2007) Surface microstructure of GeSbSe chalcogenide thin films grown at oblique angle. *Journal of Applied Physics*, **101**, 083513.

[27] Thomas, S., Al-Harthi, S.H., Ramanujan, R.V. *et al.* (2009) Surface evolution of amorphous nanocolumns of Fe-Ni grown by oblique angle deposition. *Applied Physics Letters*, **94**, 063110.

[28] González-García, L., Parra-Barranco, J., Sánchez-Valencia, J.R. *et al.* (2012) Correlation lengths, porosity and water adsorption in TiO_2 thin films prepared by glancing angle deposition. *Nanotechnology*, **23**, 205701.

[29] LaForge, J.M., Taschuk, M.T., and Brett, M.J. (2011) Glancing angle deposition of crystalline zinc oxide nanorods. *Thin Solid Films*, **519**, 3530–3537.

[30] Krause, K.M., Thommes, M., and Brett, M.J. (2011) Pore analysis of obliquely deposited nanostructures by krypton gas adsorption at 87 K. *Microporous and Mesoporous Materials*, **143**, 166–173.

[31] Beaudry, A.L., LaForge, J.M., Tucker, R.T. *et al.* (2013) Flux engineering for indium tin oxide nanotree crystal alignment and heigh-dependent branch orientation. *Crystal Growth and Design*, **13**, 212–219.

[32] LaForge, J.M., Ingram, G.L., Taschuk, M.T., and Brett, M.J. (2012) Flux engineering to control in-plane crystal and morphological orientation. *Crystal Growth and Design*, **12**, 3661–3667.

[33] Krause, K.M., Vick, D.W., Malac, M., and Brett, M.J. (2010) Taking a little off the top: nanorod array morphology and growth studied by focused ion beam tomography. *Langmuir*, **26**, 17558–17567.

[34] Djufors, B.M. (2005) Microstructural characterization of porous thin films and applications to electrochemical capacitors, PhD thesis, University of Alberta.

[35] Barabási, A.L. and Stanley, H.E. (1995) *Fractal Concepts in Surface Growth*, Cambridge University Press, Cambridge, UK.

[36] Meakin, P. (1993) The growth of rough surfaces and interfaces. *Physics Reports*, **235**, 189–289.

[37] Krug, J. (1997) Origins of scale invariance in growth processes. *Advances in Physics*, **46**, 139–282.

[38] Amar, J.G. and Family, F. (1992) Universality in surface growth: Scaling functions and amplitude ratios. *Physical Review A*, **45**, 5378–5393.

[39] Palasantzas, G. (1993) Roughness spectrum and surface width of self-affine fractal surfaces via the K-correlation model. *Physical Review B*, **48**, 14472–14478.

[40] Bubendorff, J.L., Garreau, G., Zabrocki, S. *et al.* (2009) Nanostructuring of Fe films by oblique incidence deposition on a $FeSi_2$ template onto Si(111): growth, morphology, structure and faceting. *Surface Science*, **603**, 373–379.

[41] Dolatshahi-Pirouz, A., Hovgaard, M., Rechendorff, K. *et al.* (2008) Scaling behavior of the surface roughness of platinum films grown by oblique angle deposition. *Physical Review B*, **77**, 115427.

[42] Frederick, J.R., D'Arcy-Gall, J., and Gall, D. (2006) Growth of epitaxial CrN on MgO(001): role of deposition angle on surface morphological evolution. *Thin Solid Films*, **494**, 330–335.

[43] Le Bellac, D., Niklasson, G.A., and Granqvist, C.G. (1995) Scaling of surface roughness in obliquely sputtered chromium films. *Europhysics Letters*, **32** (2), 155–159.

[44] Vauth, S., Streng, C., Mayr, S.G., and Samwer, K. (2003) Growth of vapor-deposited amorphous $Zr_{65}Al_{7.5}Cu_{27.5}$ films under oblique particle incidence investigated by experiment and simulation. *Physical Review B*, **68**, 205425.

[45] Vick, D., Smy, T., and Brett, M.J. (2002) Growth behavior of evaporated porous thin films. *Journal of Materials Research*, **17**, 2904–2911.

[46] Rechendorff, K., Hovgaard, M.B., Chevallier, J. *et al.* (2005) Tantalum films with well-controlled roughness grown by oblique incidence deposition. *Applied Physics Letters*, **87**, 073105.

[47] Álvarez, R., González-García, L., Romero-Gómez, P. *et al.* (2011) Theoretical and experimental characterization of TiO_2 thin films deposited at oblique angles. *Journal of Physics D*, **44**, 385302.

[48] Buzea, C., Kaminska, K., Beydaghyan, G. *et al.* (2005) Thickness and density evaluation for nanostructured thin films by glancing angle deposition. *Journal of Vacuum Science and Technology B*, **23**, 2545–2552.

[49] Cetinkaya, M., Malvadkar, N., and Demirel, M.C. (2008) Power-law scaling of structured poly(*p*-xylylene films deposited by oblique angle. *Journal of Polymer Science Part B*, **46**, 640–648.

[50] Dolatshahi-Pirouz, A., Sutherland, D., Foss, M., and Besenbacher, F. (2011) Growth characteristics of inclined columns produced by glancing angle deposition (GLAD) and colloidal lithography. *Applied Surface Science*, **257**, 2226–2230.

[51] Karabacak, T., Singh, J., Zhao, Y.P. *et al.* (2003) Scaling during shadowing growth of isolated nanocolumns. *Physical Review B*, **68**, 125408.

[52] Khare, C., Gerlach, J.W., Höche, T. *et al.* (2012) Effects of annealing on arrays of Ge nanocolumns formed by glancing angle deposition. *Applied Surface Science*, **258**, 9762–9769.

[53] Liu, Y.J., Chu, H.Y., and Zhao, Y.P. (2010) Silver nanorod array substrates fabricated by oblique angle deposition: morphological, optical, and SERS characterizations. *Journal of Physical Chemistry C*, **114**, 8176–8183.

[54] Main, E., Karabacak, T., and Lu, T.M. (2004) Continuum model for nanocolumn growth during oblique angle deposition. *Journal of Applied Physics*, **95**, 4346–4351.

[55] Mukherjee, S. and Gall, D. (2009) Anomalous scaling during glancing angle deposition. *Applied Physics Letters*, **95**, 173106.

[56] Smith, W., Ingram, W., and Zhao, Y. (2009) The scaling of the photocatalytic decay rate with the length of aligned TiO$_2$ nanorod arrays. *Chemical Physics Letters*, **479**, 270–273.

[57] Taschuk, M.T., Krause, K.M., Steele, J.J. *et al.* (2009) Growth scaling of metal oxide columnar thin films deposited by glancing angle depositions. *Journal of Vacuum Science and Technology B*, **27**, 2106–2111.

[58] Zhou, C.M. and Gall, D. (2008) Development of two-level porosity during glancing angle deposition. *Journal of Applied Physics*, **103**, 014307.

[59] Kardar, M., Parisi, G., and Zhang, Y.C. (1986) Dynamic scaling of growing interfaces. *Physical Review Letters*, **56**, 889–892.

[60] Mukherjee, S. and Gall, D. (2010) Power law scaling during physical vapor deposition under extreme shadowing conditions. *Journal of Applied Physics*, **107**, 084301.

[61] Mukherjee, S., Zhou, C.M., and Gall, D. (2009) Temperature-induced chaos during nanorod growth by physical vapor deposition. *Journal of Applied Physics*, **105**, 094318.

[62] Amar, J.G. and Family, F. (1995) Critical cluster size: Island morphology and size distribution in submonolayer epitaxial growth. *Physical Review Letters*, **74**, 2066–2069.

[63] Dick, B., Brett, M.J., and Smy, T. (2003) Controlled growth of periodic pillars by glancing angle deposition. *Journal of Vacuum Science and Technology B*, **21**, 23–28.

[64] Tang, X.J., Zhang, G., and Zhao, Y.P. (2006) Electrochemical characterization of silver nanorod electrodes prepared by oblique angle deposition. *Nanotechnology*, **17**, 4439–4444.

[65] Patzig, C. and Rauschenbach, B. (2008) Temperature effect on the glancing angle deposition of Si sculptured thin films. *Journal of Vacuum Science and Technology A*, **26**, 881–886.

[66] Zhou, C.M. and Gall, D. (2007) Surface patterning by nanosphere lithography for layer growth with ordered pores. *Thin Solid Films*, **516**, 433–437.

[67] Tait, R.N., Smy, T., and Brett, M.J. (1993) Modelling and characterization of columnar growth in evaporated films. *Thin Solid Films*, **226**, 196–201.

[68] Poxson, D.J., Mont, F.W., Schubert, M.F. *et al.* (2008) Quantification of porosity and deposition rate of nanoporous films grown by oblique-angle deposition. *Applied Physics Letters*, **93**, 101914.

[69] Krause, K.M., Taschuk, M.T., Harris, K.D. *et al.* (2010) Surface area characterization of obliquely deposited metal oxide nanostructured thin films. *Langmuir*, **26**, 4368–4376.

[70] Krause, K.M. (2011) Characterization and modification of obliquely deposited nanostructures, PhD thesis, University of Alberta.

[71] Robbie, K. and Brett, M.J. (1997) Sculptured thin films and glancing angle deposition: Growth mechanics and applications. *Journal of Vacuum Science and Technology A*, **15**, 1460–1465.

[72] Lintymer, J., Martin, N., Chappé, J.M. *et al.* (2004) Influence of zigzag microstructure on mechanical and electrical properties of chromium multilayered thin films. *Surface and Coatings Technology*, **180–181**, 26–32.

[73] Pandya, D.K., Rastogi, A.C., and Chopra, K.L. (1975) Obliquely deposited amorphous Ge films. I. Growth and structure. *Journal of Applied Physics*, **46**, 2966–2975.

[74] Sood, A.W., Poxson, D.J., Mont, F.W. *et al.* (2012) Experimental and theoretical study of the optical and electrical properties of nanostructured indium tin oxide fabricated by oblique-angle deposition. *Journal of Nanoscience and Nanotechnology*, **12**, 3950–3953.

[75] Gospodyn, J. and Sit, J.C. (2006) Characterization of dielectric columnar thin films by variable angle Mueller matrix and spectroscopic ellipsometry. *Optical Materials*, **29**, 318–325.

[76] Xiao, X., Dong, G., Shao, J. *et al.* (2010) Optical and electrical properties of SnO_2:Sb thin films deposited by oblique angle deposition. *Applied Surface Science*, **256**, 1636–1640.

[77] Park, Y.J., Sobahan, K.M.A., Kim, J.J., and Hwangbo, C.K. (2009) Antireflection coatings with helical SiO_2 films prepared by using glancing angle deposition. *Journal of the Korean Physical Society*, **55**, 2634–2637.

[78] Amassian, A., Kaminska, K., Suzuki, M. *et al.* (2007) Onset of shadowing-dominated growth in glancing angle deposition. *Applied Physics Letters*, **91**, 173114.

[79] Rouquerol, F., Rouquerol, J., and Sing, K. (1999) *Adsorption by Powders and Porous Solids*, Academic Press, San Diego, CA.

[80] Lowell, S., Shields, J.E., Thomas, M.A., and Thommes, M. (2006) *Characterization of Porous Solids and Powders: Surface Area, Pore Size and Density*, Springer, Dordrecht.

[81] Barton, T.J., Bull, L.M., Klemperer, W.G. *et al.* (1999) Tailored porous materials. *Chemistry of Materials*, **11**, 2633–2656.

[82] Thommes, M. (2010) Physical adsorption characterization of nanoporous materials. *Chemie Ingenieur Technik*, **82**, 1059–1073.

[83] Kiema, G.K. and Brett, M.J. (2003) Electrochemical characterization of carbon films with porous microstructures. *Journal of the Electrochemical Society*, **150**, E342–E347.

[84] Broughton, J.N. and Brett, M.J. (2002) Electrochemical capacitance in manganese thin films with chevron microstructure. *Electrochemical and Solid-State Letters*, **5**, A279–A282.

[85] Demirel, M.C. (2008) Emergent properties of spatially organized poly(*p*-xylylene) films fabricated by vapor deposition. *Colloids and Surfaces A*, **321**, 121–124.

[86] Harris, K.D., Brett, M.J., Smy, T., and Backhouse, C. (2000) Microchannel surface area enhancement using porous thin films. *Journal of the Electrochemical Society*, **147** (5), 2002–2006.

[87] Li, C.C., Huang, J.L., Lin, R.J. *et al.* (2007) Performance characterization of nonevaporable porous Ti getter films. *Journal of Vacuum Science and Technology A*, **25**, 1373–1380.

[88] Flaherty, D.W., May, R.A., Berglund, S.P. *et al.* (2010) Low temperature synthesis and characterization of nanocrystalline titanium carbide with tunable porous architectures. *Chemistry of Materials*, **22**, 319–329.

[89] Sorge, J.B., Taschuk, M.T., Wakefield, N.G. *et al.* (2012) Metal oxide morphology in argon-assisted glancing angle deposition. *Journal of Vacuum Science and Technology A*, **30**, 021507.

[90] Kim, J., Dohnálek, Z., and Kay, B.D. (2005) Structural characterization of nanoporous Pd films grown via ballistic deposition. *Surface Science*, **586**, 137–145.

[91] Flaherty, D.W., Dohnálek, Z., Dohnálková, A. *et al.* (2007) Reactive ballistic deposition of porous TiO_2 films: growth and characterization. *Journal of Physical Chemistry C*, **111**, 4765–4773.

[92] Flaherty, D.W., Hahn, N.T., Ferrer, D. *et al.* (2009) Growth and characterization of high surface area titanium carbide. *Journal of Physical Chemistry C*, **113**, 12742–12752.

[93] Dohnálek, Z., Kimmel, G.A., McCready, D.E. *et al.* (2002) Structural and chemical characterization of aligned crystalline nanoporous MgO films grown via reactive ballistic deposition. *Journal of Physical Chemistry B*, **106**, 3526–3529.

[94] Šmíd, B., Li, Z., Dohnálková *et al.* (2012) Characterization of nanoporous WO_3 films grown via ballistic deposition. *Journal of Physical Chemistry C*, **116**, 10649–10655.

[95] Kimmel, Greg A. Stevenson, K.P., Dohnálek, Z., Smith, R.S., and Kay, B.D. (2001) Control of amorphous solid water morphology using molecular beams. I. Experimental results. *Journal of Chemical Physics*, **114**, 5284–5294.

[96] Kohnálek, Z., Kimmel, G.A., Ayotte, P. *et al.* (2003) The deposition angle-dependent density of amorphous solid water films. *Journal of Chemical Physics*, **118**, 364–372.

[97] Stevenson, K., Kimmel, G.A., Dohnálek, Z. *et al.* (1999) Controlling the morphology of amorphous solid water. *Science*, **283** (5407), 1505–1507.

[98] Suzuki, M. and Taga, Y. (2001) Numerical study of the effective surface area of obliquely deposited thin films. *Journal of Applied Physics*, **90**, 5599–5605.

[99] Ma, S. and Zhou, H.C. (2010) Gas storage in porous metal-organic frameworks for clean energy applications. *Chemical Communications*, **46**, 44–53.

[100] Steele, J.J., Taschuk, M.T., and Brett, M.J. (2009) Response time of nanostructured relative humidity sensors. *Sensors and Actuators B*, **140**, 610–615.

[101] Schaming, D., Renault, C., Tucker, R.T. *et al.* (2012) Spectrochemical characterization of small hemoproteins adsorbed within nanostructured mesoporous ITO electrodes. *Langmuir*, **28**, 14065–14072.

[102] Khudhayer, W.J., Kariuki, N.N., Wang, X. *et al.* (2011) Oxygen reduction reaction electrocatalytic activity of glancing angle deposited platinum nanorod arrays. *Journal of the Electrochemical Society*, **158**, B1029–B1041.

[103] Bonakdarpour, A., Tucker, R.T., Fleischauer, M.D. *et al.* (2012) Nanopillar niobium oxides as support structures for oxygen reduction electrocatalysts. *Electrochimica Acta*, **85**, 492–500.

[104] Sing, K.S.W., Everett, D.H., Haul, R.A.W. *et al.* (1985) Reporting physisorption data for gas/solid systems. *Pure and Applied Chemistry*, **57**, 603–619.

[105] Barrett, E.P., Joyner, L.G., and Halenda, P.P. (1951) The determination of pore volume and area distributions in porous substances. I. Computations from nitrogen isotherms. *Journal of the American Chemical Society*, **73**, 373–380.

[106] Groen, J.C., Peffer, L.A.A., and Pérez-Ramírez, J. (2003) Pore size distribution in modified micro- and mesoporous materials. pitfalls and limitations in gas adsorption data analysis. *Microporous and Mesoporous Materials*, **60**, 1–17.

[107] Weidenthaler, C. (2011) Pitfalls in the characterization of nanoporous and nanosized materials. *Nanoscale*, **3**, 792–810.

[108] Karabacak, T., Mallikarjunan, A., Singh, J.P. *et al.* (2003) β-phase tungsten nanorod formation by oblique-angle sputter deposition. *Applied Physics Letters*, **83**, 3096–3098.

[109] Li, H.F., Parker, T., Tang, F. *et al.* (2008) Biaxially oriented CaF_2 films on amorphous substrates. *Journal of Crystal Growth*, **310**, 3610–3614.

[110] Xu, Y., Lei, C.H., Ma, B. *et al.* (2006) Growth of textured MgO through e-beam evaporation and inclined substrate deposition. *Superconductor Science and Technology*, **19**, 835–843.

[111] Alouach, H. and Mankey, G.J. (2004) Texture orientation of glancing angle deposited copper nanowire arrays. *Journal of Vacuum Science and Technology A*, **22**, 1379–1382.

[112] Morrow, P., Tang, F., Karabacak, T. *et al.* (2006) Texture of Ru columns grown by oblique angle sputter deposition. *Journal of Vacuum Science and Technology A*, **24**, 235–245.

[113] Okamoto, K., Hashimoto, T., Hara, K. *et al.* (1987) Columnar structure and texture of iron films prepared at various evaporation rates. *Thin Solid Films*, **147**, 299–311.

[114] Okamoto, K. and Itoh, K. (2005) Incidence angle dependences of columnar grain structure and texture in obliquely deposited iron films. *Japanese Journal of Applied Physics*, **44** (3), 1382–1388.

[115] Tucker, R.T., Beaudry, A.L., LaForge, J.M. *et al.* (2012) A little ribbing: Flux starvation engineering for rippled indium tin oxide nanotree branches. *Applied Physics Letters*, **101**, 193101.

[116] Lintymer, J., Gavoille, J., Martin, N., and Takadoum, J. (2003) Glancing angle deposition to modify microstructure and properties of sputter deposited chromium thin films. *Surface and Coatings Technology*, **174–175**, 316–323.

[117] Besnard, A., Martin, N., Carpentier, L., and Gallas, B. (2011) A theoretical model for the electrical properties of chromium thin films sputter deposited at oblique incidence. *Journal of Physics D*, **44**, 215301.

[118] Otiti, T. (2001) Optical and electrical anisotropy in obliquely evaporated Fe films. *Journal of Materials Science Letters*, **20**, 845–846.

[119] Kuwahara, K. and Shinzato, S. (1988) Resistivity anisotropy of nickel films induced by oblique incidence sputter depositions. *Thin Solid Films*, **143**, 165–168.

[120] Otiti, T., Niklasson, G.A., Svedlindh, P., and Granqvist, C.G. (1997) Anisotropic optical, magnetic, and electrical properties of obliquely evaporated Ni films. *Thin Solid Films*, **307**, 245–249.

[121] Vick, D. and Brett, M.J. (2006) Conduction anisotropy in porous thin films with chevron microstructures. *Journal of Vacuum Science and Technology A*, **24**, 156–164.

[122] Harris, K.D., van Popta, A.C., Sit, J.C. *et al.* (2008) A birefringent and transparent electrical conductor. *Advanced Functional Materials*, **18**, 2147–2153.

[123] Zhou, C.M. and Gall, D. (2006) Branched Ta nanocolumns grown by glancing angle deposition. *Applied Physics Letters*, **88**, 203117.

[124] Landauer, R. (1978) Electrical conductivity in inhomogeneous media. *AIP Conference Proceedings*, **40**, 2–45.

[125] McLachlan, D.S., Blaszkiewicz, M., and Newnham, R.E. (1990) Electrical resistivity of composites. *Journal of the American Ceramic Society*, **73**, 2187–2203.

[126] Kuwahara, K. and Hirota, H. (1974) Resistivity anisotropy in evaporated films. *Japanese Journal of Applied Physics*, **13**, 1093–1095.

[127] Van der Pauw, L.J. (1958) A method of measuring specific resistivity and Hall effect of discs of arbitrary shape. *Philips Research Reports*, **13**, 1–9.

[128] Montgomery, H.C. (1971) Method for measuring electrical resistivity of anisotropic materials. *Journal of Applied Physics*, **42**, 2971–2975.

[129] Price, W.L.V. (1972) Extension of van der Pauw's theorem for measuring specific resistivity in discs of arbitrary shape to anisotropic media. *Journal of Physics D*, **5**, 1127–1132.

[130] Bierwagen, O., Pomraenke, R., Eilers, S., and Masselink, W.T. (2004) Mobility and carrier density in materials with anisotropic conductivity revealed by van der Pauw measurements. *Physical Review B*, **70**, 165307.

[131] Dos Santos, C.A.M., de Campos, A., da Luz, M.S. *et al.* (2011) Procedure for measuring electrical resistivity of anisotropic materials: A revision of the Montgomery method. *Journal of Applied Physics*, **110**, 083703.

[132] Martin, N., Sauget, J., and Nyberg, T. (2013) Anisotropic electrical resistivity during annealing of oriented columnar titanium films. *Materials Letters*, **105**, 20–23.

[133] Pugh, E.W., Boyd, E.L., and Freedman, J.F. (1960) Angle-of-incidence anisotropy in evaporated nickel-iron films. *IBM Journal of Research and Development*, **4**, 163–171.

[134] Zhong, Y., Shin, Y.C., Kim, C.M. *et al.* (2008) Optical and electrical properties of indium tin oxide thin films with tilted and spiral microstructures prepared by oblique angle deposition. *Journal of Materials Research*, **23**, 2500–2505.

[135] Mayadas, A.F. and Shatzkes, M. (1970) Electrical-resistivity model for polycrystalline films: the case of arbitrary reflection at external surfaces. *Physical Review B*, **1**, 1382–1389.

[136] Singh, J.P., Liu, D.L., Ye, D.X. *et al.* (2004) Metal-coated Si springs: Nanoelectromechanical actuators. *Applied Physics Letters*, **84**, 3657–3659.

[137] Lalany, A., Tucker, R.T., Taschuk, M.T. *et al.* (2013) Axial resistivity measurement of a nanopillar ensemble using a cross-bridge Kelvin architecture. *Journal of Vacuum Science and Technology A*, **31**, 031502.

[138] Ohring, M. (2002) *Materials Science of Thin Films*, 2nd edn, Academic Press, San Diego, CA.

[139] Alexopoulos, P.S. and O'Sullivan, T.C. (1990) Mechanical properties of thin films. *Annual Review of Materials Science*, **20**, 391–420.

[140] Hardwick, D.A. (1987) The mechanical properties of thin films: a review. *Thin Solid Films*, **154**, 109–124.

[141] Nix, W.D. (1989) Mechanical properties of thin films. *Metallurgical Transactions A*, **20**, 2217–2245.

[142] Vinci, R.P. and Vlassak, J.J. (1996) Mechanical behavior of thin films. *Annual Reviews of Materials Science*, **26**, 431–462.

[143] Hoffman, D.W. and Thornton, J.A. (1979) Effects of substrate orientation and rotation on internal stresses in sputtered metal films. *Journal of Vacuum Science and Technology*, **16**, 134.

[144] Priest, J., Caswell, H.L., and Budo, Y. (1963) Stress anisotropy in silicon oxide films. *Journal of Applied Physics*, **34** (2), 347.

[145] Karabacak, T., Picu, C.R., Senkevich, J.J. *et al.* (2004) Stress reduction in tungsten films using nanostructured compliant layers. *Journal of Applied Physics*, **96**, 5740–5746.

[146] Robbie, K., Hnatiw, A.J.P., Brett, M.J. *et al.* (1997) Inhomogeneous thin film optical filters fabricated using glancing angle deposition. *Electronics Letters*, **33**, 1213–1214.

[147] Robbie, K., Shafai, C., and Brett, M.J. (1999) Thin films with nanometer-scale pillar microstructure. *Journal of Materials Research*, **14**, 3158–1363.

[148] Hill, A.E. and Hoffman, G.R. (1967) Stress in films of silicon monoxide. *British Journal of Applied Physics*, **18**, 13–22.

[149] Jaing, C.C., Liu, M.C., Lee, C.C. *et al.* (2008) Residual stress in obliquely deposited MgF$_2$ thin films. *Applied Optics*, **47**, C266–C270.

[150] Cuomo, J.J., Pappas, D.L., Lossy, R. *et al.* (1992) Energetic carbon deposition at oblique angles. *Journal of Vacuum Science and Technology A*, **10**, 3414–3418.

[151] Liu, F.X., Yao, K.L., and Liu, Z.L. (2007) Substrate tilting effect on structure of tetrahedral amorphous carbon films by Raman spectroscopy. *Surface and Coatings Technology*, **201**, 7235–7240.

[152] Jönsson, B. and Hogmark, S. (1984) Hardness measurements of thin films. *Thin Solid Films*, **114**, 257–269.

[153] Qasmi, M. and Delobelle, P. (2006) Influence of the average roughness R_{ms} on the precision of the Young's modulus and hardness determination using nanoindentation technique with a Berkovich indenter. *Surface and Coatings Technology*, **201**, 1191–1199.

[154] Lintymer, J., Martin, N., Chappé, J.M. *et al.* (2006) Modeling of Young's modulus, hardness and stiffness of chromium zigzag multilayers sputter deposited. *Thin Solid Films*, **503**, 177–189.

[155] Lintymer, J., Martin, N., Chappé, J.M. *et al.* (2005) Nanoindentation of chromium zigzag thin films sputter deposited. *Surface and Coatings Technology*, **200**, 269–272.

[156] Dice, G.D., Brett, M.J., Wang, D., and Buriak, J.M. (2007) Fabrication and characterization of an electrically variable, nanospring based interferometer. *Applied Physics Letters*, **90**, 253101.

[157] Seto, M.W., Dick, B., and Brett, M.J. (2001) Microsprings and microcantilevers: studies of mechanical response. *Journal of Micromechanics and Microengineering*, **11**, 582–588.

[158] Kesapragada, S.V., Victor, P., Nalamasu, O., and Gall, D. (2006) Nanospring pressure sensors grown by glancing angle deposition. *Nano Letters*, **6** (4), 854–857.

[159] Fernando, S.P., Elias, A.L., and Brett, M.J. (2006) Mechanical properties of helically perforated thin films. *Journal of Materials Research*, **21** (5), 1101–1105.

[160] Liu, D., Benstetter, G., Lodermeier, E., and Vancea, J. (2003) Influence of the incident angle of energetic carbon ions on the properties of tetrahedral amorphous carbon (ta-C) films. *Journal of Vacuum Science and Technology A*, **21**, 1665–1670.

[161] Gaire, C., Ye, D.X., Tang, F. *et al.* (2005) Mechanical testing of isolated amorphous silicon slanted nanorods. *Journal of Nanoscience and Nanotechnology*, **5** (11), 1893–1897.

[162] Gaire, C., Ye, D.X., Lu, T.M. *et al.* (2008) Deformation of amorphous silicon nanostructures subjected to monotonic and cyclic loading. *Journal of Material Research*, **23**, 328–335.

[163] Patzig, C., Khare, C., Fuhrmann, B., and Rauschenbach, B. (2010) Periodically arranged Si nanostructures by glancing angle deposition on patterned substrates. *Physica Status Solidi (b)*, **247**, 1322–1334.

[164] Seto, M.W., Robbie, K., Vick, D. *et al.* (1999) Mechanical response of thin films with helical microstructures. *Journal of Vacuum Science and Technology B*, **17**, 2172–2177.

[165] Hirakata, H., Ajioka, Y., Yonezu, A., and Minoshima, K. (2012) Fabrication and mechanical properties of column-particle nanocomposites by multiscale shape-assisted self-assembly. *Journal of Physics D*, **45**, 025302.

[166] Hirakata, H., Matsumoto, S., Takemura, M. *et al.* (2007) Anisotropic deformation of thin films comprised of helical nanosprings. *International Journal of Solids and Structures*, **44**, 4030–4038.

[167] Gaillard, Y., Rico, V.J., Jimenez-Pique, E., and González-Elipe, A.R. (2009) Nanoindentation of TiO_2 thin films with different microstructures. *Journal of Physics D*, **42**, 145305.

[168] Seto, M.W.M. (2004) Mechanical response of microspring thin films, PhD thesis, University of Alberta.

[169] Zhang, G. and Zhao, Y. (2004) Mechanical characteristics of nanoscale springs. *Journal of Applied Physics*, **95**, 267–271.

[170] Hirakata, H., Nishihira, T., Yonezu, A., and Minoshima, K. (2011) Interface strength of structured nanocolumns grown by glancing angle deposition. *Engineering Fracture Mechanics*, **78**, 2800–2808.

[171] So, E., Demirel, M.C., and Wahl, K.J. (2010) Mechanical anisotropy of nanostructured parylene films during sliding contact. *Journal of Physics D*, **43**, 045403.

[172] Hirakata, H., Nishihira, T., Yonezu, A., and Minoshima, K. (2011) Interface strength of structured nanocolumns grown by glancing angle deposition. *Engineering Fracture Mechanics*, **78**, 2800–2808.

[173] Stempflé, P., Besnard, A., Martin, N. *et al.* (2013) Accurate control of friction with nanosculptured thin coatings: Application to gripping in microscale assembly. *Tribology International*, **59**, 67–78.

[174] Malvadkar, N.A., Hancock, M.J., Sekeroglu, K. *et al.* (2010) An engineered anisotropic nanofilm with unidirectional wetting properties. *Nature Materials*, **9**, 1023–1028.

[175] Kitagawa, T., Miyauchi, K., Toyoda, N., Tsubakino, H., and Yamada, I. (2004) Optimum incidence angle of Ar cluster ion beam for superhard carbon film deposition. *Japanese Journal of Applied Physics*, **43**, 3955–3958.

[176] Shetty, R.A., Karimi, A., and Cantoni, M. (2011) Effect of deposition angle on the structure and properties of pulsed-DC magnetron sputtered TiAlN thin films. *Thin Solid Films*, **519**, 4262–4270.

[177] Starbova, K., Dikova, J., and Starbov, N. (1997) Structure related properties of obliquely deposited amorphous a-As_2S_3 thin films. *Journal of Non-Crystalline Solids*, **210**, 261–266.

5 Glancing Angle Deposition Optical Films

5.1 INTRODUCTION

The optical response of a microstructured material such as a GLAD film is intimately connected with its structure and composition. The ability to precisely control structure through GLAD therefore provides a means of manipulating the optical properties in unique ways, a fact that has motivated substantial research efforts in this area. Two main capabilities are provided by GLAD, both based on controlling microstructure to engineer film optical properties. First, using GLAD it is possible to tune the refractive index of the deposit over a wide range because the film optical constants are determined by the relative density ρ_n, which is controlled via α (Section 4.4). Second, through the specific columnar microstructure, ordinary isotropic materials can be endowed with complex optical anisotropies (e.g. birefringence, dichroism). Both of these effects can be interpreted using EMTs that relate the macroscopic optical properties of the film to the internal microstructure. These theories provide a strong foundation for GLAD film optics that has helped improve understanding and further technology development. This chapter examines several important topics in GLAD film optics. First, we describe the specification of optical properties in GLAD films and how to account for the optical anisotropy inherent in structured columnar films. We also discuss how to appropriately measure refractive index anisotropy and obtain calibration information suitable for reliable film design. This information can be used to control GLAD film optical properties and realize new optical devices; for example, single-material graded-index films and helical structures with novel polarization effects.

Understanding thin-film optical phenomena is generally a complex affair, and in the following the reader is assumed to possess a basic understanding of several topics, including thin-film optics and polarization effects. For the former, the reader is referred to introductory chapters in books by Heavens [1] and Macleod [2]. Discussions of the latter can be found in books by Born and Wolf [3], Yariv and Yeh [4] and Goldstein [5].

5.2 THE OPTICS OF STRUCTURED GLANCING ANGLE DEPOSITION FILMS

5.2.1 Optical anisotropy in columnar glancing angle deposition films

Ever since the earliest studies of oblique deposition [6–8], it has been known that the tilted columnar microstructure of obliquely deposited films generates a corresponding optical

Glancing Angle Deposition of Thin Films: Engineering the Nanoscale, First Edition.
Matthew M. Hawkeye, Michael T. Taschuk and Michael J. Brett.
© 2014 John Wiley & Sons, Ltd. Published 2014 by John Wiley & Sons, Ltd.

anisotropy, where film properties depend on the orientation between the microstructure and the optical fields. While the column material in GLAD films is generally isotropic, when shaped into an elongated, asymmetric columnar structure an anisotropic optical response is created. Light propagation through such a film exhibits polarization-sensitive properties atypical of isotropic materials, often exploited in optical devices such as polarizers and waveplates. Optical anisotropy is quite common in the natural world, with the birefringent and double refractive properties of quartz and calcite being classic examples. In these natural examples the anisotropy is intrinsic to the material, typically created by the crystal structure. In GLAD films, however, the anisotropy is a consequence of the structure and not of the material itself, and the induced anisotropy is thus frequently termed *form birefringence*.

To properly account for the inherent optical anisotropy of GLAD media, the refractive index n cannot be described by a scalar quantity. Instead, separate n values must be used to describe the index along specific orientations related to the symmetry of the film microstructure. The refractive index is then expressed in more general terms using a second-rank tensor

$$n = \begin{pmatrix} n_i & 0 & 0 \\ 0 & n_j & 0 \\ 0 & 0 & n_k \end{pmatrix}, \tag{5.1}$$

where the diagonal elements n_i, n_j and n_k are the principal refractive indices and Equation 5.1 corresponds to the tensor written in principal axis form. The principal index values may be complex to account for absorption, and wavelength dependent to account for dispersion effects. There are two basic types of anisotropy. The most general, lowest symmetry case is biaxial anisotropy, where all three principal indices are distinct ($n_1 \neq n_2 \neq n_3$). When two of the indices are equal, the anisotropy is said to be uniaxial. Whereas the optical behaviour of isotropic thin-film materials can be theoretically calculated using common 2×2 matrix techniques [2], a general description of anisotropic materials is more complex and requires the use of 4×4 matrix formalisms [9–13].

For GLAD, the orientations of the index tensor's principal axes are determined by the symmetry of the column structure. Figure 5.1a and b shows SEM views of a film with a tilted columnar microstructure and the corresponding refractive index ellipsoid. The first SEM (a) shows a side view of the columns, revealing the characteristic tilted microstructure. From this SEM, a set of *ijk*-axes is defined[1] where the *k*-axis points along the column growth direction (at an angle β from the surface normal), the *j*-axis points into the page and the *i*-axis is perpendicular to both *j* and *k*. The second SEM image (b) is a tilted top-down view obtained by looking down at the film along the $-k$-direction, revealing the asymmetric column cross-section typical of tilted column films where the columns are broader along the *j*-axis than the *i*-axis. The principal indices (Figure 5.1c) describing the optical properties of such a film are aligned with these structurally defined *ijk*-axes. Because the column structure is different along each axis, all three principal indices are unique and the film anisotropy is biaxial, with $n_i < n_j < n_k$ in general. This phenomenon can be more rigorously analysed using EMT and depolarization factors, as discussed in Section 5.2.4.

---

[1] The principal axes of GLAD films generally exhibit an orthorhombic structure where all three axes intersect at 90° angles. However, studies of certain metallic GLAD structures have shown a monoclinic symmetry, arising from intercolumn coupling effects [15, 16].

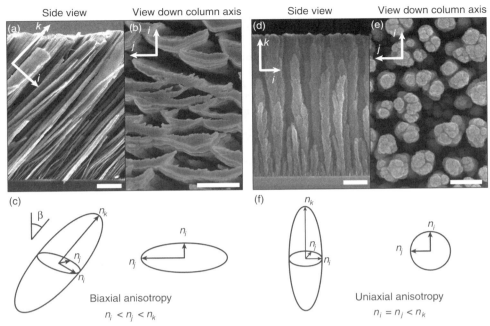

Figure 5.1 Thin-film microstructures and their corresponding optical anisotropies. (a,b) Tilted columns can be characterized by three structural axes: k points along the column axis, j is oriented along the broadening direction and i is oriented at right angles to j and k. (c) The low-symmetry structure leads to biaxial anisotropy, where the film optical properties are specified by a refractive index tensor with three distinct principal indices n_i, n_j and n_k respectively oriented along the i, j and k axes. (d,e) Vertically oriented columns present a more symmetric microstructure. As before, the k structural axis is oriented along the column axis. However, the circular column cross-section makes the i and j directions equivalent and the film is effectively isotropic in the substrate plane. (f) The increased structural symmetry leads to uniaxial anisotropy, where the refractive index tensor describing the film contains only two unique principal indices $n_i = n_j$, and n_k. The scale bars in (a) and (b) are 2 μm, and in (d) and (e) are 200 nm. Sources: (a) and (b) Reproduced from [14]. With permission from A.C. van Popta.

A more symmetric microstructure is realized by vertically oriented columnar films. This microstructure and the corresponding indices are shown in Figure 5.1d–f, and a different set of ijk-axes characterizes this microstructure. The k-axis is oriented along the column axis, which is now parallel to the surface normal, and the i- and j-axes lie in the substrate plane. However, because the average column microstructure is rotationally symmetric in the plane, there is no structural anisotropy to distinguish between the i- and j-directions as there was in the tilted column case. Consequently, n_i and n_j are equivalent and the optical anisotropy is uniaxial, being fully described by two unique indices $n_i = n_j < n_k$.

Calculating planar birefringence

When developing GLAD materials for polarization devices, a key specification is the planar birefringence, equal to the difference between the refractive index along the slow axis

(y-axis) and the fast axis (x-axis) for light at normal incidence. While this difference is zero in a uniaxial vertical-columnar film, the anisotropic cross-section of a tilted columnar GLAD film introduces significant planar birefringence (Section 5.4.2). In general, the slow axis is parallel to the column fanning direction, yielding

$$n_y = n_j. \tag{5.2}$$

The fast axis is perpendicular to the slow axis and lies in the substrate plane, thus receiving a projected contribution from both n_i and n_k according to

$$\frac{1}{n_x^2} = \frac{\cos^2 \beta}{n_i^2} + \frac{\sin^2 \beta}{n_k^2}, \tag{5.3}$$

where β is the column tilt. The planar birefringence Δn can then be calculated from the principal index values as

$$\Delta n = n_y - n_x. \tag{5.4}$$

5.2.2 Modelling glancing angle deposition films with effective medium theory

Making an exact theoretical prediction of GLAD film optical properties requires a precise and highly detailed description of the structure. Such descriptions are generally not possible in GLAD films owing to the inherent disorder; the randomized column placement, column size variations, nonuniform column structure, column surface roughness and internal columnar nanostructure conspire to create a high degree of structural uncertainty, especially over macroscopic areas. While there have been a few reports using more rigorous modelling techniques [17–19], many GLAD researchers turn to EMT to develop practical approximations, useful for explaining and predicting film properties. EMT takes an inhomogeneous composite medium, consisting of multiple material phases with a complex or unknown microstructure, and generates a representative homogeneous medium possessing (approximately) the same macroscopic optical properties. The optics of the heterogeneous medium with all its attendant complexities can thus be replaced by a homogeneous medium described by an *effective* dielectric constant ϵ_e. (The dielectric constant and the refractive index are related by $\epsilon = n^2$). The homogenization provided by an accurate EMT is very useful, as calculating measurable quantities (such as reflectance and transmittance) is significantly less complicated for a homogeneous material than for a structured one. Furthermore, with a reliable and accurate EMT approximation, it then becomes possible to quantitatively deduce important microstructural parameters, such as density, column orientation and geometry, from macroscopic optical measurements. This approach typically uses numerical techniques to vary the structural parameters and match the EMT prediction to experimental data. EMT is a well-established and very useful method for investigating the properties of composite materials. We recommend several books on the subject for further general information [20–22]. The use of EMT in thin-film technology is also common, and there are several excellent review articles on the subject [23–25].

EMT models take as basic input the dielectric constant of the constituent materials (ϵ_A and ϵ_B) and the relative amounts of each material (ρ_A and ρ_B, where $\rho_A + \rho_B = 1$). While

Table 5.1 Survey of reports using EMT descriptions of GLAD structures (HC: helical columns; PS: phisweep columns; SBD: serial bideposited columns; SP: spin–pause columns; TC: tilted columns; VC: vertical columns; ZZ: zigzag columns).

EMT model	Ref.	Material	Structure
Linear volume approximation	[26]	ITO	TC
	[27]	ITO, SiO_2	TC
Maxwell–Garnett	[28]	ZrO_2, TiO_2	TC
	[29–31]	Ag	TC
	[32]	Ag–SiO_2	TC
	[33, 34]	Al	TC
	[35]	TiO_2, Ta_2O_5, ZrO_2	TC
	[36]	Al-oxide	TC
Bruggeman	[37]	Si	TC
	[38]	Si	SBD
	[39]	TiO_2	HC
	[40]	MgF_2	HC
	[41]	MgF_2, SiO_2	TC
	[42, 43]	Si	VC
	[44, 45]	Ge	TC
	[46]	TiO_2	TC, HC
	[47]	ITO	TC
	[48, 49]	Cr	TC
	[34, 50]	Al	TC
	[51]	TiO_2, SiO_2	ZZ, SP, SBD
	[52]	Nb_2O_5	TC
	[53]	Nb_2O_5	TC
	[54]	Cu	HC, VC
	[55]	ATO	TC
	[56]	Alq_3	TC

the discussion below examines two material mixtures, certain EMTs allow consideration of mixtures of more than two materials. The columnar shape is parameterized by the depolarization factor L (Section 5.2.4). When the inclusions are nonspherical, as in the elongated columnar structures typical of GLAD, L becomes direction dependent, thus introducing anisotropy into the optical properties. A column is generally approximated as an ellipsoid, in which case three L values are used, one for each semiaxis.

Several EMT models have been used to study the optical properties of a variety of GLAD columnar structures, summarized in Table 5.1. The applicability of each EMT model is related to the specific microgeometry assumed in its derivation; where the microgeometry is in doubt, theoretical limits can be placed on the effective index value to estimate uncertainty (Section 5.2.5). The simplest EMT approach uses the linear volume approximation, where a mixture of materials A and B has an effective dielectric constant given by

$$\epsilon_e = \rho_A \epsilon_A + \rho_B \epsilon_B. \tag{5.5}$$

This equation describes the optical response of a laminar microstructure with the optical field applied parallel to the layers (Figure 5.2a). Also encountered in the literature is the classic

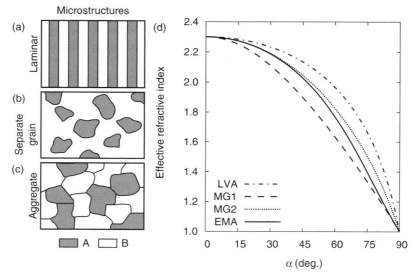

Figure 5.2 Microstructures of materials A and B used in the derivation of common EMTs. (a) The linear volume approximation (LVA) assumes a laminar structure of the two materials, with the optical field parallel to the layers. (b) The Maxwell–Garnett derivation is based on a separated grain microstructure, with inclusions of material A embedded in a continuous matrix of material B. (c) The aggregate microstructure, a random mixture of A and B, corresponds to the Bruggeman EMT. (d) Effective refractive index n_e predictions of the LVA, the Maxwell–Garnett theory with A embedded in B (MG1) and B embedded in A (MG2) and the Bruggeman approximation (EMA). The calculation is for $L = 0.5$, $n_A = 2.3$ and $n_B = 1.0$ (film–air mixture), and assumes $\rho_A = 2 \cos \alpha / (1 + \cos \alpha)$. While all four models predict a continuous decrease in n_e as the porosity increases, the predictions differ due to the different microstructures. *Source:* (b,c) adapted from [25].

Maxwell–Garnett EMT modified for elliptical geometries (also called the Bragg–Pippard equation), written as

$$\frac{\epsilon - \epsilon_B}{L\epsilon_e + (1 - L)\epsilon_B} = \rho_A \frac{\epsilon_A - \epsilon_B}{L\epsilon_A + (1 - L)\epsilon_B}. \tag{5.6}$$

The Maxwell–Garnett equation describes a separated-grain microstructure, where particles of material A are embedded within a continuous host of material B. The most common EMT approximation used to describe GLAD films is the Bruggeman model, often simply called the 'effective medium approximation' in the homogenization literature, which is given by

$$\rho_A \frac{\epsilon_A - \epsilon_e}{\epsilon_A + (L^{-1} - 1)\epsilon_e} + \rho_B \frac{\epsilon_B - \epsilon_e}{\epsilon_B + (L^{-1} - 1)\epsilon_e} = 0. \tag{5.7}$$

The Bruggeman approximation describes an aggregate microstructure (Figure 5.2c), where A and B are randomly distributed throughout the medium. The laminar and aggregate microstructures are symmetric with respect to exchanging materials A and B. This symmetry carries over to Equations 5.5 and 5.7, which predict the same effective properties whether material A is embedded in B or vice versa. However, this is not the case in the Maxwell–Garnett theory, as exchanging A and B yields an inverted microstructure and a different effective medium prediction from Equation 5.6.

To illustrate the differences between the different models, Figure 5.2d shows the effective refractive index calculated for $L = 0.5$, $n_A = 2.3$, $n_B = 1.0$, and assuming $\rho_A = 2 \cos \alpha / (1 + \cos \alpha)$ (Equation 4.11, a nominal density dependence for GLAD films). Plotted are the predictions of the LVA, the Maxwell–Garnett theory with A embedded in B (MG1) and B embedded in A (MG2), and the Bruggeman approximation (EMA). All three theories predict that the effective index continuously falls off as α increases and the film density decreases, matching experimental trends (Figure 5.4). However, the exact α dependence is different between the theories, a consequence of the microstructural differences between the models. To improve the accuracy of modelling efforts, it is therefore important to consider which microstructure best represents the actual GLAD film, often supported using experimental SEM images. Particular attention is required when examining films deposited over a range of α, owing to the pronounced microstructural differences between films deposited in different shadowing regimes. In practice, it is useful to compare the results of multiple theories and/or use the EMT limits (Section 5.2.5) to quantify the effects of microstructural uncertainty.

Finally, these EMT models are all restricted to the quasistatic regime, where the microstructural features are significantly smaller than the optical wavelength. This restriction is commonly specified as $q/\lambda < 0.1$, where q is the characteristic microstructure dimension. Some more recent theories support extending this validity to $q/\lambda < 0.3$ in certain situations [23]. Generalized EMTs incorporating multiple scattering effects are able to cope with larger structures, although the mathematical complexity is greater. In the optical spectrum (visible and infrared wavelengths), GLAD structures often satisfy the quasistatic conditions underlying most EMTs. However, one must be mindful of this assumption, particularly when working in the ultraviolet (UV) wavelengths ($\lambda < 400$ nm), with thicker films or at high α where column diameters become highly broadened. In regimes such as these, the quasistatic assumption can range from questionable to clearly violated, thus impacting the reliability of EMT models.

5.2.3 The column and void material refractive indices

In EMT calculations, GLAD films are considered as heterogeneous mixtures of columnar material and void regions, having refractive indices of n_c and n_v respectively. The void region is usually taken to be air or vacuum with $n_v = 1$, although it can be replaced by other values to model the GLAD film in different environments. For example, setting $n_v = 1.33$ can be used to model water condensation in the porous GLAD film [39]. Specifying the refractive index of the column material n_c can be done in different ways. Literature and tabulated handbook values may be used, although this is often a poor approximation as it is well known that n is highly sensitive to processing conditions (such as substrate temperature, deposition method and gas pressures). Many researchers, therefore, perform separate empirical calibrations by preparing films under nominally the same conditions but at $\alpha = 0°$. Measuring n of the resulting film thus provides an estimate of n_c more specific to the particular process.

Because the GLAD-fabricated columns may exhibit internal structure, some researchers have also used a nested EMT concept. In this approach, the columns are treated as a two-part mixture in an initial EMT calculation to produce an effective column index $n_{c,e}$. This value is then used as the column material in a second composite consisting of void and the effective column material. Le Bellac *et al.* used this concept to theoretically model aluminium oxide coatings [36], describing the columns as a composite of elongated Al particles in an aluminium oxide matrix.

When measurements are performed over a wide spectral range, it is also important to consider the optical dispersion of the material and incorporate this effect into the EMT as well. This can be done by using empirical measurements of n_c at different λ or by using an appropriate dispersion model (e.g. Cauchy dispersion polynomial, Sellemeir equation, or Tauc–Lorentz oscillator) to parameterize the wavelength dependence.

5.2.4 Modelling form birefringence via the depolarization factor

The origin of form birefringence in columnar microstructures can be modelled by parameterizing the column shape through the depolarization factor L, which describes the electrostatic polarizability of the microstructure. The larger the depolarization factor, the smaller the resulting refractive index. In the columnar architectures typical of GLAD fabrication, the reduced structural symmetry creates an anisotropic polarizability, which is incorporated into EMT modelling by specifying a depolarization factor for each principal axis; that is, $\{L_i, L_j, L_k\}$, where $L_i + L_j + L_k = 1$. While arbitrary structures require numerical calculation of L [57, 58], ellipsoidal geometries permit useful analytical solutions and provide a first-order approximation of GLAD column structures.

One such geometry is the prolate spheroid, obtained by rotating an ellipse about its semimajor axis to produce a cigar-like shape. This geometry provides a first-order approximation of a vertically oriented column of length a and width b (Figure 5.3). Two L values parameterize

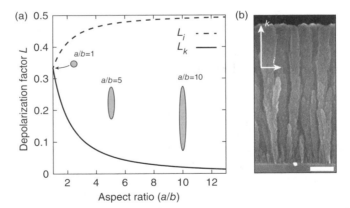

Figure 5.3   (a) Relationship between the depolarization factors L_i and L_k and the microstructural aspect ratio (a/b) in a prolate spheroid (cigar shape). For $a/b = 1$, the structure is spherical and $L_i = L_k = 1/3$. The structure becomes more elongated as a/b increases, leading to a monotonic increase in L_i and monotonic decrease in L_k. In high aspect ratio structures $(a/b > 10)$, L_i and L_k asymptotically approach 0.5 and 0 respectively. (b) Vertical column microstructures can exhibit large aspect ratios, leading to saturated depolarization factors and maximized differences between n_k and n_i. Scale bar corresponds to 200 nm.

such a structure, one parallel to the column axis given by

$$L_k = \frac{1 - e^2}{2e^3} \left[\ln \left(\frac{1 + e}{1 - e} \right) - 2e \right]$$

(5.8)

and the second along the equator given by

$$L_i = L_j = \frac{1}{2}(1 - L_z),$$

(5.9)

where $e = [1 - (b/a)^2]^{1/2}$ is the eccentricity. Figure 5.3a plots Equations 5.8 and 5.9 for structures of different aspect ratio (a/b), alongside an SEM image of vertical columnar film for comparison (Figure 5.3b). In the spherical case $(a/b = 1)$ the polarization response is isotropic and L_k and L_i are both equal to 1/3. Increasing the aspect ratio breaks the symmetry and generates anisotropy with L_k decreasing and L_i increasing. In the high aspect ratio limit, corresponding to long and thin vertical columns, $L_k \rightarrow 0$ and $L_i \rightarrow 1/2$. The depolarization factor thus captures the different polarizability of the columnar microstructure for light polarized parallel and perpendicular to the column axis. Because $L_i = L_j > L_k$, the in-plane refractive index of a vertical column structure is less than the index along the substrate normal, a prediction consistent with experimental findings.

Modelling tilted column GLAD microstructures requires accounting for the cross-sectional fanning that occurs during column growth (Section 2.6.4). This microstructure can be approximated by a general ellipsoid having semi-axes a, b and c, where a is oriented along the column axis (k-axis), b is oriented parallel to the fanning direction (j-axis) and c is orthogonal to a and b (i-axis). While crude, this approximation is practical and very common as the depolarization factors of a general ellipsoid can be determined analytically. However, the computation is decidedly more complex than the prolate spheroid case. For $a > b > c$, the depolarization factors of the ellipsoid are given by [22]

$$L_a = \frac{abc[F(k, \phi) - E(k, \phi)]}{(a^2 - b^2)\sqrt{(a^2 - c^2)}},$$

(5.10)

$$L_b = 1 - L_a - L_c,$$

(5.11)

$$L_c = \frac{b}{b^2 - c^2} \left[b - \frac{ac}{\sqrt{a^2 - c^2}} E(\phi, k) \right],$$

(5.12)

where the functions $E(\phi, k)$ and $F(\phi, k)$ are incomplete elliptic integrals of the first and second kind defined as

$$F(k, \phi) = \int_0^\phi \frac{d\psi}{\sqrt{1 - k^2 \sin^2 \psi}},$$

(5.13)

$$E(k, \phi) = \int_0^\phi d\psi \sqrt{1 - k^2 \sin^2 \psi}.$$

(5.14)

where $k^2 = (a^2 - b^2)/(a^2 - c^2)$ and $\sin^2 \phi = 1 - c^2/a^2$. Tabulated values for these integrals can be found in mathematical handbooks and are often implemented in software packages such as Matlab and Mathematica. While these expressions are complex, they provide a link between the measured principal index values and the underlying structural anisotropy of the column, enabling optical measurement of microstructural parameters.

5.2.5 Dealing with microstructural uncertainty: bounds on the effective dielectric function

Since it is often the case that the microscopic geometry is not known precisely, it is not always clear which EMT approximation should be used. Fortunately, there exist a series of theoretical bounds that can be used to place physical limits on the possible ϵ_e values. These bounds are useful for estimating the impact of microstructural uncertainty on ϵ_e and corresponding structural parameters. The first and widest are the Wiener bounds [23], given by

$$\epsilon_1 = \rho_a \epsilon_a + \rho_b \epsilon_b \tag{5.15}$$

$$\epsilon_2^{-1} = \rho_a \epsilon_a^{-1} + \rho_b \epsilon_b^{-1}. \tag{5.16}$$

More stringent restrictions on the effective properties were developed by Hashin and Shtrikman [59], who showed that, when ρ_a and ρ_b are known, ϵ_e will lie between

$$\epsilon_1 = \epsilon_b + \frac{\rho_a \epsilon_b (\epsilon_a - \epsilon_b)}{\epsilon_b + L\rho_b(\epsilon_a - \epsilon_b)} \tag{5.17}$$

and

$$\epsilon_2 = \epsilon_a + \frac{\rho_b \epsilon_a (\epsilon_b - \epsilon_a)}{\epsilon_a + L\rho_a(\epsilon_b - \epsilon_a)}. \tag{5.18}$$

These limits are equivalent to the Maxwell–Garnett expression, the first with ϵ_b embedded in ϵ_a and the second with ϵ_a embedded in ϵ_b.

5.3 CALIBRATING OPTICAL PROPERTIES OF GLANCING ANGLE DEPOSITION FILMS

Obtaining an accurate experimental calibration of film optical constants is an important prerequisite of optical design. Many techniques have been developed to measure optical constants in conventional thin films, and excellent references on the topic exist [1,60–62]. In general, the experimental methodologies involve performing a set of optical measurements and then estimating the optical constants using a theoretical model. Certain simple cases permit algebraic analysis, whereas more complex measurements and models require numerical fitting routines that match model predictions to experimental data and extract the desired parameters. Compared with conventional thin-film characterization, the optical anisotropy inherent in GLAD films is an additional complicating factor that must be factored into the characterization methodology. The measurement reliability is not merely determined by instrumental precision but is also limited by how accurately the model represents the GLAD film. Special consideration is thus required when designing an experimental approach as not all methods are equal. Some methods are simpler but only provide limited information or are restricted to certain experimental configurations. Other methods require complex instrumentation and sophisticated analysis but are more general in their applicability. It is therefore up to the practitioner to select a method that provides the information required and to the necessary accuracy. The following discussion outlines several methods that have been used to calibrate the optical properties of GLAD films, presented in order of increasing capability and complexity. These techniques have been used in the analysis of many different materials and GLAD microstructures, which are surveyed in Table 5.2.

5.3.1 Basic measurements: isotropic approximations

The easiest way to deal with optical anisotropy as a complicating factor is to ignore it. At first, this may seem like bad advice given all the prior discussions of anisotropy in GLAD films. However, approximating the structured GLAD film as an isotropic medium may be sufficiently accurate as, in certain cases, the difference between the principal indices may only be a few percent. An isotropic measurement technique can provide a scalar n value that generally falls within the range of the actual principal indices of the anisotropic medium. In addition to providing a rough estimate, the analysis required is significantly simpler as the theories used to model isotropic media are less complex than their anisotropic counterparts. This leads to faster implementation and greater ease of use. Consequently, the isotropic approach is often sufficient for the purposes of basic calibration, and applications include quick parameter measurement, assessing the impact of deposition conditions, simple process monitoring and providing starting estimates for subsequent complicated calibrations.

One approach employed by several researchers is the envelope method, which is also commonly used for routine characterization of conventional films [63–66]. In this straightforward analysis technique, the normal-incidence transmission spectrum of the film is measured, producing characteristic thin-film interference fringes. Envelopes around the fringe maxima and minima are then constructed, from which n can be directly calculated at every wavelength. The envelope technique has been used by several researchers to obtain scalar n approximations of GLAD structured films of As_2S_3 [67,68], As_2Se_3 [67], GeS_2 [69], Nb_2O_5 [52], ATO [55], TiO_2 [70] and ZrO_2 [53].

Spectrophotometric methods are similar to envelope methods but use slightly more complex analysis and instrumentation. The spectrophotometric approach involves measuring the reflectance and/or transmittance spectrum, often at multiple angles of incidence, and then numerically fitting the experimental data with a simple optical thin-film model to extract n. While the use of numerical fitting routines adds to the analysis complexity compared with the envelope technique, the modelling approach provides greater flexibility [71], allowing incorporation of different optical-dispersion models, extension to multilayer films, and consideration of inhomogeneities and surface roughness effects. Spectrophotometric determination of n in GLAD films has been performed by various groups, examining materials such as Cu [54], Ge [44,72], GeSbSe [73] and SiO_2 [74].

A third, standard optical analysis technique is ellipsometry, a widely used analytical technique for thin-film characterization, capable of excellent instrumental precision and compatible with many material systems [75,76]. In ellipsometry, one measures optical polarization changes after reflection from or transmission through the film. In standard ellipsometric

Table 5.2 Material survey of reported GLAD film optical constants. HC: helical columns; PS: phisweep columns; SBD: serial bideposited columns; SP: spin-pause columns; TC: tilted columns; VC: vertical columns; ZZ: zigzag columns. Alq$_3$: tris-(8-hydroxyquinoline)aluminium; ATIR: attenuated total internal reflection; ATO: antimony-doped tin oxide; ITO: indium-doped tin oxide.

Mater.	Ref.	Structure	Parameters	Technique
Ag	[30]	TC	Principal indices	Spectrophotometry
Co	[87]	TC	Principal indices	Ellipsometry
Cr	[15]	TC	Principal indices	Ellipsometry
Cu	[54]	TC	Scalar n	Spectrophotometry
Ge	[44]	TC	Scalar n	Spectrophotometry
	[45]	TC	Scalar n	Ellipsometry
	[72]	TC	Scalar n	Spectrophotometry
Si	[103]	TC	Scalar n	Ellipsometry
	[43]	VC	Principal indices	Ellipsometry
	[42]	VC	Principal indices	Ellipsometry
	[37]	SBD	Principal indices	Ellipsometry
Ti	[16]	TC	Principal indices	Ellipsometry
Alq$_3$	[56]	TC	Principal indices	Ellipsometry
As$_2$S$_3$	[67]	TC	Scalar n	Envelope method
	[68]	TC	Scalar n	Envelope method
As$_2$Se$_3$	[67]	TC	Scalar n	Envelope method
ATO	[55]	TC	Scalar n, linear biref.	Envelope method
GeSbSe	[73]	TC	Scalar n	Spectrophotometry
GeSe$_2$	[69]	TC	Scalar n	Envelope method
HfO$_2$	[88]	TC	Linear biref.	Ellipsometry
	[99]	TC	Principal indices	Prism coupler
ITO	[77]	TC	Scalar n	Ellipsometry
	[46]	TC, HC	Scalar n	Ellipsometry
	[47]	TC	Scalar n	Ellipsometry
	[106]	SBD	Linear biref.	Ellipsometry
MgF$_2$	[81]	SBD	Linear biref.	Envelope method
	[41]	TC	Principal indices	Ellipsometry
	[96]	TC	Principal indices	Pol. conversion
	[97]	PS	Principal indices	Pol. conversion
MoO$_3$	[107]	VC	Principal indices	Ellipsometry

Mater.	Ref.	Structure	Parameters	Technique
SiO$_2$	[74]	TC	Scalar n	Multiple
	[78]	TC	Scalar n	Ellipsometry
	[94]	TC	Principal indices	Spectrophotometry
	[101]	TC	Principal indices	ATIR
	[41]	TC	Principal indices	Ellipsometry
	[102]	ZZ, SP, SBD	Principal indices	Ellipsometry
TiO$_2$	[70]	TC	Scalar n, linear biref.	Envelope method
	[80]	VC	Planar n	Spectrophotometry
	[82]	SBD	Linear biref.	Ellipsometry
	[92]	SBD	Linear biref.	Ellipsometry
	[93]	TC	Principal indices	Spectrophotometry
	[98]	TC	Principal indices	Prism coupling
	[94]	TC	Principal indices	Spectrophotometry
	[91]	TC	Principal indices	Ellipsometry
	[104]	TC	Principal indices	Ellipsometry
	[35]	TC	Principal indices	Multiple
Ta$_2$O$_5$	[102]	ZZ, SP, SBD	Principal indices	Ellipsometry
	[82]	SBD	Linear biref.	Ellipsometry
	[95]	TC	Principal indices	Spectrophotometry
	[91]	TC	Principal indices	Ellipsometry
	[35]	TC	Principal indices	Multiple
WO$_3$	[105]	TC	Principal indices	Ellipsometry
ZnS	[83]	TC	Scalar n, linear biref.	Spectrophotometry
ZrO$_2$	[53]	TC	Scalar n, Linear biref.	Envelope method
	[82]	SBD	Linear biref.	Ellipsometry
	[93]	TC	Principal indices	Spectrophotometry
	[91]	TC	Principal indices	Ellipsometry
	[35]	TC	Principal indices	Multiple

analysis, these polarization changes are interpreted via models based on light propagation in isotropic media to yield the optical constants. (Note that generalized ellipsometry techniques exist that are able to characterize complex, anisotropic media such as GLAD films, a topic discussed below.) Standard ellipsometry techniques have been used to provide isotropic n approximations of GLAD films in a number of studies, examining materials such as Ge [45], ITO [46, 47, 77], SiO_2 [78] and TiO_2 [78].

As a final note, it is important to re-emphasize that the isotropic assumption is generally inaccurate and cannot fully characterize the GLAD film. When possible, the validity of the approximation should be assessed to provide confidence in the approach. This can be done, for example, by comparing the results of basic isotropic models against the predictions of anisotropic calculations assuming reasonable values of anisotropy.

5.3.2 Calibrating anisotropy with polarization-sensitive measurements

In the design of anisotropic optical devices using GLAD, it is critical to obtain an accurate characterization of the relevant anisotropy parameters. The isotropic approaches are ill-suited to the measurement of these parameters and the characterization of more complex film structures that exploit these anisotropies, such as thin-film waveplates and helical microstructures. A full characterization of anisotropy in GLAD structures requires intricate measurement and analysis methods, which are discussed in Section 5.3.3. However, measurements of certain parameters, such as planar birefringence and retardance, can be obtained by performing polarization-resolved versions of the basic envelope and spectrophotometric methods. Such experimental refinements also provide a more accurate measurement of GLAD film optical properties, and are useful for improved calibration procedures.

While more complicated experimentally owing to the requirement of polarization optics, certain alignments between the medium's principal axes and the optical polarizations allow simplified analysis [79]. For example, vertical columnar microstructures exhibit uniaxial anisotropy and behave isotropically for light polarized in the surface plane. Therefore, normally incident light of any polarization 'sees' an isotropic medium and the isotropic assumption can be made without reservation. For obliquely incident light, the electric field of the s-polarization state (also called transverse-electric (TE) polarization) always lies in the plane and therefore also sees a regular isotropic medium. The planar refractive index of vertical columnar structures can therefore be accurately obtained in an s-polarized spectrophometric measurement, which has been performed for TiO_2 GLAD films [80].

Similar high-symmetry configurations exist in the case of biaxial, tilted columnar structures, allowing accurate characterization via simpler analysis. By aligning the polarizations along high-symmetry directions, the polarization states are decoupled and the optical response can be examined using isotropic analysis. This can be used to characterize the linear birefringence of the biaxial film in a straightforward manner. A common approach involves measuring the normal-incidence transmittance spectra twice, first with the optical field polarized normal to the deposition plane and second with the field polarized in the deposition plane. Applying the envelope or model-fitting analysis method to these transmittance spectra provides the two planar indices and thus the linear birefringence. This method has been used to calibrate the linear birefringence in biaxial films of MgF_2 [81], Nb_2O_5 [52], ATO) [55], Ta_2O_5 [35, 82], TiO_2 [35, 70, 82], ZrO_2 [35, 53, 82] and ZnS [83]. To ensure accurate measurement, it is important that the optical polarization and the film optical axes are aligned as well as possible. To assist alignment, it is helpful to place the film between cross-polarizers and align the film optical axes to the polarizer axes by minimizing the transmitted signal.

5.3.3 In-depth characterization with generalized techniques

When even greater detail is required from the optical calibration, researchers turn to more general methodologies capable of measuring each principal index of the anisotropic medium as well as their spatial orientation. Knowledge of these parameters provides a sophisticated description of GLAD film optical properties, one that is better suited for high-accuracy design, characterization of complex films and microstructural studies (via EMT). Although much more is learned from these calibrations, the measurement apparatus and analysis techniques needed to obtain a generalized characterization are more complex than previously discussed techniques.

While basic ellipsometric techniques are commonly used for isotropic film analysis as previously discussed, the generalized ellipsometry approach provides a broader scope of analysis and is often used to characterize complex anisotropic materials [84–86]. These advanced ellipsometry measurements incorporate a more complete description of anisotropy in their analysis, thus providing the framework to extract additional information from polarization-change measurements. The generalized ellipsometry approach has become a predominant technique for advanced characterization of optical GLAD films, being used to measure principal indices in several material systems, including Alq_3 [56], Co [87], Cr [15], HfO_2 [88], MgF_2 [41], Si [38, 42, 89, 90], SiO_2 [41, 51], Ta_2O_5 [91], Ti [16], TiO_2 [51, 91, 92] and ZrO_2 [91]. Additionally, because ellipsometric measurements are sensitive to nanometre-scale thickness changes, several reports have adapted the technique for in-situ growth characterization [90, 91].

Several groups have also extended the spectrophotometric approach to the case of anisotropic materials. Here, one first measures polarization-resolved reflectance and/or transmittance spectra and then extracts the desired principal-index parameters using an anisotropic optical model [30, 93–95]. Another approach involves measuring cross-polarized reflectance spectra, where the sample is illuminated with one polarization state and the detector measures the amount converted into the orthogonal polarization after reflection of the sample [96, 97]. This conversion is determined by the sample anisotropy, and the principal index values can thus be calculated.

Another standard technique for optical characterization is prism coupling, where incident light is evanescently coupled to thin-film waveguide modes. By modifying the analysis to consider anisotropic film effects, these approaches become suitable for GLAD film characterization and extraction of principal indices [98, 99].

5.3.4 Additional factors

The preceding discussions have focused on methods to incorporate optical anisotropy into the measurement technique in order to provide a progressively more accurate and complex description of the structured GLAD film. However, it is important to note that, while anisotropy effects are particularly relevant in GLAD film characterization, the analysis may be further complicated by additional factors, such as the presence of unintentional index gradients and surface roughness. Fortunately, these further complications are also present in conventional optical coatings, and analysis methods have been developed to deal with their effects; for example, index gradients in Ref. [71] and roughness in Ref. [100].

5.4 CONTROLLING GLANCING ANGLE DEPOSITION FILM OPTICAL PROPERTIES

In conventional film fabrication, the optical properties are sensitive to several processing variables, such as substrate temperature, gas pressures, and deposition rate. These sensitivities generally allow properties to be tuned over a limited range only; the goal then becomes minimizing process variations to fabricate a consistent deposit from run to run. However, optical properties can be controlled over a very wide range using GLAD technology by manipulating film microstructure. Furthermore, the general aspects of this control are material independent and the techniques have been demonstrated across a wide range of materials (Table 5.2).

As understood through EMT, GLAD refractive index engineering is accomplished in two ways, which may be combined. The first approach introduces porosity in the film microstructure: by manipulating the relative amounts of column and void regions in the microstructure, the effective refractive index of the film can be controlled. This control is achieved by using α to alter the shadowing length. The second approach involves realizing specific optical anisotropies by altering the shape and orientation of the columnar microstructures. Both α and φ motions are often required to reach the design goal, depending on the optical effect and accuracy required. The following examines first a preliminary approach to refractive index engineering, focused on using α to control film density and structural anisotropy in a rudimentary, yet powerful way. Adding φ-rotation to the fabrication process enables more sophisticated control over column structure, which directly translates to sophisticated control over the resulting form birefringence.

5.4.1 Basic refractive index engineering with α

Figure 5.4a shows measured refractive indices for TiO_2 films deposited at different α. The data shown are compiled from several sources and include measurements reported by multiple groups, using a range of measurements, and examining several microstructures. Values correspond to wavelengths near 600 nm, and the n_y value was selected in cases where optical anisotropy was reported. These results clearly demonstrate the refractive index engineering capability provided by GLAD technology. By controlling only a *single* parameter, the refractive index of the deposited film can be continuously tuned over a large range. For smaller α ($<50°$), the film remains dense, with the relatively short shadow lengths introducing minimal porosity to the microstructure. As α increases and the shadow length grows longer, greater portions of the microstructure are occupied by void regions (Figure 5.4b–d) and the index falls off commensurately. At very glancing angles, a highly porous film is produced with a film microstructure dominated by isolated columnar features surrounded by void regions (Figure 5.4e). The refractive index drops off quickly under these conditions, approaching $n = 1$ in the $\alpha \rightarrow 90°$ limit. By tuning α between these two extremes, any intermediate n value can achieved.

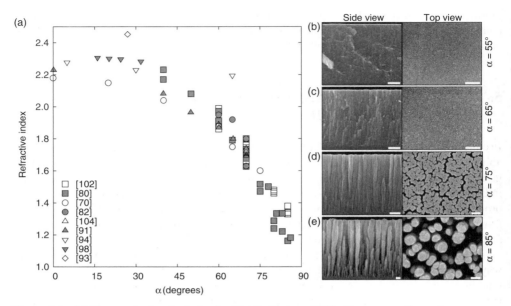

Figure 5.4   (a) Measured refractive index for GLAD-fabricated TiO$_2$ films reported by various groups: [70, 80, 82, 91, 93, 94, 98, 102, 104]. All indices were measured for wavelengths near 600 nm. (b–e) Side and top SEM views of vertical columnar film microstructures deposited at several α, illustrating the change in film porosity and microstructure over a wide α range. All scale bars indicate 200 nm. (b)–(e) Reproduced from [108]. With permission from K.M. Krause.

The observed variation over the combined data set is consistent with the known sensitivity of TiO$_2$ films to normal processing conditions (e.g. substrate temperature, oxygen pressure). In amorphous, thin-film form the index of deposited TiO$_2$ typically varies from 2.2 to 2.4, depending on the preparation method [109]. Measurements on GLAD films deposited at near-normal incidence are consistent with this range, and the variability carries through to other α. Moreover, several of the individual data sets exhibit less variability, presumably related to the specific degree of process control achieved in each report. It should therefore be possible to achieve precise fabrication of target n values through process optimization and quality control.

As a final point, this discussion examined TiO$_2$ as the conclusions reached are strengthened by the large number of available data sets. However, the $n(\alpha)$ behaviour presented is general, and similar trends are seen in different materials. Material-specific features aside, this speaks to the generality of the underlying mechanism. The refractive index is controlled by manipulating the film microstructure via ballistic shadowing, which affects many materials in similar ways, making GLAD a powerful technique in optical coating engineering.

5.4.2 Controlling planar birefringence with α

The planar birefringence Δn is a key design parameter for polarization optics fabricated with GLAD technology. Unlike the simpler, monotonic relationship between n and α, Δn exhibits

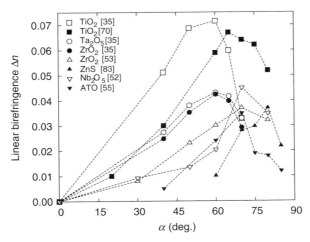

Figure 5.5 Linear birefringence values measured for several materials, showing the complex α-dependence of the birefringence. Across the multiple data sets, the general empirical trends are consistent and the birefringence is typically maximized near 70°. However, sensitivity to material and processing conditions leads to significant variability ($\sim 10°$), and accurate optimization requires experimental calibration.

a more complex α scaling. Empirical studies have examined several materials, including ATO [55], Bi_2O_3 [110], HfO_2 [88], Nb_2O_5 [52], Ta_2O_5 [35, 110], TiO_2 [35, 70], WO_3 [110], ZrO_2 [35, 53] and ZnS, and Figure 5.5 shows a representative selection of the reported data. These investigations have consistently reported similar α dependencies: Δn is negligible for $\alpha \sim 0$, rises gradually with increasing α, reaches a material-dependent peak value at an intermediate α, and then decreases quickly as α approaches 90°. For most materials, the peak Δn values can greatly surpass the birefringence of many common natural materials, such as quartz ($\Delta n = 0.009$) and MgF_2 ($\Delta n = 0.006$). To be competitive with highly anisotropic materials such as calcite ($\Delta n = 0.172$) or nematic liquid crystals, more advanced GLAD motions must be used, as discussed in Section 5.4.3. The birefringence maximum typically occurs at $70° \pm 10°$, although the exact shape of the curve is determined by the specific α dependencies of the column tilt and the principal indices (Equation 5.4). An accurate determination of the optimal α value requires experimental calibration as the birefrigence is both material and process specific. For example, the birefringence peaks of the two TiO_2 data sets presented in Figure 5.5 reach similar Δn values but are offset by $\sim 10°$, indicating the sensitivity of the process to experimental conditions. A similar conclusion is reached by examining the ZrO_2 data in Figure 5.5 as well.

5.4.3 Optimizing birefringence with serial bideposition

Optimizing form birefringence involves calibrating an optimal α; depending on material, this approach can be used to achieve Δn values on the order of 0.04–0.07, as shown in Figure 5.5. Although these values are already larger than many common natural materials, GLAD film birefringence can be further increased by using advanced substrate motions to realize column

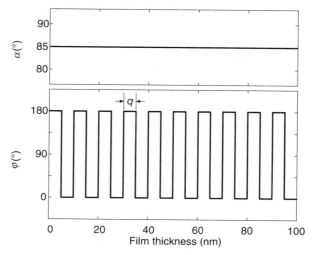

Figure 5.6   Substrate motion algorithm for the SBD technique. The substrate tilt angle α is held constant throughout the deposition. The substrate is rotated quickly between $\varphi = 0°$ and $180°$ after every growth interval q.

geometries with even greater structural anisotropies, and thus enhanced form birefringence properties. This is the objective of the SBD technique, developed by Hodgkinson *et al.* [112]. The SBD substrate algorithm, depicted in Figure 5.6, involves holding the substrate at a fixed α and implementing step-like $180°$ φ rotations after every growth interval q, which is typically sub-10 nm. Note that φ can either rotate back and forth between two angles separated by $180°$ as shown, or step through a series of angles $180°$ apart (e.g. $0°$, $180°$, $360°$, $540°$).

The SBD motion is similar to that used to form a zigzag microstructure (Figure 5.7a,b) where the alternating growth direction produces a microstructure with distinct columnar arms. However, the SBD technique produces different columnar features by drastically reducing the growth interval between the φ rotations. By setting q less than the characteristic column width, distinct columnar arms are unable to form before the subsequent φ rotation (Figure 5.7c) and the zigzag microstructure degenerates into a vertically oriented column (Figure 5.7d). This microstructural change is similar to the helix-to-vertical-column transition observed when moving from slow to rapid φ rotation (Section 2.5.2). However, the SBD vertical columns are strikingly different than those produced by rapid, continuous φ rotation. In the latter case, the vapour flux arrives equally from all directions to establish a symmetric shadowing environment (Figure 5.8b). The resulting columns have circular planar cross-sections and exhibit uniaxial optical anisotropy (Figure 5.8b). In the SBD case, however, the vapour flux direction alternates between two opposing directions, creating a strictly one-dimensional shadowing environment where lateral column-fanning remains (Figure 5.8c). SBD, therefore, produces vertically oriented columns with a pronounced elliptical cross-section and a corresponding biaxial optical anisotopy (Figure 5.8d).

The SBD technique has been used to enhance the linear birefringence of anisotropic Si [37], ITO [106], Ta_2O_5 [82], TiO_2 [82,92,104] and ZrO_2 [82] microstructures. By optimizing

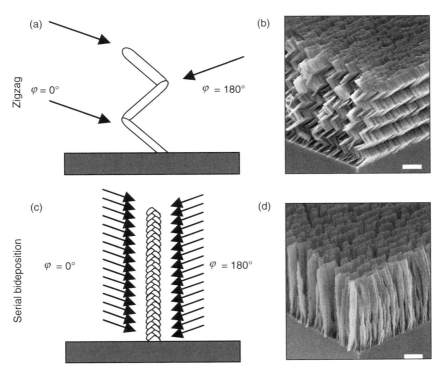

Figure 5.7   In the typical zigzag growth procedure, (a) the φ is periodically rotated from 0° to 180° to (b) produce a series of column arms with alternating growth direction. For SBD, (c) the growth interval between φ rotations is much shorter, causing the columnar growth sections to (d) degenerate into a single vertically oriented column. Both scale bars indicate 1000 nm. (a) and (c) Reproduced from [14]. With permission from A.C. van Popta. (b) and (d) Reprinted with permission from [111]. © 2010, American Chemical Society.

the α and q process variables the form birefringence of SBD-fabricated films can be greatly increased compared with an equivalent tilted column film. As with basic, obliquely deposited tilted columns, the birefringence of SBD-fabricated structures exhibits a strong dependence on α, with a peak value obtained between $\alpha = 60$ and 70° (Figure 5.9a). However, the sub-deposit thickness q also influences the resulting birefringence, as shown by Hodgkinson et al. [82] in studies of SBD-deposited TiO_2 (Figure 5.9b). For $q > 20$ nm, the SBD motion has little effect and the linear birefringence of the deposited film is equivalent to that of a tilted columnar microstructure. However, the birefringence quickly increases as q is reduced below 20 nm, and for $q < 5$ nm the birefringence is double the basic value. Therefore, when implementing the SBD algorithm it is recommended that q be only a few nanometres to maximize the effect.

Post-deposition treatments can also be used to further increase the birefringence properties of SBD structures (see Section 5.7.1 for more details). For example, van Popta et al. used thermal annealing treatments to crystallize SBD-deposited TiO_2 structures and achieve an additional twofold increase of Δn [92].

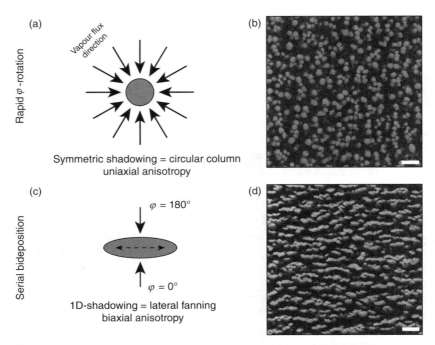

Figure 5.8 Rapid φ rotation produces (a) a uniform, 2D shadowing environment that yields (b) vertically oriented, circular column cross-sections. By depositing from opposing sides, the SBD technique creates (c) an asymmetric shadowing environment that produces (d) vertically oriented columns with highly elongated cross-sections. The scale bars indicate 1000 nm. (b) and (d) Reprinted with permission from [111]. © 2010, American Chemical Society.

5.4.4 Modulating birefringence with complex φ motions

While the φ rotations of the SBD technique provide a useful route to maximize form birefringence in deposited structures, refined control over the birefringence can be obtained by further engineering the $\varphi(t)$ motion profile. The vertical- and SBD-column microstructures shown in Figure 5.8b and d can be seen as two limiting cases: the former uses linear, continuous $\varphi(t)$ motion to produce a uniaxial microstructure with $\Delta n = 0$, whereas the latter implements discrete, stepwise $\varphi(t)$ motion to increase structural anisotropy and maximize Δn. To smoothly interpolate between the two extremes, van Popta *et al.* created a family of $\varphi(t)$ motions where the φ rotation-speed as a function of film thickness t is given by [113]

$$\frac{d\varphi}{dt} = N\frac{1 - \cos(2\pi t/q)}{1 + e\cos(2\pi t/q)}. \tag{5.19}$$

Here, N is a normalizing constant set so that $\phi = 360°$ when $t = 2q$. The parameter e is the primary control variable, set on the interval $[-1, 1]$ to determine the exact shape of the motion profile. Figure 5.10 shows the $\varphi(t)$ motion profile for different values of e. The limiting cases described above are achieved when $e = -1$, yielding linear $\varphi(t)$ motion, and

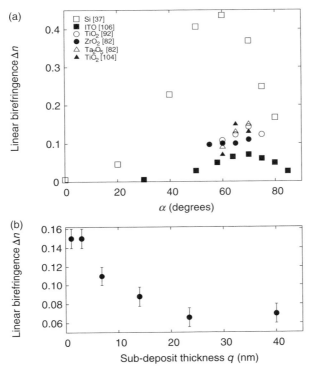

Figure 5.9 (a) α scaling of linear birefringence in SBD-fabricated films of various materials. (b) Effect of varying the sub-deposit thickness q on the SBD film linear birefringence (TiO$_2$ at $\alpha = 70°$). For $q > 20$ nm, the birefringence is unaffected. Reducing q below the 20 nm threshold progressively increases the linear birefringence, reaching a 100% increase for $q < 5$ nm. Adapted from [82].

$e = 1$, producing the step-like SBD motion. As e is increased from -1 to $+1$, the $\varphi(t)$ motion becomes increasingly nonlinear and gradually approaches the step-like SBD motion.

The symmetry of the shadowing environment is controlled by the e parameter, enabling interpolation between the high 2D-uniformity of Figure 5.8a and the one-dimensional anisotropic shadowing of Figure 5.8c. Consequently, the anisotropy of the fabricated microstructure is strongly influenced by the specific e parameter. Figure 5.11a–d shows top-down SEM images of a series of TiO$_2$ films deposited at $\alpha = 70°$ and different e values. Although the microstructure is relatively dense at this α, making it difficult to resolve columnar features, the structural anisotropy can be more easily inferred from the orientation of intercolumn void regions. For $e = -1$, the slit-like void structures exhibit no preferred orientation, indicating a high degree of structural symmetry. As e is increased and the shadowing environment becomes more nonuniform, the structural anisotropy also increases, as shown by the void regions becoming progressively more aligned. These observations are supported by linear birefringence measurements in these films (Figure 5.11e), which demonstrate a strong positive correlation between Δn and e. Modulating e during deposition allows Δn to be controlled through the film thickness, a technique used to realize complex birefringence gradients and advanced helical film designs (Section 5.6.6).

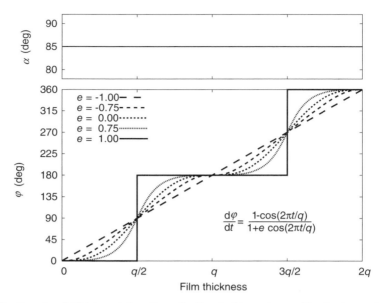

Figure 5.10   Equation 5.19 (shown) describes a family of $\varphi(t)$ rotation profiles that interpolate between vertical column and SBD motion [113]. The key control parameter is e, which varies between -1 and $+1$ and specifies the shape of the φ motion. For $e \to -1$ the motion is highly linear (~vertical column) while as $e \to 1$ the motion becomes increasingly step-like (~SBD).

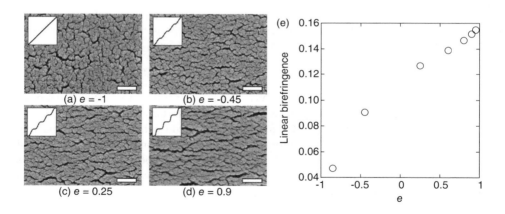

Figure 5.11   (a–d) Top-down SEMs of the film microstructure reveal the increasingly anisotropic microstructure developed as e is increased and the substrate φ motion becomes more SBD-like. The scale bars represent 200 nm. (e) The tunable microstructural anisotropy enables control over the linear birefringence of the deposited film. When the e parameter approaches -1, the symmetric microstructure leads to minimal birefringence, whereas e values near $+1$ produce highly anisotropic structures with maximal birefringence. Tuning e between these limits allows realization of tailored birefringence levels in the deposited film. Reproduced with permission from [113]. © 2007, The Optical Society.

5.4.5 Controlling n with advanced glancing angle deposition motions

While α provides basic tuning of GLAD film structure and the corresponding optical properties, several advanced motion algorithms with specific φ rotations provide significantly greater control over the column growth. The SBD technique and van Popta's modification thereof provide excellent control over film birefringence via engineered φ rotations. However, there is a whole suite of advanced GLAD motion algorithms that can be used to further manipulate the column microstructure, such as spin–pause (Section 2.7.2) and phisweep (Section 2.7.3). These techniques achieve sophisticated control over shadowing-growth effects through engineered φ rotations, thereby providing control over various structural parameters, such as ρ_n, β and column fanning. Because these parameters collectively influence the principal indices of the GLAD film, the advanced GLAD algorithms can be used to realize specific columnar morpholgies with tailored anisotropies. Wakefield *et al.* devised a unifying treatment of these advanced GLAD substrate motions, the columnar microstructures they produce and the resulting optical anisotropy [51]. Their approach predicts the optical properties generated by a given motion algorithm, enabling the rational design of columnar microstructures to target specific optical effects not produced by basic GLAD algorithms. For instance, by manipulating the column tilt and the degree of column fanning using the phisweep technique, it is possible to carefully tune the planar column microstructure and the resulting principal indices to access different regimes of anisotropy. By selecting growth conditions that produce a matched n_x and n_y, a uniaxial anisotropy can be produced in a tilted microstructure. Another possibility is to use a technique such as phisweep to suppress column fanning and create films exhibiting negative birefringence.

5.5 GRADED-INDEX COATINGS: DESIGN AND FABRICATION

While the GLAD technique allows the refractive index of a single deposited layer to be controlled as desired, it is possible to extend the concept to the fabrication of multiple layers. By changing α during deposition, controlled density variations can be introduced into the film. These density changes produce an equivalent modulation of the film refractive index, enabling the fabrication of single-material films with engineered refractive-index profiles, so-called graded-index films. Because the index is determined by the relative amounts of column-to-void(air) material, a wide index modulation may be achieved. Whereas conventional graded-index fabrication methods, such as co-evaporation [114] or plasma-enhanced chemical vapour deposition [115], are typically limited in the choice of materials and the accessible refractive index ranges, the GLAD approach is compatible with a diverse range of materials and can achieve any refractive index between the bulk material value and air, thus offering very powerful design and fabrication capabilities.

This graded-index approach has been used to fabricate various graded-index optical coatings in a wide range of materials, summarized in Table 5.3. In the following sections we describe a general procedure that can be used to achieve desired $n(t)$ profiles from calibrated measurements of $n(\alpha)$. Specific examples for generating important index profiles are also discussed.

Table 5.3   Survey of graded-index optical coatings fabricated using the GLAD technique (ITO: indium-doped tin oxide).

Index profile	Ref.	Material	Notes
Antireflection	[116]	SiO_2	Gaussian (continuous)
	[78]	TiO_2, SiO_2	Quintic, 5-layer
	[117]	TiO_2, SiO_2	Quintic, 7-layer
	[118]	ITO	Quintic, 6-layer
	[119]	TiO_2, SiO_2	Optimized, 3-layer
	[120]	SiO_2	High–low stack, 4-layer
	[121]	TiO_2	Quintic, 5-layer
	[122]	SiO_2	Optimized, 4-layer
	[123]	Ge	Linear, Gaussian, quintic, 3-layer
Rugate	[124]	MgF_2	
	[125]	ZrO_2	Birefringent rugate
	[42]	Si	Regular and apodized profiles
	[126]	Si	Regular and apodized profiles
	[127]	TiO_2	Microcavity
	[128]	TiO_2	Birefringent microcavity
	[129]	SiO_2	Index-matched layer
	[130]	TiO_2	Microcavity
Distributed Bragg reflector	[43]	Si	
	[77]	ITO	
	[131]	Y_2O_3:Eu	Photoluminescent
	[132]	ITO	Conductive
	[133]	Si	
	[134]	TiO_2	Vapour sensing
	[135]	TiO_2	Ultrasound transducer

5.5.1 General design method for glancing angle deposition graded-index coatings

The design problem is to develop a set of α and ϕ substrate motions that will realize a desired refractive index profile $n(t)$. This index profile can assume virtually any shape, limited only by the range of accessible $n(\alpha)$ values and the capabilities of the α and ϕ motors. As an illustrative case, consider the example shown in Figure 5.12.

The first step is to discretize a target index profile $n(t)$ into a series of N sublayers with thickness dt_i and refractive index n_i, as depicted in Figure 5.12. While dt_i should ideally be infinitesimal, it is limited by the resolution of the substrate motion as determined by the motor hardware. Thus, in practice, the discretization should be sufficiently fine to reproduce the optical properties $n(t)$ to the desired level of accuracy. Tolerances can be as assessed via numerical modelling of the optical properties to gauge the effect of dt_i. Note that the sublayer thickness need not be uniform, but can be adapted to provide a finer discretization where dn/dt is large. Once discretized, the deposition angle α_i needed to realize n_i in each sublayer can be determined from the $n(\alpha)$ calibration data, thereby transforming the $n(t)$ design into a corresponding $\alpha(t)$ substrate motion trajectory. Curve fits of the $n(\alpha)$ data are useful here to estimate values over the entire α range. Fitting functions can be constructed from EMT models discussed above or using empirical polynomials.

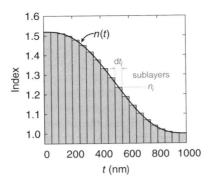

Figure 5.12 Discretization of a continuous refractive index profile $n(t)$ into a series of constant-index sublayers, with index n_i and thickness dt_i. Here, $n(t)$ corresponds to a quintic refractive index profile, a high-performance antireflection coating design [136].

Reproducing the given $n(t)$ design requires not only accurately depositing layers of the target n_i, but also of the desired thickness, which plays an equally important role in determining the optical response of the overall film. Accurate thickness calibration of the GLAD process is therefore required, which demands special attention when depositing films with variable α motion because the vertical growth rate of the film is α dependent. As discussed in Section 7.3, this dependence is acquired from both a geometric factor and because of the α-dependent film density. Because the GLAD motion is generally controlled relative to a monitored deposition rate, which is α independent, it is important to calibrate the deposition ratio $D(\alpha)$, a quantity equal to the ratio of the film growth rate to the monitored deposition rate. The $D(\alpha)$ calibration can be obtained while determining $n(\alpha)$ by also measuring the film thickness, which is usually required in the optical modelling. Numerical curve fits of the $D(\alpha)$ data can be performed, typically using empirical polynomials or via Equation 7.6. Once the $D(\alpha)$ relationship is established, the ith-sublayer thickness can be converted to a thickness deposited at the rate monitor according to

$$dt'_i = \frac{dt_i}{D(\alpha_i)}. \tag{5.20}$$

The sublayer thicknesses can be accurately controlled using this expression, a key requirement for realizing the desired optical index profile.

5.5.2 Designing φ motions for high-accuracy graded-index coatings

This discussion has thus far focused on controlling the α motion to achieve a target index profile as α is the main control parameter for ρ_n. However, an appropriate φ motion must be implemented in the design as well. The principal effect of φ is to determine the orientation and magnitude of optical anisotropies in the deposited film. Because common optical coating designs, including antireflection (AR) layers and interference filters, are specified for isotropic materials, it is important to consider the optical anisotropies inherent to GLAD-fabricated films and construct appropriate φ motions for high-accuracy fabrication.

The simplest method for achieving a target index profile is to ignore the optical anisotropy and hold φ fixed during the deposition while the $\alpha(t)$ trajectory is followed. This substrate motion will construct a series of slanted-column sublayers, producing a film with a variable-β microstructure. However, because of the slanted columnar microstructure, the film will also exhibit a corresponding biaxial optical anisotropy associated with column fanning. The fixed-φ approach will therefore only yield a first-order approximation to the intended film because the presence of the optical anisotropy will cause discrepancies between the optical properties of the fabricated film and the isotropic design. While this error may be acceptable depending on the application-specific tolerance, achieving greater accuracy requires implementing φ motion to realize a vertically oriented microstructure. Because such a geometry exhibits uniaxial anisotropy with a vertical optical axis, the microstructured film behaves isotropically for normally incident light (and obliquely incident s-polarized light). An added benefit of φ rotation is an improvement in film uniformity across the wafer (Section 7.4).

The vertical column microstructure is thus more ideally suited to realizing equivalents of common isotropic optical coating designs. As discussed in Section 2.5.2, this microstructure is created by continuous φ rotation. The angular velocity of the substrate must be sufficiently large such that the growth pitch p, equal to the amount of vertical film growth per revolution, is smaller than the column width w. While w is a material- and process-dependent parameter (Section 4.3.2) and requires calibration, setting $p = 10$–20 nm will generally produce a vertical-column film. This p value also makes a good starting point for further growth optimization. Because the vertical growth rate changes with α, the φ rotation rate must change as well to compensate and maintain a constant pitch. Over the ith growth interval dt_i, the substrate must rotate through an angle $d\varphi_i$ equal to

$$d\varphi_i = \frac{dt_i}{p_i} \times 360° \tag{5.21}$$

in order to realize a specified growth pitch p_i. In terms of growth at the rate monitor and the deposition ratio (Equation 5.20), this yields

$$d\varphi_i = \frac{D \cdot dt_i'}{p_i} \times 360°. \tag{5.22}$$

Using these relations, a $\varphi(t)$ trajectory can be developed to accompany a given $\alpha(t)$ motion and realize the vertically oriented columnar microstructure. In general, p_i would be constant for every sublayer in the view of fabricating a consistent microstructure. However, it could be varied to optimize the growth of each sublayer.

One final comment is worth emphasizing. The vertical columnar microstructure generally provides the most accurate reproduction of isotropic coating designs and the preceding $\varphi(t)$ design recipe is thus recommended. However, manipulating $\varphi(t)$ can be used to engineer optical anisotropies and create unconventional graded-index coatings with novel polarization-selective responses [18,19,43,125,128]. Because manipulating optical anisotropy is a unique capability, there is little design precedence in the optical coating industry, and thus much research and development opportunity for the enterprising engineer.

5.5.3 Specific examples

Section 5.5.2 has provided a framework for constructing $\alpha(t)$ and $\varphi(t)$ substrate motion algorithms to realize a specified $n(t)$ profile target using calibrated $n(\alpha)$ and $D(\alpha)$ values. Here, this framework is implemented to create AR and graded-index interference coatings, two designs selected for their technological importance.

5.5.4 Antireflection coatings

AR layers are a common application of optical coatings, used to eliminate surface reflection losses from elements in optical systems. One approach to AR coating design uses graded-index layers, where $n(t)$ smoothly varies between a value matched to the substrate index n_s and one matched to the index of the incident medium n_i (usually air). Several groups have reported on graded-index AR coatings fabricated using GLAD technology (Table 5.3), using complex $\alpha(t)$ and $\varphi(t)$ motions to achieve the smoothly varying index profile required.

Southwell described a high-performance graded-index AR coating using a fifth-order polynomial [136], commonly called a quintic matching layer, given by

$$n(t) = n_i + (n_s - n_i)(10x^3 - 15x^4 + 6x^5), \tag{5.23}$$

where $x = 1 - t/L$ and L is the thickness of the matching layer. This index profile is shown in Figure 5.13a, with parameters of $n_i = 1$ and $n_s = 1.52$, indices corresponding to air and typical glass, and $L = 1000$ nm. As can be seen, the refractive index profile smoothly interpolates between the high-index substrate and the low-index air, leading to a significant reduction in the Fresnel reflection losses.

For the design example, a representative set of artificial calibration curves was constructed. Figure 5.13b shows $n(\alpha)$ calibration data, generated by using Equation 5.7 and assuming a dispersionless column-material index of 1.5 (typical value for SiO_2) and an α-dependent ρ_n given by Equation 4.11. A depolarization factor $L = 0.5$ was used, representative of light normally incident upon high aspect ratio vertical columns. By controlling α, the refractive index of the deposit can be tuned over a wide range, from 1.5 at $\alpha = 0$, corresponding to a dense deposit, to ~ 1.0, corresponding to a predominantly void-filled film microstructure. Figure 5.13c shows $D(\alpha)$ calibration data, obtained from Equation 7.6 with $F_0 = 1$ and assuming the same α-dependent ρ_n curve used to generate $n(\alpha)$. In the case where no experimental values are available, generating artificial calibration curves from universal equations in this manner can be useful in estimating preliminary values. However, accurate designs should always be constructed from experimental calibrations.

The first step is to invert the $n(\alpha)$ data and construct the necessary $\alpha(t)$ profile. This profile is then converted to $\alpha(t')$ using the nonlinear correction given by the $D(\alpha)$ curve (Equation 5.20). The resulting α motion is shown in Figure 5.13c. As the first layer is required to initially match the higher index substrate, the substrate is initially positioned at $\alpha = 0°$, thus depositing a matching layer. To realize the slow index variation, α is systematically increased following the calibration curves. The α increase is relatively rapid at first, but this rate of change decreases steadily and tapers off as α approaches $90°$ to produce a highly rarefied microstructure with n approaching unity. The $\varphi(t')$ motion required to achieve a constant

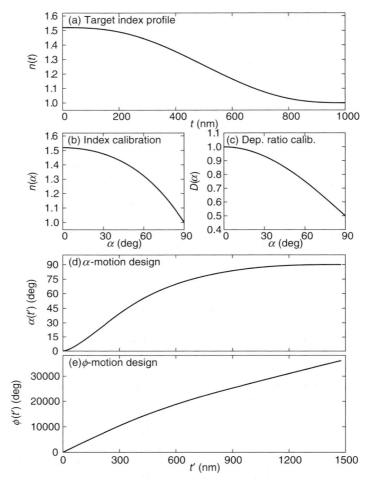

Figure 5.13 Designing a substrate motion algorithm to realize a graded-index AR coating. (a) The design target consists of a quintic $n(t)$ profile providing index matching between a glass substrate ($n = 1.52$) and air ($n = 1$). Artificial (b) refractive index and (c) deposition ratio calibration curves generated for this design example (details in text). Using the procedure described in Section 5.5.1, α and φ motions were generated to realize the design target. The α position is initially near $0°$ to deposit a dense layer with an index equal the glass substrate. As deposition continues, α is slowly increased to reduce the film index according to the design target. The substrate undergoes continuous φ rotation to produce vertically oriented, uniaxial structures. However, the φ rotation rate decreases during deposition to maintain a constant growth pitch.

$p = 10$ nm throughout the deposition is shown in Figure 5.13d, calculated using Equation 5.22 and the $D(\alpha)$ calibration curve. Because film growth relative to the rate-monitor thickness decreases as α is increased, the angular velocity of the φ rotation decreases to compensate, which is reflected in the slow and continuous decrease in the slope of $\varphi(t')$. Additional features of the substrate motion profiles are also worth pointing out. The final thickness deposited at the rate monitor is nearly 50% greater than that deposited at the substrate

$(t'_f = 1477$ nm compared with $t_f = 1000$ nm). This difference occurs because film growth is slower at the substrate than at the rate monitor, as indicated by $D < 1$ for all α. Also note that $n(t)$ is symmetric about its midpoint $(t = L/2)$, with $n(t) > (n_s + n_i)/2$ for $t < L/2$ and $n(t) < (n_s + n_i)/2$ for $t > L/2$. However, this symmetry is not present in the $\alpha(t')$ motion because the $n(\alpha)$ and $D(\alpha)$ calibration curves are nonlinear. This loss of symmetry is a general feature of complex motion algorithms.

Experimental results

As summarized in Table 5.3, several groups have synthesized various graded-index AR coatings using the GLAD technique. An example is presented in Figure 5.14, showing the results of Kennedy and Brett, who were the first to fabricate GLAD AR coatings [116]. Vertically oriented SiO$_2$ structures with the target graded-index profile were realized by slowly increasing α during film growth, while rotating the substrate in φ (Figure 5.14a). The exact α substrate motion profile was designed to create a Gaussian index profile that closely approximates the ideal quintic profile of Equation 5.23 and the φ rotation ensures the columnar structures are vertically oriented and isotropic in the plane. SEM inspection of the film (side view in Figure 5.14b and oblique view in Figure 5.14c) reveals the graded microstructure produced by this deposition process. The initially deposited film is highly dense due to the near-normal deposition geometry. As α increases, ballistic shadowing introduces greater amounts of porosity into the film and the typical columnar microstructure emerges. During the final growth stages, larger, isolated columns are formed as α approaches $90°$, and low film density is realized. The density profile results in an index matching effect, reducing the Fresnel reflection of the glass substrate. Figure 5.14d shows transmittance spectra measured at different incidence angles $(0–30°)$ for both TE and TM polarizations. At normal incidence, greater than 99% transmittance is measured over a broad wavelength range $(400–1000$ nm$)$, demonstrating the excellent AR characteristics of the GLAD coating. Note also that the uniaxial anisotropy ensures that the TE and TM results are identical at normal incidence owing to the lack of planar birefringence in the film. The AR performance for both polarizations is maintained as the angle of incidence is increased to $30°$. Also shown on the spectra are the results of optical simulations (dashed line), which accurately reproduced the experimental results for both TE and TM polarizations and all measured angles, although the discrepancy is increased at $30°$. The agreement between experiment and theory indicates that the target index profile is achieved and validates the graded-index GLAD approach to AR coating fabrication.

5.5.5 Rugate interference filters

Another important graded-index coating is the rugate optical interference filter [137], having a sinusoidal index profile given by

$$n(t) = n_a + 0.5n_p \sin(2\pi t/P), \qquad (5.24)$$

where n_a is the average film index, n_p is the index variation amplitude and P is the period of the index variation. This refractive index profile is plotted in Figure 5.15a for parameters of $n_a = 1.6, n_p = 1.0$ and $P = 200$. The periodicity of the refractive index profile has a dramatic

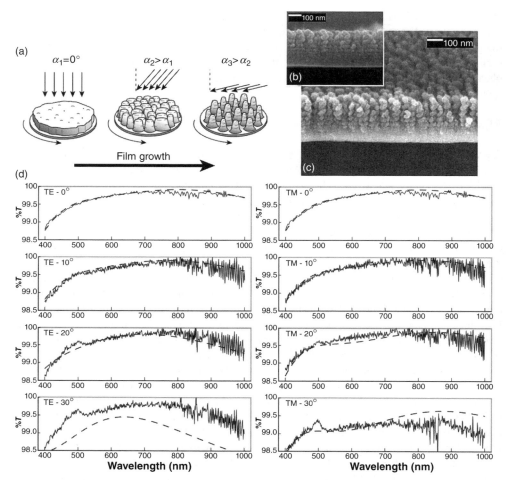

Figure 5.14 Experimental results on graded-index SiO$_2$ AR coatings fabricated using GLAD [116]. (a) During film growth α is gradually increased to realize a tapered column nanostructure with the desired index gradient. Viewing the film microstructure (b) at an oblique angle and (c) from the side reveals the gradual microstructural change that accompanies the α increase. The film is highly dense towards the substrate, whereas the upper regions of the film consist of isolated columnar microstructure with a low effective index. (d) Measured (solid line) and calculated (dashed line) transmittance spectra for TE and transverse-magnetic (TM) polarized light at different angles of incidence. Greater than 99% transmittance is observed over a wide range of wavelength and viewing angles, demonstrating the excellent AR performance of the coating. Reproduced with permission from [116]. © 2007, The Optical Society.

effect on the optical properties. Light incident on the film experiences constructive and destructive interference effects that produce a spectral reflection band centred at $\lambda = 2n_a P$, an effect exploited in many optical filtering applications.

Following the same set of procedures as described in the previous design example, an artificial set of $n(\alpha)$ and $D(\alpha)$ calibration curves was constructed (Figure 5.15b,c). However, in this instance, a column material refractive index of 2.3 was used for the EMT calculation,

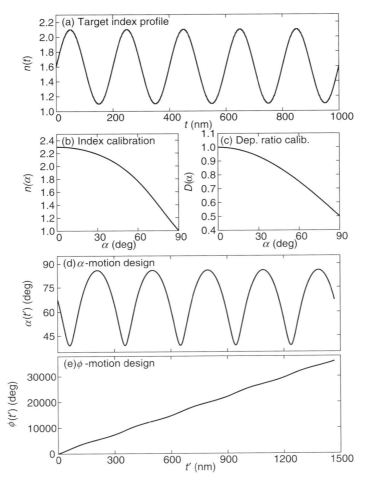

Figure 5.15 Designing a substrate motion algorithm to realize a rugate interference coatings. (a) The design target consists of a sinusoidal $n(t)$ profile (Equation 5.24) with $n_a = 1.6$, $n_p = 1.0$ and $P = 200$. Artificial (b) refractive index and (c) deposition ratio calibration curves generated for this design example (details in text). Using the procedure described in Section 5.5.1, α and φ motions were generated to realize the design target. The α position oscillates between 86° and 39° to produce the desired index contrast, and between these limits α follows a non-sinusoidal trajectory due to the nonlinear $n(\alpha)$ and $D(\alpha)$ variation. The φ rotation rate modulates to compensate for the growth-rate α dependence and recover a constant growth pitch.

an approximate value for TiO$_2$ material. As TiO$_2$ has a large bulk index, a wide range of index values is accessible, and achieving the index modulation specified by the rugate design does not require using the full α range. The upper and lower bounds of the index profile bounds ($n = 2.1$ and 1.1) correspond to α limits of 39° and 86°.

The $\alpha(t')$ and $\varphi(t')$ motions calculated from the calibration curves are presented in Figure 5.15d and e. The periodicity of the $n(t)$ design is carried over to the substrate motion profiles, and the substrate cycles between high and low α values to realize the desired index modulation. Similarly, the φ rotation velocity is periodically modulated in order to consistently realize a 10 nm growth pitch. However, as in the AR design example, the exact shapes are distorted by the nonlinear calibration curves, an effect that is particularly evident in the $\alpha(t')$ motion. Because the growth rate is reduced at higher α, the substrate spends a greater portion of each cycle at high α than at low α. The $\alpha(t')$ motion, therefore, exhibits a skewed duty cycle in order to reproduce the symmetric variation of $n(t)$.

Experimental results

Rugate interference filters have been fabricated using GLAD by multiple researchers, as summarized in Table 5.3. Figure 5.16 presents a collection of results on TiO_2 rugate filters, assembled from Ref. [138]. The SEM image in Figure 5.16a shows the film microstructure, revealing the density modulation created throughout the film by the periodic α motion. Throughout the film growth, α was varied between $30°$ and $80°$ for a total of 16 periods and the motion was designed to realize a structural period of $P = 181$ nm. The continuous φ rotation during growth, designed according to the discussion of Section 5.5.2, generates

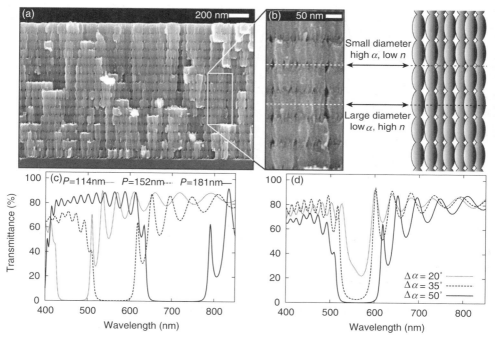

Figure 5.16   Microstructure and optical properties of GLAD-fabricated rugate filters. (a) SEM view of a 16-period rugate film, showing the periodic density modulation achieved in the microstructure. (b) Closer examination of the film reveals the microstructure consists of vertically oriented columns with a periodically varying diameter. The large-diameter regions are created by low α deposition and correspond to the high n regions, whereas the smaller diameter portions are deposited at high α and exhibit low n values. (c) Measured transmittance spectra of rugate filters with different spatial periods P, showing that the optical stopband (region of low transmittance) can be spectrally tuned by controlling P. (d) The stopband width is determined by the index contrast in the film, which can be tailored by adjusting the limits of the periodic α motion. Reproduced from [138]. With permission from M.M. Hawkeye.

vertically oriented columnar structures with a uniaxially anisotropic optical response. Figure 5.16b presents a closer examination of the film alongside a representative illustration of the columnar features, showing the microstructural variation over a single period. The column diameter and film density vary periodically, tracking the α change during deposition. At low α, the diameter is largest and the film density is at its peak, whereas at high α the diameter is smaller and the film is highly porous. In between these two α endpoints the column diameter smoothly varies following the prescribed α motion to produce the desired sinusoidal refractive index modulation.

The periodic index modulation leads to destructive interference of incident light, creating a stopband of drastically reduced transmittance. The optical characteristics of the stopband are readily tuned by modifying the substrate motion algorithm. The spectral location of the stopband can be controlled by adjusting P, as demonstrated by the transmittance spectra in Figure 5.16c, which were measured for three 16-period rugate films with different P. As P is increased, the optical stopband redshifts proportionally: for these three films with $P = 114$ nm, $P = 152$ nm, and $P = 181$ nm, the stopband occurs at 467 nm, 564 nm and 714 nm respectively, creating a set of optical filters that reflect blue-, green- and red-coloured light. The bandwidth of the stopband is determined by the amplitude of the index variation, which is controlled by changing the upper and lower limits of the substrate motion. The difference between the α limits (Δn) directly impacts the bandwidth, as can be seen in Figure 5.16d, which shows transmittance spectra measured for three rugate films having identical P and n_a parameters but with $\Delta\alpha = 20°, 35°$ and $50°$. As the α modulation increases, the index variation in the film increases commensurately, producing a wider and deeper optical stopband at the same wavelength.

5.5.6 Avoiding high-α growth instabilities in graded-index films

Realizing index profiles with highly porous, low-index layers often requires implementing substrate motions that enter into the high-α, extreme-shadowing regime. As this regime is characterized by greatly increased column competition, evolutionary growth, and column broadening phenomena, high-α growth requires extra consideration. In a combined experimental and theoretical investigation, Kaminska *et al.* examined how these extreme-shadowing growth instabilities affect film optical properties by depositing rugate filters with upper and lower α bounds of 86° and 63° [126]. Their results indicate that although the first several periods correctly follow the design, latter periods exhibit a slow degradation of the index modulation due to shadowing-related structure broadening. This effect was only observed when portions of the α motion entered into the extreme shadowing regime as it was not observed in films fabricated with an upper α limit of 76°.

Different methods can be used to compensate for or even avoid these negative growth effects. In their report, Kaminska *et al.* modified the substrate motion algorithm to slowly increase the α modulation-amplitude over the course of the deposition [126]. This modification corrected for the broadening-induced loss of index modulation and recovered the intended index profile. Alternatively, broadening phenomena can be reduced by changing the initial design to one with less extreme α values. For example, switching to a material with a smaller intrinsic refractive index allows low-index layers to be deposited without resorting to extreme α. Other options include synthesizing an alternative $n(t)$ without low-index values that still realizes the desired optical response. The conventional optical coating literature offers many such design synthesis techniques (e.g. see Ref. [139]).

5.6 DESIGNING HELICAL STRUCTURES FOR CIRCULAR POLARIZATION OPTICS

Helically structured columns, formed by oblique deposition with slow substrate rotation (Section 2.5.1), possess a unique polarization-sensitive optical response. However, unlike regular columnar films where the optical properties depend on linear polarization, helical columns exhibit properties specific to *circular* polarization. These effects were first reported by Young and Kowal in 1959 [140], who studied MgF_2 films deposited with slow substrate rotation and observed optical activity (rotation of the polarization plane), an effect associated with chiral media. Although this report was essentially an early precursor of GLAD technology, little work followed until the 1990s, when a series of key studies uncovered the basic physical and structural mechanisms behind the chiral optical phenomenon [141–144]. These reports generated much subsequent interest in the use of helically structured GLAD films for polarization optics. This section begins with an introduction to chiral optics before moving to basic and advanced fabrication and design examples.

5.6.1 Optics of chiral glancing angle deposition media

Structurally, chiral media exhibit a rich set of polarization-sensitive optical phenomena, and general predictions of optical properties require sophisticated computation utilizing 4×4 matrix or dyadic formalisms. For greater detail on these topics, the interested reader is referred to treatments of polarization optics [5] and the electromagnetics of complex, anisotropic media [9, 13, 145]. Here, we provide a utilitarian discussion of optical rotation and circular Bragg reflection, the two effects most often examined in helical GLAD structures.

The problem of light propagation in a helical medium can be solved exactly for light at normal incidence (z-axis propagation) [145]. In such a configuration, incident light sees the planar birefringence of the GLAD film, quantified by the in-plane indices n_o and n_e. Along the thickness of the film the orientation of these indices slowly rotates with a periodicity P determined by the helical structure pitch. Light propagation in such a medium is characterized by Bloch modes of the form

$$\begin{pmatrix} E_x + iE_y \\ E_x - iE_y \end{pmatrix} = \begin{pmatrix} ae^{i2\pi z/P} \\ be^{-i2\pi z/P} \end{pmatrix} e^{-ikz}, \tag{5.25}$$

where E_x and E_y are the electric field vectors in the x- and y-directions, a and b are the field amplitudes, and k is the mode wavenumber. Solving Maxwell's equations yields four eigenmodes with wavenumbers given by

$$\pm k_1 = \frac{1}{2}\left(n_e^2 + n_o^2\right)k_0^2 + \left(\frac{2\pi}{P}\right)^2 - \sqrt{2\left(\frac{2\pi}{P}\right)^2\left(n_e^2 + n_o^2\right)k_0^2 + \frac{1}{4}\left(n_e^2 + n_o^2\right)k_0^2} \tag{5.26}$$

$$\pm k_2 = \frac{1}{2}\left(n_e^2 + n_o^2\right)k_0^2 + \left(\frac{2\pi}{P}\right)^2 + \sqrt{2\left(\frac{2\pi}{P}\right)^2\left(n_e^2 + n_o^2\right)k_0^2 + \frac{1}{4}\left(n_e^2 + n_o^2\right)k_0^2}, \tag{5.27}$$

where the $\pm$ denotes forward and backward wave-propagation and $k_0 = 2\pi/\lambda_0$ is the free-space wavenumber of the field. In general, these modes are elliptically polarized, with a handedness defined by the chirality of the structure. In a right-handed medium, the $\pm k_1$ modes are left handed and the $\pm k_2$ modes are right handed. The situation is reversed in a left-handed medium, wherein $\pm k_1$ is right handed and $\pm k_2$ is left handed. It follows that left- and right-handed media typically exhibit enantiomorphic properties: what a right-handed medium does to left-handed polarization states, a left-handed medium does to right-handed states.

The general solutions can be categorized into one of four regimes, depending on the optical wavelength, the linear birefringence $\Delta n = n_e - n_o$ and P. In the Maugin regime ($\lambda_0 \ll 0.5\Delta nP$), the structural variation occurs so slowly that the normal modes are essentially linearly polarized, one polarized parallel and the other perpendicular to the local optical axis. However, this regime is uncommon in GLAD structures as it requires P on the order of hundreds of micrometres for typical situations such as $\lambda \sim 500$ nm (visible spectrum) and $\Delta n \sim 0.1$. More common are the short-wavelength ($0.5\Delta nP \ll \lambda_0 \ll P$) and long-wavelength ($n_e P \ll \lambda_0$) circular regimes where the polarizations of the normal modes are approximately circular. These regimes are noted for their optical activity, related to a phenomenon known as circular birefringence, where the left-circularly polarized (LCP) and right-circularly polarized (RCP) modes propagate through the medium with different phase velocities. Linearly polarized light can be viewed as a superposition of LCP and RCP light, and the angle of the linear polarization plane is determined by the phase between the LCP and RCP states. Propagation in the circularly birefringent medium alters this phase difference and, after travelling a distance z, the polarization plane is rotated by an angle

$$\Delta\psi = \frac{1}{2}(k_L - k_R)z, \qquad (5.28)$$

where k_L and k_R are the wavenumbers of the LCP mode and RCP mode determined from Equations 5.26 and 5.27. In the short-wavelength regime, the optical rotation is given by

$$\Delta\psi = \frac{1}{2}(k_1 - k_2)z = -\frac{\pi(\Delta n)^2 P}{4\lambda_0^2} z. \qquad (5.29)$$

Being negative, the optical rotation occurs in the direction opposite to the handedness of the medium; that is, a right-handed helix induces a left-handed rotation. In the long-wavelength regime, the optical rotation is given by

$$\Delta\psi = \frac{1}{2}(-k_1 - k_2)z = -\frac{\pi n^2(\Delta n)^2 P^3}{4\lambda_0^4} z, \qquad (5.30)$$

where $n^2 = 0.5(n_e^2 + n_o^2)$. In this case, the optical rotation direction follows the medium chirality; that is, a right-handed helix induces a right-handed rotation. In both regimes, the optical rotation is proportional to the square of the linear birefringence and increases linearly as the light travels through the medium. However, the two regimes are distinguished by their wavelength scaling and dependence on the structural pitch. Note that these simplified relationships only apply when the inequalities defining the regimes are comfortably satisfied.

The fourth regime is known as the circular Bragg regime ($n_o P < \lambda_0 < n_e P$), the most well-studied case in the GLAD literature. In this regime, the first mode (k_1) is characterized by a complex wavenumber, yielding an evanescently propagating mode. Consequently, incident light of this polarization is Bragg reflected by the medium. Conversely, the second mode (k_2) remains real-valued, indicating that the mode travels freely through the medium and incident light of this polarization does not experience Bragg reflection. This is the circular Bragg effect, where the helical medium selectively reflects circular polarizations of equivalent chirality: a left-handed helix selectively reflects LCP light, whereas a right-handed helix selectively reflects RCP light. The Bragg effect only occurs over a specific wavelength range, producing a reflection band at the wavelength satisfying the Bragg condition

$$\lambda_{cBr} = nP \tag{5.31}$$

and with a spectral bandwidth related to the linear birefringence

$$\frac{\Delta\lambda_{cBr}}{\lambda_{cBr}} = \frac{\Delta n}{n}. \tag{5.32}$$

While optical rotation also occurs in the Bragg regime, the wavelength dependence of the effect is complex owing to the strong dispersion of the Bragg resonance.

5.6.2 Engineering basic helical structures

From Section 5.6.1, the optical properties of the helical GLAD structure are determined by several parameters, particularly Δn, n and P. Each of these parameters may be directly accessed using the GLAD technique, allowing the optical response to be engineered. Tailored circular polarization effects can thus be achieved via straightforward modifications of the basic helical column growth algorithm (Figure 2.12) [146, 147].

Basic modifications to the φ rotation provide control over the resulting helical microstructure (Figure 5.17a). The handedness of the helical columns is determined by the direction of φ rotation, allowing selective Bragg reflection of either RCP or LCP light (Figure 5.17b) and control over the optical rotation direction (Figure 5.17c). The enantiomorphic symmetry of the optical effects follows from the mirror symmetry of the helical structures. Reversing the structural chirality switches the respective roles of LCP and RCP light. Also note the anomalous wavelength dispersion observed in the optical rotation spectra due to operating in the circular Bragg regime. The spectral location of the circular reflection band can be tuned by adjusting the structural pitch P, set by the substrate φ rotation speed relative to the vertical growth rate (Figure 5.17d). Adjusting P will also affect the optical rotatory power; in the circular regimes, the rotation scales following Equation 5.29 or 5.30.

The linear birefringence is a key determinant of both optical rotation and the circular Bragg effect. In addition to being material specific, the linear birefringence exhibits a strong α dependence, as discussed in Section 5.4.2, thereby providing a means of tuning these polarization effects in any material. Figure 5.18a shows the maximum selective transmittance of circularly polarized light in helical structures deposited at different α. The peak selectivity observed near $\alpha = 65°$ corresponds to the maximization of the circular Bragg effect due to optimal linear birefringence conditions (Figure 5.5). The second rise in selectivity found

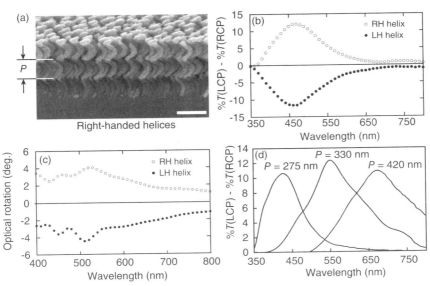

Figure 5.17 Optical properties of basic helical structures. (a) The helical microstructure is defined by its handedness (here, right-handed, RH) and structural pitch P, which are respectively controlled by the direction and speed of the substrate φ rotation during growth. (b) The circular Bragg effect: the helical film selectively reflects circularly polarized light of equivalent handedness. An RH helix preferentially transmits LCP light, whereas the left-handed (LH) helix transmits more RCP light. (c) Linearly polarized light incident on the helical film experiences optical rotation of the plane of polarization in a direction determined by the structural handedness. (d) The spectral position of the circular Bragg effect is controlled by tuning P, enabling straightforward fabrication of circular polarization filters at different wavelengths. Scale bar represents 500 nm. (a) Reproduced from [14]. With permission from A.C. van Popta. (b), (c), (d) Reproduced with permission from [146]. © 2004, The Optical Society.

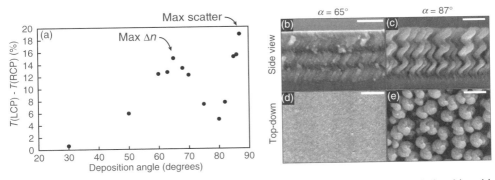

Figure 5.18 (a) The magnitude of the circular Bragg effect exhibits a complex relationship with the deposition angle. At low α the effect is small owing to the limited linear birefringence (Δn) that develops in such films. Increasing α produces larger Δn and the circular polarization discrimination increases commensurately, peaking between $\alpha = 60°$ and $70°$ at a point corresponding to maximum Δn. The magnitude of the Bragg effect initially falls off as α is increased further. However, a second rise in polarization selectivity is observed due to the onset of polarization-sensitive scattering that accompanies the transition to an open microstructure. The film microstructure associated with (b,d) the maximum Δn regime and (c,e) maximum scattering regime. All scale bars represent 500 nm. Reproduced with permission from [147]. © 2004, The Optical Society.

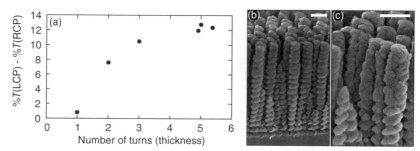

Figure 5.19 (a) In helical structures deposited at $\alpha = 85°$, the magnitude of the circular Bragg effect saturates after several turns due to (b,c) a broadening-induced structural degradation and consequent loss of form birefringence in very thick films. Scale bar represents 1 μm. (a) Reproduced with permission from [146]. © 2004, The Optical Society. (b) Reproduced from [14]. With permission from A.C. van Popta.

for $\alpha > 80°$ has a different origin; namely, a rise in polarization-sensitive scattering losses associated with high-α deposition. While the chiral morphology can be observed in films at all α, as revealed by side-view SEM images of the helical films (Figure 5.18b and d), films deposited at high α produce larger, isolated helical columns that preferentially scatter light of equivalent handedness (Figure 5.18c and e). Optical coating fabrication near the first peak region (60–70°) is thus preferred, as then Δn is maximized and scattering losses are suppressed. However, such films do not benefit from the stress reduction effects that accompany high-α deposition (Section 4.8.1).

While the circular reflection band can be strengthened by adding more turns to the helix, this becomes problematic when fabricating helices at high α owing to broadening-associated growth instabilities that gradually degrade the helical structure, as shown in Figure 5.19. In extreme cases the degradation can lead to a complete loss of form birefringence after several turns. The structural broadening has a negative effect on the circular Bragg phenomenon, offsetting the gains associated with more helical turns and saturating the reflectance peak (Figure 5.19). A similar saturation is seen in the optical rotatory power. Routes to mitigating the broadening effects include depositing at lower α, where column widths are smaller, increasing P to help retain the helical structure, employing an advanced GLAD motion (e.g. phisweep), and optimization of process variables (e.g. temperature, pressure, flux collimation). A remarkable exception to this broadening effect is seen in helically deposited Alq_3 films, where many helical turns can be added to achieve large selective reflection effects [148].

5.6.3 Polygonal helical structures

While circular helical structures have received much attention, polygonal helices also exhibit an unusual and complex interaction with polarized light [149]. Polygonal helices are realized by replacing the continuous, smooth φ rotation with discrete rotations (Section 2.5.3): an m-sided polygonal helix with pitch P is realized by rotating φ through an angle $\xi = \pm 2\pi/m$ after each P/m growth interval. (The $\pm$ denotes a right- versus left-handed helix.) The structural chirality of the polygonal helix produces two circular Bragg resonances, one

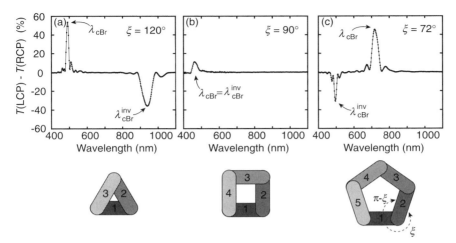

Figure 5.20 Selective transmittance of circularly polarized light in right-handed polygonal helical structures. (a) An 18-turn, 245 nm pitch triangle helix; (b) a 7-turn, 240 nm pitch square helix; (c) a 10-turn, 375 nm pitch pentagon structure. The column orientation, and hence the orientation of the linear birefringence, in the sublayers is shown as well. Reproduced with permission from [149]. © 2005, AIP Publishing LLC.

for each circular polarization. In a right-handed helix,[2] the birefringence rotates by ξ in a clockwise direction for incident RCP light with a period P. This produces an RCP Bragg resonance at $\lambda_{cBr} = nP$, the same condition present in circular helices. For LCP incident on the structure, the birefringence is seen to rotate in the opposite direction by an angle $\pi - \xi$ with a different period equal to $2P/(m-2)$. An LCP Bragg resonance is thus generated at

$$\lambda_{cBr}^{inv} = \frac{2}{m-2}nP = \frac{\xi}{\pi - \xi}\lambda_{cBr}. \tag{5.33}$$

Van Popta *et al.* confirmed these effects in a combined experimental and theoretical study of TiO_2 polygonal helices [149]. Figure 5.20 shows the selective transmittance spectra of a right-handed (a) triangle helix ($\xi = 120°$), (b) square helix ($\xi = 90°$) and (c) pentagonal helix ($\xi = 72°$). The relative alignment of the corresponding column sublayers is shown as well, illustrating the different birefringence rotations seen by LCP and RCP light. The positive peak corresponds to the λ_{cBr} condition, where RCP light is strongly reflected by the chiral structure. The inverted peak appears at a different spectral location, tuning according to Equation 5.33. In the square-helix case, the two Bragg resonances occur at the same wavelengths and counteract one another. The selective transmittance is greatly reduced, although a residual peak remains due to the greater strength of the RCP peak.

---

[2] The following properties are also enantiomorphic. For a left-handed structure, the behaviours of RCP and LCP light will be reversed.

5.6.4 Optimization of circular bragg phenomena with serial bideposition

The polarization effects observed in chiral GLAD media are closely linked to the linear birefringence of the helical microstructure, with the circular-reflection bandwidth proportional to Δn and the optical rotation proportional to $(\Delta n)^2$ (see Section 5.6.1). Increasing Δn to optimize these effects is consequently a common fabrication target. SBD-fabricated nanostructures (Section 5.4.3) exhibit the highest reported Δn values owing to the maximization of form birefringence, and Hodgkinson et al. [150] have extended the SBD concept to the fabrication of helical nanostructures. In this substrate motion algorithm, the rapid, 180° φ rotations characteristic of SBD are superimposed on a slower φ rotation, as shown in Figure 5.21. To produce a slow helical-turn of pitch P, the forward and backward steps of the SBD motion must be unbalanced, producing a forward step of $\Delta\varphi = 180° + \delta\varphi$ and a backward step of $\Delta\varphi = 180° - \delta\varphi$. The quantity $\delta\varphi$ is determined by P and q according to

$$\delta\varphi = \pm\frac{q}{P} \times 360°. \tag{5.34}$$

When $\delta\varphi$ is positive, the forward step is greater than the backward step and the substrate rotates in the positive φ direction. Conversely, a negative $\delta\varphi$ produces a net backward motion and a negative φ rotation. Note that any point of the φ motion can be incremented or decremented by 360° without affecting the algorithm. Because of this, the backward step can actually be replaced by another forward step to end up at the same φ point, albeit displaced by 360°.

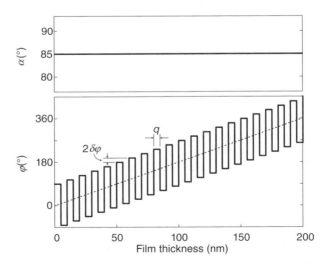

Figure 5.21 Substrate motion algorithm for producing SBD helical columns. The substrate tilt angle α is held constant throughout the deposition. Rapid $\Delta\varphi = 180°$ rotations (occurring after every growth interval q) are superimposed on a slower φ rotation (with a pitch P) to produce a helical microstructure with maximized linear birefringence.

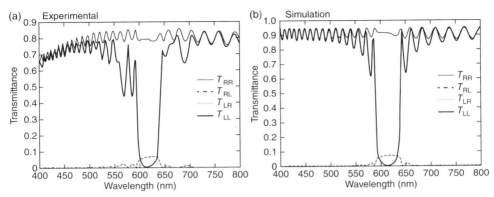

Figure 5.22   (a) Polarization-resolved transmittance spectra for an SBD-fabricated (left-handed, $\alpha = 70°$, 15-turn) chiral medium, measured at normal incidence. (b) Theoretical transmittance spectra calculated using anisotropic matrix-methods. Reproduced from [151]. With permission from SPIE.

By progressively stepping the basic SBD motion in this manner, the orientation of the birefringent sublayers is made to rotate about the substrate normal, thus inducing a helical structure to the film. The associated chiral optical phenomena benefit from the increased Δn provided by the SBD microstructure, and very strong polarization effects can be obtained. Figure 5.22 shows (a) measured and (b) simulated polarization-resolved transmittance spectra for a left-handed, 15-turn TiO_2 helical microstructure fabricated using the SBD-helix algorithm [151]. Note that, by depositing at $\alpha = 70°$, structures with many helical turns can be fabricated without column broadening effects destroying the planar form birefringence. The left-handed helical film strongly reflects LCP light, with $T_{LL} < 0.5\%$ at the circular Bragg resonance ($\lambda_{cBr} = 613$ nm), whereas incident RCP light is transmitted (T_{RR}) through the film. Cross-polarization components T_{RL} and T_{LR}, representing polarization conversion from one circular polarization state to the other, are created by reflections at the air–film and film–substrate interfaces and may be minimized with index matching [151].

5.6.5 Microcavity design in helical structures

Microcavities are frequently encountered devices in optical technology, created by confining light into a wavelength-scale cavity between two reflecting surfaces [153]. The classic example of a microcavity is the Fabry–Pérot etalon, and the concept has been realized in optical coatings, photonic crystals and whispering-gallery resonators. Microcavities find use in many applications, including optical filtration, laser technology, spectroscopy, and sensing, leveraging the narrow bandwidth and resonant enhancement of confined light. These effects can also be realized in GLAD-fabricated helical structures by engineering structural features ('defects') to act as cavity resonators. However, the microcavity effects become sensitive to circular polarization as the reflectivity of the helical structure arises from the circular Bragg process.

Two design approaches have been demonstrated, both having a similar spectral effect. The first involved modifying the helical motion algorithm to fabricate a constant-index spacer layer in the centre of the film (Figure 5.23a) [149, 154]. This spacer was formed

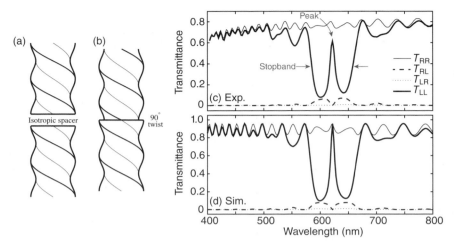

Figure 5.23 Chiral microcavity resonators are fabricated by introducing a defect into the helical microstructure. The defect can be achieved either by (a) adding an isotropic spacer layer (spacer defect) or (b) discontinuously rotating φ by 90° during growth (twist defect). Reprinted from [152], with permission from Elsevier.

by depositing an $\alpha = 0°$ layer having a half-wave optical thickness. The second design is spacerless, creating a cavity by introducing an extra 90° φ rotation at the midpoint of the helix-growth algorithm [149, 152]. This extra rotation creates a twist defect in the helical structure and a microcavity is formed at the interface (Figure 5.23b). The optical effect of helical defects is presented in Figure 5.23, showing (c) experimental and (d) simulated transmittance spectra for a left-handed chiral structure with a twist defect. The circular Bragg effect strongly reflects incident LCP light, leading to a decreased T_{LL} over the width of the stopband. However, the presence of the defect leads to the resonant confinement and enhanced transmittance of LCP light over a narrow spectral band in the centre of the stopband and a narrow peak in the T_{LL} spectrum is produced.

5.6.6 Fabricating graded-birefringence thin-film designs

Examination of Figure 5.22 shows that helically structured films exhibit sidelobes occurring on either side of the Bragg reflection band. To achieve a more ideal filter profile, conventional optical coating designs suppress sidelobe formation by apodizing the refractive index profile $n(t)$ with an envelope function $w(t)$ [155], thus producing a filter with an index profile $w(t) \cdot n(t)$. An equivalent sidelobe reduction can be induced in helically structured films by applying this apodization to the linear birefringence; that is, realizing a film with a linear birefringence that varies as $w(t) \cdot \Delta n(t)$. While Δn can be controlled via α, thus suggesting one possible approach, α changes introduce n modulations that impact the optical spectra. However, the desired effect can be achieved at fixed α by using the advanced φ motions developed by van Popta *et al.* (Section 5.4.4).

Figure 5.24a compares a regular Δn profile and a profile apodized by a Gaussian envelope function. In the regular profile, Δn is constant through the film, whereas in the

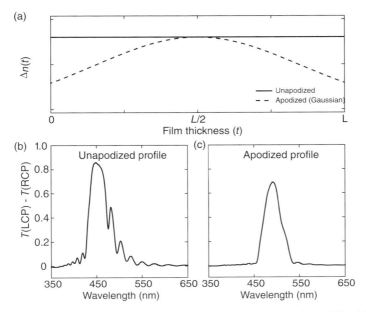

Figure 5.24   (a) In a typical helical film the linear birefringence as a function of film thickness $\Delta n(t)$ is constant (unapodized). Applying a Gaussian apodization function tapers the $\Delta n(t)$ profile towards the film endpoints, producing a graded birefringence transition that reduces interface reflections and sidelobe effects. Experimentally measured selective transmittance of LCP and RCP light in (b) an unapodized helical film and (c) a helical film with an apodized $\Delta n(t)$ profile, implemented using advanced φ rotation control. Applying the apodization removes the sidelobes from the circular Bragg peak and produces a smoother optical filter spectrum. Reproduced with permission from [113]. © 2007, The Optical Society.

apodized version the birefringence peaks at the film midpoint $(t = L/2)$ and tapers to the $1/e$-point at the film end points $(t = 0, L)$ following the Gaussian envelope. Van Popta et al. fabricated filters with both profiles, using the SBD-helix algorithm to realize the regular Δn profile. The apodized version was fabricated by using Equation 5.19 and $\Delta n(e)$ calibration data (Figure 5.11) to design a complex φ rotation profile that deposits sublayers with the prescribed birefringence variation [113]. Figure 5.24b and c shows the selective transmittance of the unapodized and apodized filters respectively. The unapodized filter exhibits a circular Bragg peak with accompanying sidelobes on both sides of the band. Applying the apodization profile successfully removes these sidelobes from the spectrum, thus demonstrating the effectiveness of the design technique. As observed in conventional coatings, apodization also decreases the magnitude and bandwidth due to the overall decrease in Δn in the medium.

Engineering the linear birefringence profile in this manner provides a means of incorporating many of the design synthesis methods seen in conventional optical coatings into a helical optics counterpart. This presents the opportunity to mimic designs of important optical devices, such as antireflection layers and narrow-bandpass and edge filters, in a circular polarization context.

5.7 PRACTICAL INFORMATION AND ISSUES

5.7.1 Post-deposition tuning

Several techniques have been developed to modify the optical properties of GLAD films after the deposition process, useful for fine-tuning filter spectra, for correcting fabrication errors and for property optimization. In general, the approaches involve a post-deposition processing step to modify the columnar microstructure, change the material density and/or alter the chemical composition of the column material. Each of these changes affects the optical properties, as can be interpreted using the previously developed EMT framework.

Thermal annealing treatments

Several researchers have investigated thermal annealing as a means to alter the optical properties of as-deposited structures. Thermal annealing impacts the optical properties via multiple mechanisms. At sufficiently high temperatures, densification and crystallization of the column material generally occur, which increase the column refractive index and act to increase the overall effective index. However, the densification causes a decrease in column volume, thus increasing the void fraction in the film and reducing the effective index. The net impact on the effective index is thus determined by the relative strengths of these competing effects, and experimental findings on TiO_2 films have found that thermal annealing can either redshift [156] or blueshift [157] the optical spectrum. However, these two reports considered different microstructures deposited at different α, indicating the specific experimental conditions dictate which mechanism will be dominant.

Van Popta *et al.* investigated the use of thermal annealing treatments to increase the linear birefringence of TiO_2 SBD-fabricated films [92]. Annealing as-deposited films at 400 °C for 10^4 s yielded optimal results, increasing the linear birefringence by 40–100%, depending on deposition angle. The increased birefringence is attributed to an increased column index due to crystallization of the column material and an increase in the microstructural anisotropy produced by densification. Higher temperature anneals (>700 °C) severely roughened the films and led to significant scattering losses.

Thermal annealing processes can also be applied to chemically transform the column composition, and produce a corresponding change in the optical properties. Robbie *et al.* used high-temperature (>600 °C) treatments in an oxygen environment to convert graded-index microstructures fabricated from Si, which is opaque in the visible spectrum, into SiO_2 replicas, which are transparent [129]. This oxidation process can thus be used to completely alter intrinsic material properties and realize filters suited for different spectral regions. Note that such an oxidation treatment will also yield structural changes associated with the different densities of Si and SiO_2, most notably altering the volume fraction ρ_n of the film.

Column thinning via chemical etching

Another route to post-deposition tuning of the optical properties uses chemical etch treatments to modify the columnar microstructure (Section 6.2.3). Chemical etchants can be used to controllably remove column material and thin-down the column structures, thus decreasing the volume fraction and the corresponding effective index. In comparison with the densification process that occurs in thermal annealing treatments, chemical etch processes can leave the column material relatively unaffected. However, some modification of the column surface and microporosity (Section 4.5.2) would be expected due to penetration of the etchant into the intracolumn structure. Pursel *et al.* have investigated using diluted HCl to etch TiO_2 [158] microstructures and blueshift the optical spectra. They observed a 0.3 nm shift per minute immersed in the chemical etchant, indicating that fine control over the spectral features is achievable.

Structural modification with ion milling

Ion-milling techniques use high-energy ion bombardment to modify columnar structure via resputtering effects (discussed in Section 6.2.4). These ion-milling-induced structural modification are accompanied by a corresponding change in the effective optical properties of the structured film. Huang *et al.* have used this effect to transform as-deposited vertical columnar films into tapered structures with an increased density at the substrate [159]. After ion milling, the structures possess a density gradient that is greater at the substrate and then tapering off at the air interface, producing an index profile similar to the graded-index AR coating designs discussed in Section 5.5.1. The ion-milled structures thus exhibit greatly improved AR capabilities when compared with the as-deposited structures.

5.7.2 Environmental sensitivity

Capillary condensation of water vapour into the porous columnar film microstructures is a well-known problem, one that affects even comparatively dense optical coatings fabricated using conventional PVD approaches. As moisture condenses into a porous GLAD film the effective optical properties change as the void index increases, causing a corresponding redshift of spectral features. Localized pockets of condensed moisture in the film microstructure can also increase optical scattering losses [110, 160]. Together, these effects lead to an inherent environmental sensitivity in GLAD-fabricated optical coatings.

The environmental sensitivity can be considered from two viewpoints. The first is that such sensitivity is undesirable and must be minimized for stable device operation. The structured GLAD film must be either protected from the environment using an encapsulation technique (Section 6.5.2) or chemically functionalized with hydrophobic surface groups to inhibit water condensation (Section 6.2.5). The second viewpoint is that the heightened environmental sensitivity can be exploited for the development of optical sensor devices. Quantification of fluids trapped in the porous microstructure can be done by measuring shifts of the optical spectra, a route explored by several researchers in proof-of-concept experiments [39, 161] and optimized designs [134].

5.7.3 Optical scattering

The microstructure of GLAD-fabricated structures is generally subwavelength, and they are therefore accurately approximated as continuous and homogeneous effective media. However, this homogenization is still always an approximation, and a portion of the incident light is diffusely scattered by microstructural inhomogeneities in the film. While scattering is generally a second-order effect and only important in high-precision measurements, scattering can cause significant losses when the microstructural features approach the wavelength scale. Scattering worsens with increasing film thickness and/or α increases, owing to increased column diameters and surface roughening.

In addition to being an important consideration for high-precision optical coating design, scattering measurements also provide a useful characterization of internal film microstructure [162, 163]. In studies of obliquely deposited Ta_2O_5 ($\alpha = 70°$, thickness 3 μm), Motohiro and Taga measured total diffuse scattering losses between 0.0 and 8.3% depending on the method of preparation and environmental conditions (moisture) [110]. Minimal scattering conditions were achieved by minimizing deposition pressure, and surface roughening was identified as the primary scattering source. A series of studies on tilted columnar films ($\alpha < 60°$, thickness 2 μm) examined the spatial distribution of the scattered light, finding highly anisotropic scatter patterns correlated to the orientation of the columnar microstructure [164, 165]. In a study of helical columnar films (TiO_2, thickness 1000 nm), Sorge *et al.* found a rise in scattering effects for $\alpha > 80°$ related to increased dimensions of microstructural features in the high-α regime [147].

As a rough guideline, scattering losses of GLAD optical coatings can be minimized by avoiding a combination of thick films and high-α depositions when possible and optimizing process conditions to reduce structural broadening and surface roughness. Preventing water penetration into the film microstructure (e.g. by encapsulation) will also help to reduce scattering effects.

REFERENCES

[1] Heavens, O.S. (1965) *Optical Properties of Thin Solid Films*, Dover.

[2] Macleod, H.A. (2001) *Thin-Film Optical Filters*, Institute of Physics Publishing, Bristol, UK.

[3] Born, M. and Wolf, E. (1999) *Principles of Optics*, 7th edn, Cambridge University Press, Cambridge, UK.

[4] Yariv, A. and Yeh, P. (1984) *Optical Waves in Crystals: Propagation and Control of Laser Radiation*, Wiley, Cambridge, UK.

[5] Goldstein, D.H. (2010) *Polarized Light*, 3rd edn, CRC Press.

[6] Holland, L. (1953) The effect of vapor incidence on the structure of evaporated aluminum films. *Journal of the Optical Society of America*, **43**, 376–380.

[7] Smith, D.O., Cohen, M.S., and Weiss, G.P. (1960) Oblique-incidence anisotropy in evaporated permalloy films. *Journal of Applied Physics*, **31**, 1755–1762.

[8] Cohen, M.S. (1961) Anisotropy in permalloy films evaporated at grazing incidence. *Journal of Applied Physics*, **32**, S87–S88.

[9] Berreman, D.W. (1972) Optics in stratified and anisotropic media: 4×4 matrix formulation. *Journal of the Optical Society of America*, **62**, 502–510.

[10] Teitler, S. and Henvis, B.W. (1970) Refraction in stratified, anisotropic media. *Journal of the Optical Society of America*, **60**, 830–834.

[11] Yeh, P. (1979) Electromagnetic propagation in birefringent layered media. *Journal of the Optical Society of America*, **69**, 742–756.

[12] Schubert, M. (1996) Polarization-dependent optical parameters of arbitrarily anisotropic homogeneous layered systems. *Physical Review B*, **53**, 4265–4274.

[13] Lakhtakia, A. and Messier, R. (2005) *Sculptured Thin Films: Nanoengineered Morphology and Optics*, SPIE Press, Bellingham, WA.

[14] Van Popta, A.C. (2009) Optical materials and devices fabricated by glancing angle deposition, PhD thesis, University of Alberta.

[15] Schmidt, D., Booso, B., Hofmann, T. *et al.* (2009) Generalized ellipsometry for monoclinic absorbing materials: determination of optical constants of Cr columnar thin films. *Optics Letters*, **34**, 992–994.

[16] Schmidt, D., Booso, B., Hofmann, T. *et al.* (2009) Monoclinic optical constants, birefringence, and dichroism of slanted titanium nanocolumns determined by generalized ellipsometry. *Applied Physics Letters*, **94**, 011914.

[17] Tabunshchyk, K.V., Hawkeye, M.M., Kovalenko, A., and Brett, M.J. (2007) Three-dimensional simulation of periodically structured thin films with uniaxial symmetry. *Journal of Physics D*, **40**, 4936–4942.

[18] Leontyev, V., Wakefield, N.G., Tabunshchyk, K. *et al.* (2008) Selective transmittance of linearly polarized light in thin films rationally designed by FDTD and FDFD theories and fabricated by glancing angle deposition. *Journal of Applied Physics*, **104**, 104302.

[19] Leontyev, V., Hawkeye, M., Kovalenko, A., and Brett, M.J. (2012) Omnidirectional reflection from nanocolumnar TiO$_2$ films. *Journal of Applied Physics*, **112**, 084317.

[20] Milton, G.W. (2002) *The Theory of Composites*, 1st edn, Cambridge University Press, Cambridge, UK.

[21] Choy, T.C. (1999) *Effective Medium Theory: Principles and Applications*, Oxford University Press, New York.

[22] Sihvola, A. (1999) *Electromagntic Mixing Formulas and Applications*, 1st edn, The Institution of Engineering and Technology, London, UK.

[23] Aspnes, D.E. (1982) Optical properties of thin films. *Thin Solid Films*, **89**, 249–262.

[24] Granqvist, C.G. and Hunderi, O. (1978) Optical properties of Ag-SiO$_2$ cermet films: a comparison of effective-medium theories. *Physical Review B*, **18**, 2897–2906.

[25] Niklasson, G.A., Granqvist, C.G., and Hunderi, O. (1981) Effective medium models for the optical properties of inhomogeneous materials. *Applied Optics*, **20**, 26–30.

[26] Sood, A.W., Poxson, D.J., Mont, F.W. *et al.* (2012) Experimental and theoretical study of the optical and electrical properties of nanostructured indium tin oxide fabricated by oblique-angle deposition. *Journal of Nanoscience and Nanotechnology*, **12**, 3950–3953.

[27] Poxson, D.J., Mont, F.W., Schubert, M.F. *et al.* (2008) Quantification of porosity and deposition rate of nanoporous films grown by oblique-angle deposition. *Applied Physics Letters*, **93**, 101914.

[28] Abeysuriya, K. and Hodgkinson, I.J. (1988) Plate-void model for thin-film form birefringence. *Journal of the Optical Society of America A*, **5**, 1549–1553.

[29] Ng, M.W., Smith, G.B., and Dligatch, S. (1995) Spectral switching of the preferred transmission direction in absorbing anisotropic composites. *Journal of Physics D*, **28**, 2578–2584.

[30] Wang, X.J., Abell, J.L., Zhao, Y.P., and Zhang, Z.M. (2012) Angle-resolved reflectance of obliquely aligned silver nanorods. *Applied Optics*, **51**, 1521–1531.

[31] Fu, J., Park, B., and Zhao, Y. (2009) Nanorod-mediated surface plasmon resonance sensor based on effective medium theory. *Applied Optics*, **48**, 4637–4649.

[32] Suzuki, M. and Taga, Y. (1992) Anisotropy in the optical absorption of Ag–SiO$_2$ thin films with oblique columnar structures. *Journal of Applied Physics*, **71**, 2848–2854.

[33] Ditchburn, R.J. and Smith, G.B. (1992) A model for the optical response of obliquely deposited thin films. *Journal of Physics D*, **25**, 334–337.

[34] Smith, G.B. (1990) Theory of angular selective transmittance in oblique columnar thin films containing metal and voids. *Applied Optics*, **29**, 3685–3693.

[35] Hodgkinson, I., Wu, Q.H., and Collett, S. (2001) Dispersion equations for vacuum-deposited tilted-columnar biaxial media. *Applied Optics*, **40**, 452–457.

[36] Le Bellac, D., Niklasson, G.A., and Granqvist, C.G. (1995) Angular-selective optical transmittance of highly transparent Al-oxide-based films made by oblique-angle sputtering. *Journal of Applied Physics*, **78**, 2894–2896.

[37] Beydaghyan, G., Kaminska, K., Brown, T., and Robbie, K. (2004) Enhanced birefringence in vacuum evaporated silicon thin films. *Applied Optics*, **43**, 5343–5349.

[38] Beydaghyan, G., Buzea, C., Cui, Y. *et al.* (2005) Ex situ ellipsometric investigation of nanocolumns inclination angle of obliquely evaporated silicon thin films. *Applied Physics Letters*, **87**, 153103.

[39] Lakhtakia, A., McCall, M.W., Sherwin, J.A. *et al.* (2001) Sculptured-thin-film spectral holes for optical sensing of fluids. *Optics Communications*, **194**, 33–46.

[40] Sunal, P.D., Lakhtakia, A., and Messier, R. (1998) Simple model for dielectric thin-film helicoidal bianisotropic media. *Optics Communications*, **158**, 119–126.

[41] Gospodyn, J. and Sit, J.C. (2006) Characterization of dielectric columnar thin films by variable angle Mueller matrix and spectroscopic ellipsometry. *Optical Materials*, **29**, 318–325.

[42] Kaminska, K., Brown, T., Beydaghyan, G., and Robbie, K. (2003) Vacuum evaporated porous silicon photonic interference filters. *Applied Optics*, **42**, 4212–4219.

[43] Kaminska, K. and Robbie, K. (2004) Birefringent omnidirectional reflector. *Applied Optics*, **43**, 1570–1576.

[44] Leem, J.W. and Yu, J.S. (2013) Design and fabrication of amorphous germanium thin film-based single-material distributed Bragg reflectors operating near 2.2 μm for long wavelength applications. *Journal of the Optical Society of America B*, **30** (4), 838–842.

[45] Leem, J.W. and Yu, J.S. (2012) Broadband and wide-angle distributed Bragg reflectors based on amorphous germanium films by glancing angle deposition. *Optics Express*, **20**, 20576–20581.

[46] Zhong, Y., Shin, Y.C., Kim, C.M. *et al.* (2008) Optical and electrical properties of indium tin oxide thin films with tilted and spiral microstructures prepared by oblique angle deposition. *Journal of Materials Research*, **23**, 2500–2505.

[47] Kim, J.K., Gessmann, T., Schubert, E.F. *et al.* (2006) GaInN light-emitting diode with conductive omnidirectional reflector having a low-refractive-index indium–tin oxide layer. *Applied Physics Letters*, **88**, 013501.

[48] Le Bellac, D., Niklasson, G.A., and Granqvist, C.G. (1995) Angular-selective optical transmittance of anisotropic inhomogeneous Cr based films made by sputtering. *Journal of Applied Physics*, **77**, 6145–6151.

[49] Mbise, G.W., Niklasson, G.A., Granqvist, C.G., and Palmer, S. (1996) Angular-selective optical transmittance through obliquely evaporated Cr films: experiments and theory. *Journal of Applied Physics*, **80**, 5361–5364.

[50] Ditchburn, R.J. and Smith, G.B. (1991) Useful angular selectivity in oblique columnar aluminum. *Journal of Applied Physics*, **69**, 3769–3771.

[51] Wakefield, N.G., Sorge, J.B., Taschuk, M.T. *et al.* (2011) Control of the principal refractive indices in biaxial metal oxide films. *Journal of the Optical Society of America A*, **28**, 1830–1840.

[52] Xiao, X., Dong, G., Xu, C. *et al.* (2008) Structure and optical properties of Nb_2O_5 sculptured thin films by glancing angle deposition. *Applied Surface Science*, **255**, 2192–2195.

[53] Wang, S., Xia, G., Fu, X. *et al.* (2007) Preparation and characterization of nanostructured ZrO_2 thin films by glancing angle deposition. *Thin Solid Films*, **515**, 3352–3355.

[54] Savaloni, H., Babaei, F., Song, S., and Placido, F. (2009) Characteristics of sculptured Cu thin films and their optical properties as a function of deposition rate. *Applied Surface Science*, **255**, 8041–8047.

[55] Xiao, X., Dong, G., Shao, J. *et al.* (2010) Optical and electrical properties of SnO_2:Sb thin films deposited by oblique angle deposition. *Applied Surface Science*, **256**, 1636–1640.

[56] Szeto, B., Hrudey, P.C.P., Gospodyn, J. *et al.* (2007) Obliquely deposited tris(8-hydroxyquinoline) aluminium (Alq_3) biaxial thin films with negative in-plane birefringence. *Journal of Optics A*, **9**, 457–462.

[57] Sareni, B., Krähenbühl, L., Beroual, A., and Brosseau, C. (1997) Effective dielectric constant of random composite materials. *Journal of Applied Physics*, **81**, 2375–2383.

[58] Mejdoubi, A. and Brosseau, C. (2006) Finite-element simulation of the depolarization factor of arbitrarily shaped inclusions. *Physical Review E*, **74**, 031405.

[59] Hashin, Z. and Shtrikman, S. (1962) A variational approach to the theory of the effective magnetic permeability of multiphase materials. *Journal of Applied Physics*, **33**, 3125–3131.

[60] Heavens, O.S. (1960) Optical properties of thin films. *Reports on Progress in Physics*, **23**, 1–65.

[61] Palik, E.D. (ed.) (1998) *Handbook of Optical Constants of Solids*, 2nd edn, Academic Press, San Diego, CA.

[62] Abelès, F. (1963) Methods for determining optical parameters of thin films. *Progress in Optics*, **2**, 249–288.

[63] Manifacier, J.C., Gasiot, J., and Fillard, J.P. (1976) A simple method for the determination of the optical constants n, k and the thickness of a weakly absorbing thin film. *Journal of Physics E*, **9**, 1002–1004.

[64] Swanepoel, R. (1983) Determination of the thickness and optical constants of amorphous silicon. *Journal of Physics E*, **16**, 1214–1222.

[65] Swanepoel, R. (1984) Determination of surface roughness and optical constants of inhomogeneous amorphous silicon films. *Journal of Physics E*, **17**, 896–903.

[66] Poelman, D. and Smet, P.F. (2003) Methods for the determination of the optical constants of thin films from single transmission measurements: a critical review. *Journal of Physics D*, **36**, 1850–1857.

[67] Bhardwaj, P., Shishodia, P.K., and Mehra, R.M. (2003) Photo-induced changes in optical properties of As_2S_3 and As_2Se_3 films deposited at normal and oblique incidence. *Journal of Materials Science*, **38**, 937–940.

[68] Starbova, K., Dikova, J., and Starbov, N. (1997) Structure related properties of obliquely deposited amorphous a-As_2S_3 thin films. *Journal of Non-Crystalline Solids*, **210**, 261–266.

[69] Bhardwaj, P., Shishodia, P.K., and Mehra, R.M. (2007) Optical and electrical properties of obliquely deposited a-$GeSe_2$ films. *Journal of Materials Science*, **42**, 1196–1201.

[70] Wang, S., Xia, G., He, H. *et al.* (2007) Structural and optical properties of nanostructured TiO_2 thin films fabricated by glancing angle deposition. *Journal of Alloys and Compounds*, **431**, 287–291.

[71] Borgogno, J.P., Lazarides, B., and Pelletier, E. (1982) Automatic determination of the optical constants of inhomogeneous thin films. *Applied Optics*, **21**, 4020–4029.

[72] Pandya, D.K. and Chopra, K.L. (1976) Obliquely deposited amorphous Ge films. I. Optical properties. *Physica Status Solidi (a)*, **35**, 725–734.

[73] Martín-Palma, R.J., Ryan, J.V., and Pantano, C.G. (2007) Spectral behavior of the optical constants in the visible/near infrared of GeSbSe chalcogenide thin films grown at glancing angle. *Journal of Vacuum Science and Technology A*, **25**, 587–591.

[74] Xi, J.Q., Kim, J.K., and Schubert, E.F. (2005) Silica nanorod-array films with very low refractive indices. *Nano Letters*, **5**, 1385–1387.

[75] Azzam, R.M.A. and Bashara, N.M. (1977) *Ellipsometry and Polarized Light*, North-Holland Publishing Co., Amsterdam.

[76] Aspnes, D.E. (1997) The accurate determination of optical properties by ellipsometry, in *Handbook of Optical Constants of Solids* (ed. E.D. Palik), vol. 1, Academic Press, San Diego, CA, pp. 89–112.

[77] Schubert, M.F., Xi, J.Q., Kim, J.K., and Schubert, E.F. (2007) Distributed Bragg reflector consisting of high- and low-refractive-index thin film layers made of the same material. *Applied Physics Letters*, **90**, 141115.

[78] Xi, J.Q., Schubert, M.F., Kim, J.K. *et al.* (2007) Optical thin-film materials with low refractive index for broadband elimination of Fresnel reflection. *Nature Photonics*, **1**, 176–179.

[79] Dignam, M.J., Moskovits, M., and Stobie, R.W. (1971) Specular reflectance and ellipsometric spectroscopy of oriented molecular layers. *Transactions of the Faraday Society*, **67**, 3306–3317.

[80] Hawkeye, M.M. and Brett, M.J. (2009) Controlling the optical properties of nanostructured TiO_2 thin films. *Physica Status Solidi a*, **206**, 940–943.

[81] Liu, M.C., Lee, C.C., Chiang, C.J., and Jaing, C.C. (2009) Enhanced birefringence of MgF_2 thin film at 193 nm by serial bideposition. *Optical Review*, **16**, 562–565.

[82] Hodgkinson, I.J. and Wu, Q.H. (1999) Serial bideposition of anisotropic thin films with enhanced linear birefringence. *Applied Optics*, **38**, 3621–3625.

[83] Wang, S., Fu, X., Xia, G. *et al.* (2006) Structure and optical properties of ZnS thin films grown by glancing angle deposition. *Applied Surface Science*, **252**, 8734–8737.

[84] Azzam, R.M.A. and Bashara, N.M. (1974) Application of generalized ellipsometry to anisotropic crystals. *Journal of the Optical Society of America*, **64**, 128–133.

[85] Schubert, M. (1998) Generalized ellipsometry and complex optical systems. *Thin Solid Films*, **313–314**, 323–332.

[86] Jellison, Jr, G.E. (2004) Generalized ellipsometry for materials characterization. *Thin Solid Films*, **450**, 42–50.

[87] Schmidt, D., Kjerstad, A.C., Hofmann, T. *et al.* (2009) Optical, structural, and magnetic properties of cobalt nanostructure thin films. *Journal of Applied Physics*, **105**, 113508.

[88] Zuber, A., Jänchen, H., and Kaiser, N. (1996) Perpendicular-incidence photometric ellipsometry of biaxial anisotropic thin films. *Applied Optics*, **35**, 5553–5556.

[89] Kaminska, K., Amassian, A., Martinu, L., and Robbie, K. (2005) Growth of vacuum evaporated ultraporous silicon studied with spectroscopic ellipsometry and scanning electron microscopy. *Journal of Applied Physics*, **97**, 013511.

[90] Amassian, A., Kaminska, K., Suzuki, M. *et al.* (2007) Onset of shadowing-dominated growth in glancing angle deposition. *Applied Physics Letters*, **91**, 173114.

[91] Hodgkinson, I., Wu, Q.H., and Hazel, J. (1998) Empirical equations for the principal refractive indices and column angle of obliquely deposited films of tantalum oxide, titanium oxide, and zirconium oxide. *Applied Optics*, **37**, 2653–2659.

[92] Van Popta, A.C., Cheng, J., Sit, J.C., and Brett, M.J. (2007) Birefringence enhancement in annealed TiO_2 thin films. *Journal of Applied Physics*, **102**, 013517.

[93] Hodgkinson, I.J., Horowitz, F., Macleod, H.A. *et al.* (1985) Measurement of the principal refractive indices of thin films deposited at oblique incidence. *Journal of the Optical Society of America*, **2**, 1693–1697.

[94] Wang, H. (1995) Assessment of optical constants of multilayer thin films with columnar-structure-induced anisotropy. *Journal of Physics D*, **28**, 571–575.

[95] Qi, H., Xiao, X., He, H. *et al.* (2009) Optical properties and microstructure of Ta_2O_5 biaxial film. *Applied Optics*, **48**, 127–133.

[96] Jen, Y.J., Peng, C.Y., and Chang, H.H. (2007) Optical constant determination of an anisotropic thin film via polarization conversion. *Optics Express*, **15**, 4445–4451.

[97] Jen, Y. and Lin, C.F. (2008) Anisotropic optical thin films finely sculptured by substrate sweep technology. *Optics Express*, **16**, 5372–5377.

[98] Flory, F., Endelema, D., Pelletier, E., and Hodgkinson, I. (1993) Anisotropy in thin films: modeling and measurement of guided and nonguided optical properties – application to TiO_2 films. *Applied Optics*, **32**, 5649–5659.

[99] Jänchen, H., Endelema, D., Kaiser, N., and Flory, F. (1996) Determination of the refractive indices of highly biaxial anisotropic coatings using guided modes. *Pure and Applied Optics*, **5**, 405–415.

[100] Aspnes, D.E., Theeten, J.B., and Hottier, F. (1979) Investigation of effective-medium models of microscopic surface roughness by spectroscopic ellipsometry. *Physical Review B*, **20**, 3292–3302.

[101] Jen, Y., Hsieh, C.H., and Lo, T.S. (2005) Optical constant determination of an anisotropic thin film via surface plasmon resonance: analyzed by sensitivity calculation. *Optics Communications*, **244**, 269–277.

[102] Wakefield, N.G. and Sit, J.C. (2011) On the uniformity of films fabricated by glancing angle deposition. *Journal of Applied Physics*, **109**, 084332.

[103] Jang, S.J., Song, Y.M., Choi, H.J. *et al.* (2010) Structural and optical properties of silicon by tilted angle evaporation. *Surface Coatings Technology*, **205**, S447–S450.

[104] Hodgkinson, I.J. and Wu, Q.H. (1999) Vacuum deposited biaxial thin films with all principal axes inclined to the substrate. *Applied Optics*, **17**, 2928–2932.

[105] Beydaghyan, G., Bader, G., and Ashrit, P.V. (2008) Electrochromic and morphological investigation of dry-lithiated nanostructured tungsten trioxide thin films. *Thin Solid Films*, **516**, 1646–1650.

[106] Harris, K.D., van Popta, A.C., Sit, J.C. *et al.* (2008) A birefringent and transparent electrical conductor. *Advanced Functional Materials*, **18**, 2147–2153.

[107] Beydaghyan, G., Doiron, S., Haché, A., and Ashrit, P.V. (2009) Enhanced photochromism in nanostructured molybdenum trioxide films. *Applied Physics Letters*, **95**, 051917.

[108] Krause, K.M. (2011) Characterization and modification of obliquely deposited nanostructures, PhD thesis, University of Alberta.

[109] Bennett, J.M., Pelletier, E., Albrand, G. *et al.* (1989) Comparison of the properties of titanium dioxide films prepared by various techniques. *Applied Optics*, **28**, 3303–3317.

[110] Motohiro, T. and Taga, Y. (1989) Thin film retardation plate by oblique deposition. *Applied Optics*, **28**, 2466–2482.

[111] Jim, S.R., Taschuk, M.T., Morlock, G.E. *et al.* (2010) Engineered anisotropic microstructures for ultrathin-layer chromatography. *Analytical Chemistry*, **82**, 5349–5356.

[112] Hodgkinson, I.J., Wu, Q.H., Brett, M.J., and Robbie, K. (1998) Vacuum deposition of biaxial films with surface-aligned principal axes and large birefringence Δn, in *Optical Interference Coatings*, vol. 9 of *OSA Technical Digest Series*, Optical Society of America, Washington, DC, pp. 104–106.

[113] Van Popta, A.C., van Popta, K.R., Sit, J.C., and Brett, M.J. (2007) Sidelobe suppression in chiral optical filters by apodization of the local form birefringence. *Journal of the Optical Society of America A*, **24**, 3140–3149.

[114] Ouellette, M.F., Lang, R.V., Yan, K.L. *et al.* (1991) Experimental studies of inhomogeneous coatings for optical applications. *Journal of Vacuum Science and Technology A*, **9**, 1188–1192.

[115] Martinu, L. and Poitras, D. (2000) Plasma deposition of optical films and coatings: a review. *Journal of Vacuum Science and Technology A*, **18**, 2619–2645.

[116] Kennedy, S.R. and Brett, M.J. (2003) Porous broadband antireflection coating by glancing angle deposition. *Applied Optics*, **42**, 4573–4579.

[117] Kuo, M.L., Poxson, D.J., Kim, Y.S. *et al.* (2008) Realization of a near-perfect antireflection coating for silicon solar energy utilization. *Optics Letters*, **33**, 2527–2529.

[118] Kim, J.K., Chhajed, S., Schubert, M.F. *et al.* (2008) Light-extraction enhancement of GaInN light-emitting diodes by graded-refractive-index indium tin oxide anti-reflection contact. *Advanced Materials*, **20**, 801–804.

[119] Poxson, D.J., Schubert, M.F., Mont, F.W. *et al.* (2009) Broadband omnidirectional antireflection coatings optimized by genetic algorithm. *Optics Letters*, **34**, 728–730.

[120] Sobahan, K.M.A., Park, Y.J., Kim, J.J., and Hwangbo, C.K. (2011) Nanostructured porous SiO_2 films for antireflection coating. *Optics Communications*, **284**, 873–876.

[121] Jayasinghe, R.C., Perera, A.G.U., Zhu, H., and Zhao, Y. (2012) Optical properties of nanostructured TiO_2 thin films and their application as antireflection coatings on infrared detectors. *Optics Letters*, **37**, 4302–4304.

[122] Yan, X., Poxson, D.J., Cho, J. *et al.* (2013) Enhanced omnidirectional photovoltaic performance of solar cells using multiple-discrete-layer tailored- and low-refractive index anti-reflection coatings. *Advanced Functional Materials*, **23**, 583–590.

[123] Leem, J.W. and Yu, J.S. (2013) Multi-functional antireflective surface-relief structures based on nanoscale porous germanium with graded refractive index profiles. *Nanoscale*, **5**, 2520–2526.

[124] Robbie, K., Hnatiw, A.J.P., Brett, M.J. *et al.* (1997) Inhomogeneous thin film optical filters fabricated using glancing angle deposition. *Electronics Letters*, **33**, 1213–1214.

[125] McPhun, A.J., Wu, Q.H., and Hodgkinson, I.J. (1998) Birefringent rugate filters. *Electronics Letters*, **34**, 360–361.

[126] Kaminska, K., Suzuki, M., Kimura, K. *et al.* (2004) Simulating structure and optical response of vacuum evaporated porous rugate filters. *Journal of Applied Physics*, **95**, 3055–3062.

[127] Van Popta, A.C., Hawkeye, M.M., Sit, J.C., and Brett, M.J. (2004) Gradient-index narrow-bandpass filter fabricated with glancing-angle deposition. *Optics Letters*, **29**, 2545–2547.

[128] Hawkeye, M.M. and Brett, M.J. (2006) Narrow bandpass optical filters fabricated with one-dimensionally periodic inhomogeneous thin films. *Journal of Applied Physics*, **100**, 044322.

[129] Robbie, K., Cui, Y., Elliott, C., and Kaminska, K. (2006) Oxidation of evaporated porous silicon rugate filters. *Applied Optics*, **45**, 8298–8303.

[130] Hawkeye, M.M., Joseph, R., Sit, J.C., and Brett, M.J. (2010) Coupled defects in one-dimensional photonic crystal films fabricated with glancing angle deposition. *Optics Express*, **18**, 13220–13226.

[131] Gospodyn, J., Taschuk, M.T., Hrudey, P.C. *et al.* (2008) Photoluminescence emission profiles of Y_2O_3:Eu films composed of high-low density stacks produced by glancing angle deposition. *Applied Optics*, **47**, 2798–2805.

[132] Yan, X., Mont, F.W., Poxson, D.J. *et al.* (2011) Electrically conductive thin-film color filters made of single-material indium-tin-oxide. *Journal of Applied Physics*, **109**, 103113.

[133] Jang, S.J., Song, Y.M., Yeo, C.I. *et al.* (2011) Highly tolerant a-Si distributed Bragg reflector fabricated by oblique angle deposition. *Optical Materials Express*, **1**, 451–457.

[134] Hawkeye, M.M. and Brett, M.J. (2011) Optimized colorimetric photonic-crystal humidity sensor fabricated using glancing angle deposition. *Advanced Functional Materials*, **21**, 3652–3658.

[135] Hajireza, P., Krause, K., Brett, M., and Zemp, R. (2013) Glancing angle deposited nanostructured film Fabry–Perot etalons for optical detection of ultrasound. *Optics Express*, **21**, 6391–6400.

[136] Southwell, W.H. (1983) Gradient-index antirefflecton coatings. *Optics Letters*, **8** (11), 584–586.

[137] Southwell, W.H. (1988) Spectral response calculations of rugate filters using coupled-wave theory. *Journal of the Optical Society of America*, **5**, 1558–1564.

[138] Hawkeye, M.M. (2011) Engineering optical nanomaterials using glancing angle deposition, PhD thesis, University of Alberta.

[139] Tikhonravov, A.V., Trubetskov, M.K., and DeBell, G.W. (2007) Optical coating design approaches based on the needle optimization technique. *Applied Optics*, **46**, 704–710.

[140] Young, N.O. and Kowal, J. (1959) Optically active fluorite films. *Nature*, **183**, 104–105.

[141] Azzam, R.M.A. (1992) Chiral thin solid films: Method of deposition and applications. *Applied Physics Letters*, **61**, 3118–3120.

[142] Robbie, K., Brett, M.J., and Lakhtakia, A. (1995) First thin film realization of a helicoidal bianisotropic medium. *Journal of Vacuum Science and Technology A*, **13**, 2991–2993.

[143] Robbie, K., Brett, M.J., and Lakhtakia, A. (1996) Chiral sculptured thin films. *Nature*, **384**, 616.

[144] Lakhtakia, A. and Weiglhofer, W.S. (1995) On light propagation in helicoidal bianisotropic mediums. *Proceedings of the Royal Society of London A*, **448**, 419–437.

[145] Yeh, P. and Gu, C. (2010) *Optical of Liquid Crystal Displays*, 2nd edn, Wiley.

[146] Van Popta, A.C., Sit, J.C., and Brett, M.J. (2004) Optical properties of porous helical thin films. *Applied Optics*, **43**, 3632–3639.

[147] Sorge, J.B., van Popta, A.C., Sit, J.C., and Brett, M.J. (2006) Circular birefringence dependence on chiral film porosity. *Optics Express*, **14**, 10550–10577.

[148] Hrudey, P.C.P., Westra, K.L., and Brett, M.J. (2006) Highly ordered organic Alq₃ chiral luminescent thin films fabricated by glancing-angle deposition. *Advanced Materials*, **18**, 224–228.

[149] Van Popta, A.C., Brett, M.J., and Sit, J.C. (2005) Double-handed circular Bragg phenomena in polygonal helix thin films. *Journal of Applied Physics*, **98**, 083517.

[150] Hodgkinson, I., Wu, Q.H., De Silva, L., and Arnold, M. (2004) Inorganic positive uniaxial films fabricated by serial bideposition. *Optics Express*, **12** (16), 3840Ű7.

[151] Wu, Q., Hodgkinson, I.J., and Lakhtakia, A. (2000) Circular polarization filters made of chiral sculptured thin films: experimental and simulation results. *Optical Engineering*, **39**, 1863–1868.

[152] Hodgkinson, I.J., Wu, Q.H., Thorn, K.E. *et al.* (2000) Spacerless circular-polarization spectral-hole filters using chiral sculptured thin films: theory and experiment. *Optics Communications*, **184**, 57–66.

[153] Vahala, K.J. (2003) Optical microcavities. *Nature*, **424**, 839–846.

[154] Hodgkinson, I.J., Wu, Q.H., Lakhtakia, A., and McCall, M.W. (2000) Spectral-hole filter fabricated using sculptured thin-film technology. *Optics Communications*, **177**, 79–84.

[155] Southwell, W.H. (1989) Using apodization functions to reduce sidelobes in rugate filters. *Applied Optics*, **28**, 5091–5094.

[156] Van Popta, A.C., Sit, J.C., and Brett, M.J. (2004) Optical properties of porous helical thin films and the effects of post-deposition annealing, in *Organic Optoelectronics and Photonics* (eds P.L. Heremans, M. Muccini, and H. Hofstraat), vol. 5464 of *Proceedings of SPIE*, SPIE Press, Bellingham, WA, pp. 198–208.

[157] Pursel, S.M., Horn, M.W., and Lakhtakia, A. (2006) Blue-shifting of circular Bragg phenomenon by annealing of chiral sculptured thin films. *Optics Express*, **14**, 8001–8012.

[158] Pursel, S.M., Horn, M.W., and Lakhtakia, A. (2007) Tuning of sculptured-thin-film spectral-hole filters by postdeposition etching. *Optical Engineering*, **46**, 040507.

[159] Huang, Z., Hawkeye, M.M., and Brett, M.J. (2012) Enhancement in broadband and quasi-omnidirectional antireflection of nanopillar arrays by ion milling. *Nanotechnology*, **23** (27), 275703.

[160] Gledhill, R.F., Hodgkinson, I.J., and Wilson, P.W. (1986) Anisotropic optical scatter from moisture patches in thin films deposited obliquely. *Journal of Applied Physics*, **59**, 1453–1455.

[161] Steele, J.J., van Popta, A.C., Hawkeye, M.M. *et al.* (2006) Nanostructured gradient index optical filter for high-speed humidity sensing. *Sensors and Actuators B: Chemical*, **120**, 213–219.

[162] Amra, C. (1993) From light scattering to the microstructure of thin-film multilayers. *Applied Optics*, **32**, 5481–5491.

[163] Kassam, S., Hodgkinson, I.J., Wu, Q., and Cloughley, S.C. (1995) Light scattering from thin films with an oblique columnar structure and with granular inclusions. *Journal of the Optical Society of America A*, **12**, 2009–2021.

[164] Hodgkinson, I., Bowmar, P.I., Wu, Q.H., and Kassam, S. (1995) Scatter from tilted-columnar birefringent thin films: observation and measurement of anisotropic scatter distributions. *Applied Optics*, **34**, 163–168.

[165] Hodgkinson, I., Cloughley, S., Wu, Q.H., and Kassam, S. (1996) Anisotropic scatter patterns and anomalous birefringence of obliquely deposited cerium oxide films. *Applied Optics*, **35** (28), 5563–5568.

6 Post-Deposition Processing and Device Integration

6.1 INTRODUCTION

Microstructured GLAD films have proven to be robust to diverse post-deposition techniques, making it possible to establish experimental procedures and recipes to modify and control film properties. The development of a post-deposition 'toolbox' presents an important technological advantage, allowing GLAD film functionality to be altered/improved and enabling the integration of GLAD layers into microdevice designs. This chapter surveys the many strategies that have been developed for post-deposition modification of GLAD films, covering common microprocessing processes, including annealing, oxidation, lithographic patterning and etching, in addition to examining several issues specific to working with GLAD films, such as deposition onto nonplanar surfaces, electrode integration, film behaviour in liquid environments and template fabrication approaches.

6.2 POST-DEPOSITION STRUCTURAL CONTROL

6.2.1 Annealing

GLAD microstructures are often amorphous or exhibit poor crystallinity owing to the low-temperature deposition conditions required to maximize the ballistic shadowing effect. Consequently, thermal annealing treatments of as-deposited films are commonly used to promote the crystallization of the column material, typically corroborated by an accompanying measurement of crystal structure, such as electron or X-ray diffraction, or Raman spectroscopy.

Care must be taken when designing annealing processes for GLAD films as thermal treatments above a material-dependent threshold temperature can lead to severe degradation of the column microstructure due to the onset of diffusion processes and melting of the column material. The temperatures associated with these thermal effects are lower than observed in bulk material or dense-film form due to the enhanced surface-to-volume ratios of the microstructured GLAD films, a phenomenon common to nanostructured materials in general. Microstructural degradation due to thermal annealing has been examined by several groups, typically as part of a broader process-optimization study. These reports are summarized in Table 6.1, grouped by material. When developing annealing treatments at elevated temperatures, it is important to consider these effects as the processing window can be quite narrow and temperature changes as small as 50 °C can lead to a loss of structural definition. Note that these values should be treated as rough guides and can be expected to vary strongly with experimental conditions. For example, changing α and φ will modify the

Glancing Angle Deposition of Thin Films: Engineering the Nanoscale, First Edition.
Matthew M. Hawkeye, Michael T. Taschuk and Michael J. Brett.
© 2014 John Wiley & Sons, Ltd. Published 2014 by John Wiley & Sons, Ltd.

Table 6.1 Temperature threshold T_{th} for significant microstructural degradation as reported in different materials. The bulk material melting point T_m is shown as well (sp: sublimation point); taken from Ref. [13].

Material	Ref.	T_{th} (°C)	T_m (°C)
Cu	[1]	300	1085
	[2]	400	1085
WO$_3$	[3]	500	1473
Ge	[4]	600	938
	[5]	600	938
TiO$_2$	[6]	700	1843
	[7]	700	1843
	[8]	700	1843
Y$_2$O$_3$:Eu	[9]	900	2439
C	[10]	900	3825 (sp)
TiC	[11]	1000	3067
Si	[12]	>1000	1414

surface-to-volume ratio of the microstructure, thus altering the impact of surface diffusion effects and affecting the temperature threshold for microstructure degradation.

The typical impact of annealing at temperatures near or greater than the structural degradation threshold is seen in Figure 6.1. The SEM images show TiO$_2$ helical columns before annealing, and after a 60 min annealing treatment at 700 °C and 1000 °C. For context, the melting point of bulk TiO$_2$ is 1843 °C [13]. Minor microstructural changes are observed in the lower temperature case, attributable to the anatase crystallization of the TiO$_2$ material and a small degree of surface smoothening from surface diffusion. At higher temperatures, however, the columnar morphology is dramatically changed (Figure 6.1c). Thermally activated surface- and bulk-diffusion cause significant mass transport processes, causing columns to coalesce and merge throughout the film. The microstructural changes impact the optical properties of the helical film, as shown in Figure 6.1d. As the annealing temperature is increased above the threshold and the helical microstructure degrades, the circular Bragg effect (Section 5.5) created by the helical structure diminishes as well. The increase in microstructural disorder, seen by comparing Figure 6.1a and d, increases optical scattering losses and decreases the overall film transmittance. The normally clear films take on a milky visual appearance due to the increased diffuse scattering. Generally, severe structural degradation leads to an undesirable loss of function, and such high-temperature treatments should be avoided. However, Karabacak et al. point out that the reduced temperature required for structural coalescence could be exploited for low-temperature soldering or wafer bonding applications [2].

The majority of thermal annealing processes in GLAD technology are performed below the degradation threshold to preserve the columnar microstructure of the deposited film as best as possible. Annealing treatments are generally conducted to improve the crystallinity of as-deposited column material and optimize film properties for specific applications. A short list of examples includes:

• optimizing planar birefringence in SBD-fabricated TiO$_2$ nanostructures [8];
• spectral tuning of TiO$_2$ helical columns [15];

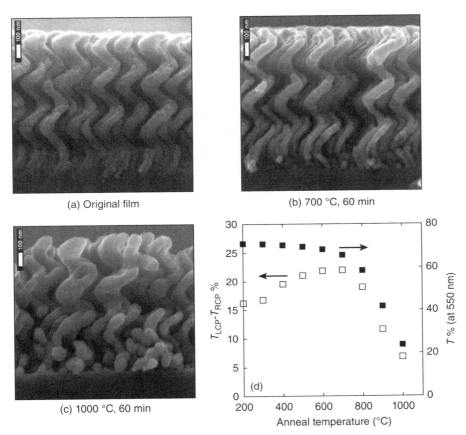

Figure 6.1 Effect of thermal annealing temperature on TiO_2 helical columns, deposited at 85°. (a) The original three-turn helical film. (b) After annealing at 700 °C the film retains a helical morphology, although the surface roughness of the columns is reduced. (c) After annealing at 1000 °C the diffusion processes have degraded the helical microstructure, with significant coalescence and column merging evident. (d) The loss of structure influences the optical properties of the film. When annealed at temperatures above 700 °C the circular polarization selectivity (open squares) decreases due to the loss of helical structure, and scattering losses associated with the increased structural disorder decrease the transmittance through the film (filled squares). Reproduced from [7]. With permission from SPIE.

- improving the electrochemical reactivity of structured C electrodes [10];
- increasing short-circuit currents in TiO_2-based dye-sensitized solar cells [14];
- enhancing the separation performance of SiO_2 chromatography plates [16];
- increasing photoconversion efficiency in α-Fe_2O_3 electrodes for water oxidation [17];
- enhancing the photocatalytic activity of TiO_2 nanostructures [18]; and
- improving photoluminescence intensity of Y_2O_3:Eu structured films [19].

Certain microstructural changes often accompany thermal processing even at sub-degradation temperatures. While large-scale mass transport does not occur and the columnar

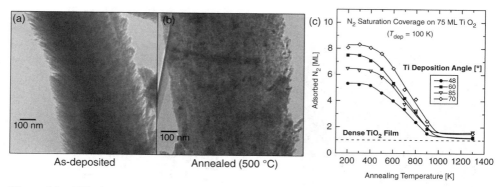

Figure 6.2   TEM images of TiO_2 nanocolumns (a) before and (b) after a 500 °C annealing treatment showing a change in the internal nanostructure of the column. (c) Surface area measurements of annealed TiO_2 microstructures show a reduction in the total surface area when annealed at temperatures greater than 400 °C, related to the collapse of micropore structures internal to the column. (a) and (b) Reprinted from [14], with permission from Elsevier, (c) Reprinted with permission from [6]. © 2007, American Chemical Society.

geometry remains largely unaffected, gentler annealing treatments can provide sufficient thermal energy to affect intracolumn structural features. This effect can be seen in the TEM images of Figure 6.2 showing TiO_2 nanocolumns before and after a 500 °C annealing process [14]. The internal structure of the as-deposited column is characterized by bundled nanofibres, giving the column a feathered appearance. Annealing the sample induces anatase crystallization of the column material and a conglomeration of the fibre bundles, significantly altering the fine details of the column micro/mesopore structure. The observed collapse of the internal porosity is supported by surface area measurements of TiO_2 structures annealed at different temperatures (Figure 6.2c), showing a steady decrease in surface area beginning at temperatures much below the degradation threshold [6]. While annealing can prove beneficial for improving intrinsic column-material properties, it generally decreases total film porosity. Annealing process development can therefore require careful calibration for device applications leveraging the high surface area of GLAD films.

Localizing annealing treatments

While thermal process are typically performed on the wafer-scale and the entire film is treated simultaneously, tightly focused lasers have recently been used to selectively anneal localized film areas. Absorbing column material in the laser-beam focus is heated by the incident light, inducing microstructural changes similar to those described above. This approach has been used to locally enhance dichroic properties of Ag films and create defined patterns by scanning the laser focus across the substrate [20, 21]. As with regular furnace-based annealing, the authors found increasing the laser intensity beyond a threshold had a negative impact on column structure, leading to a loss of dichroism. The technique has currently been demonstrated for beam diameters on the order of 60 μm. Using higher numerical aperture optics, however, the beam spot could be reduced to wavelength dimensions, allowing selective annealing of only a few columns at a time.

6.2.2 Chemical composition control

In addition to inducing column material crystallization and densification, annealing treatments can also be used to alter the chemical composition of column material by adding process gases to induce chemical reactions. While GLAD is broadly compatible with many materials, certain materials exhibit superior deposition characteristics, such as lower adatom mobility, more stable growth rates or lower processing pressures. Post-deposition chemical modification, therefore, allows high-quality nanostructures to be realized in difficult-to-deposit materials. For example, Dang *et al.* converted as-deposited Ta_2O_5 nanostructures into Ta_3N_5 via thermal nitridation in an ammonia environment at temperatures from 700 to 850 °C [22]. The converted nanostructures were subsequently used as a high-surface-area photocatalyst for water oxidation applications. Chemical reactions can also be used to change the composition, thus altering or correcting the intrinsic material properties. For example, Robbie *et al.* converted Si columnar structures into SiO_2 using both wet and dry thermal oxidation processing at 600 °C and 650 °C [23]. Whereas the original structures were opaque owing to strong Si absorption of visible and near-infrared light, the converted SiO_2 structures were highly transparent and more appropriate for use as optical filters over these wavelengths. Lower temperature oxidation processes can also be used to correct stoichiometric deficiencies in materials such as TiO_2, which often deposit in sub-stoichiometric forms $(TiO_{x<2})$ [24].

As well as altering chemical composition, the oxidation process affects the structural dimensions of the columnar microstructure owing to changes in the intrinsic material density upon reaction, an effect well known from thermal oxide growth in silicon microfabrication. Summers and Brett used this effect to modify the composition and structural filling factor of periodic, Si square-spiral columns [25]. Figure 6.3 shows the as-deposited film microstructure alongside the microstructure after wet oxidation at 700 °C. Oxidation of the column material causes a volumetric expansion of the columnar structures, increasing the film thickness and the filling fraction (Figure 6.3c). Complete stoichiometric conversion to SiO_2 and a maximum structural expansion were observed at 700 °C. Increasing the processing temperature leads to collapse of the internal column micropore structures, causing a densification of the film and a corresponding structural contraction. Altering the microstructure density in this manner is useful for preparing templates (Section 6.8.1) with fill fractions different from the natural equilibrium density of the as-deposited films.

6.2.3 Microstructural control via chemical etching

A few reports have detailed the use of chemical etch processes to selectively remove column material, thereby thinning individual columns, increasing film porosity and modifying the overall film morphology. Controlling film microstructure through etch processing can thus be used to tune physical properties and improve device characteristics. Pursel *et al.* used dilute HCl solutions to increase the porosity of TiO_2 columns, causing a well-controlled blueshift of optical properties [27] (Section 5.6.1). However, wet etch techniques are ill-suited to processing of films deposited at high α (typically $\alpha > 80$–$85°$, depending on film thickness) where the film microstructure is characterized by isolated columnar features. Under these conditions, capillary forces can cause column bending and stiction effects when

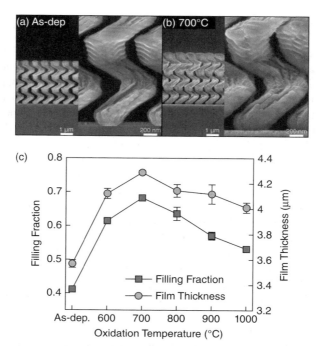

Figure 6.3 Periodic Si square-spiral columns (a) before and (b) after wet thermal oxidation. The conversion of the column material from Si to SiO$_2$ causes a volumetric expansion of the column structures, which (c) increases the film thickness and fill fraction. Reprinted from [25], with permission from Elsevier.

the films are placed into liquids. After liquid exposure and drying, columns are clumped together and the microstructure is irreversibly altered (the nanocarpet effect, Section 6.7). To avoid these effects Kupsta *et al.* developed a dry reactive ion etching (RIE) process using CF$_4$ to alter the microstructure of TiO$_2$ GLAD films [26]. The authors obtained greater control over the etch process by placing a PTFE spacer between the sample and RIE electrode to absorb free fluorine radicals generated by the plasma. Figure 6.4 shows the effect of different etch durations on the column morphology. The unetched film (Figure 6.4a) shows the typical vertical column microstructure, with extinct columns and an even, disordered column distribution. After a 60 s etch (Figure 6.4b), the film microstructure is visibly altered: extinct columns are removed by the etch process and channel-like void regions with widths on the order of a column diameter start to form. These microstructural changes are exacerbated with increased processing time, and after 120 s and 180 s etch durations (Figure 6.4b and c) both the channel region width and column clumping progressively increase. The microstructural impact of the RIE process has been used to improve device performance in different applications. The channel formation increases the diffusivity of water vapour in and out of the film, which Kupsta *et al.* exploited to improve the adsorption and desorption times of capacitive humidity sensors based on the porous GLAD film (response times shown in Table 6.2) [26]. Jim *et al.* also developed a similar RIE process to improve the separation capability of GLAD-fabricated chromatography plates [16].

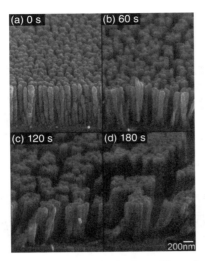

Figure 6.4 Microstructural changes in TiO_2 vertical column films after different RIE process times: (a) as-deposited film; after etch treatments for (b) 60 s, (c) 120 s and (d) 180 s. Multiple structural changes are observed with increasing etch time: channel-like void regions form between columns; smaller, extinct columns are completely removed; and the degree of column clumping increases. The measured water vapour adsorption and desorption times decrease with increased etch duration (Table 6.2), indicating that channel formation improves diffusivity and sensor speed. © 2009 IEEE. Reprinted with permission from [26].

6.2.4 Ion-milling structural modification

Another processing technique used to modify as-deposited GLAD nanostructures is ion milling, where the nanostructured sample is bombarded with high-energy ions (~1 keV). The ion bombardment etches column material via sputtering at a rate determined by the ion current, the column material and the local angle of incidence, as has been found from studies of ion-milled GLAD nanostructures composed of Si [28, 30], SiO_2 [28, 29] and TiO_2 [28]. Ion milling causes a nonuniform erosion of the film microstructure, as shown in Figure 6.5. The as-deposited film structure (Figure 6.5a) is transformed by several mechanisms (Figure 6.5b). Because sputter yield is increased for off-normal ion incidence, the etch rates are greater for ions impacting the edges of the curved column apexes. This nonuniform etch rate leads to a progressive sharpening of the column tips during ion bombardment. The material

Table 6.2   Effect of RIE etch on sensor adsorption and desorption response times.

Etch time (s)	Adsorption (ms)	Desorption (ms)
0	151 ± 4	129 ± 15
15	53 ± 5	67 ± 14
30	42 ± 3	50 ± 3
60	49 ± 8	52 ± 8

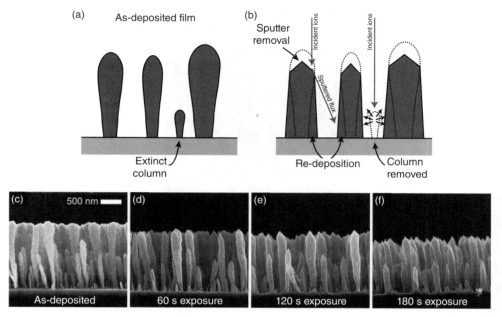

Figure 6.5   Structural modifications occurring during ion milling of GLAD-fabricated nanostructures. (a,b) Incident ions sputter etch the initially rounded columns into sharpened, tapered structures and can completely remove smaller, extinct columns. SEM images of SiO_2 vertical column structures (c) before ion milling and after (d) 60 s, (e) 120 s and (f) 180 s exposure. The observed microstructural changes occur gradually, indicating that fine morphology control can be obtained. (a) and (b) Reproduced from [28]. © IOP Publishing. Reproduced by permission of IOP Publishing. All rights reserved. (c)–(f) Reproduced from [29]. © IOP Publishing. Reproduced by permission of IOP Publishing. All rights reserved.

removed by the sputtering process is predominantly ejected downwards into the film and is subsequently redeposited farther down along neighbouring columns, thus increasing the base diameter of the column structures. The ion milling also removes smaller, extinct columns, which are resputtered onto neighbouring columns and the substrate. The microstructural modification associated with ion milling can be seen in Figure 6.5c–f, which shows SiO_2 vertical column structures after progressively longer ion-milling durations. The column sharpening is clearly seen, as is the increase of base column diameter and removal of extinct column structures.

The structural modifications induced by ion milling have been investigated for multiple applications. The decreased radius of curvature of the sharpened columns improves the field emission properties of the structures [30]. The increased base diameter of the column mechanically stiffens the structures against bending, making the modified structures less susceptible to the capillary-force-induced nanocarpet effect in liquid environments [29]. In a third application, the tapering of the ion-milled columns creates a graded refractive-index profile that improves substrate index-matching and enhances the antireflection properties of the structured layer [28].

6.2.5 Column surface modifications

Surface chemistry tuning

The surface properties of GLAD films are also a common target for post-deposition process-ing, useful for achieving specific chemical functionality (e.g. passivation). A series of papers reports the tailoring of the surface chemical properties of various film materials using both solution- [31, 32] and vapour-phase [32, 33] siloxane-based chemistries. Characterizations of the treated films indicate that the GLAD structures survive the chemical processes and that the entire film is functionalized. A number of chemical moieties have been covalently bonded to the oxide surface, demonstrating chemical tunability and the realization of super-hydrophobic surfaces. For example, Figure 6.6a and b shows contact-angle measurements of water on as-deposited and siloxane-functionalized SiO_2 films. The untreated surface is hydrophilic, with high wettablity and an ∼0° contact angle, whereas the chemically modified surface becomes superhydrophobic, with a contact angle >150°. Such surface treatments have proven useful in reducing the environmental sensitivity of GLAD-fabricated porous optical interference coatings (Section 5.6.2). As-deposited structures readily take up water vapour, which affects the optical spectrum of the porous coating. Vapour-phase functional-ization rendered the films hydrophobic and led to a sixfold reduction in humidity sensitivity, thus increasing the operational stability of the coating [34].

Conformal sub-nanometre surface coatings with atomic layer deposition

Atomic layer deposition (ALD) is a gas-phase chemical process able to conformally coat high-aspect-ratio structures with a number of different materials [36]. The self-limiting nature of the reaction steps involved in ALD leads to a sequential, monolayer-by-monolayer growth process that provides sub-nanometre thickness control. ALD has been used to conformally coat GLAD columns with ultrathin material layers, realizing various core–shell nanostructure combinations. Using two separate ALD processes, Albrecht *et al.* coated Si nanocolumns

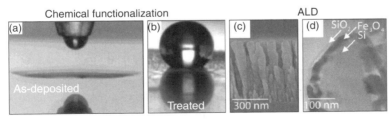

Figure 6.6 Surface chemical modification of GLAD microstructures. By using chemical function-alization treatments, the surface energy of the GLAD film can be tuned. Here, SiO_2 columns have been functionalized with hydrophobic siloxane molecules, rendering the (a) hydrophilic as-deposited surface into (b) a superhydrophobic surface, as demonstrated by these contact-angle measurements. The atomic layer deposition (ALD) process can be used to conformally coat (c) GLAD microstructures with (d) nanometre-thick layers of a different material, providing another route to tailoring the surface characteristics. (a) and (b) Reprinted with permission from [32]. © 2006, American Chemical Society. (c) and (d) Reproduced with permission from [35]. Copyright © 2010 Wiley–VCH Verlag GmbH & Co. KGaA, Weinheim.

with a 20 nm Fe_3O_4 layer to provide magnetic functionality, followed by a protective 5 nm SiO_2 layer [35]. The resulting columnar microstructure is shown in Figure 6.6c and d, where the SiO_2 and Fe_3O_4 conformal shell structures can be seen via high-resolution SEM and TEM imaging. Taschuk *et al.* have used ALD to coat TiO_2 and Si nanocolumns with 0.3–4 nm thick TiO_2 layers [37]. In the case of Si columns, adding the ALD layer transfers the photocatalytic activity of the TiO_2 surface onto a different material scaffold. For the TiO_2 columns, the highly conformal ALD layer was shown to alter the intracolumn micropore structure, which altered the humidity sensitivity of the capacitive response. This work demonstrates a decoupling of surface functionality and nanoscale structure, enabling greater design freedom in gas-sensing devices.

6.3 DEPOSITION ONTO NONPLANAR GEOMETRIES

While GLAD is typically performed on featureless, flat substrates to provide uniform starting conditions for nucleation and ballistic shadowing processes, depositing onto pre-patterned, nonplanar substrates provides additional design freedom and can be used to realize different microstructure geometries. As was discussed in Chapter 3, depositing onto sub-micrometre seed structures has received much attention as it enables fabrication of high-uniformity nanostructure arrays. Although combining GLAD with larger substrate features is less-well examined, it has led to the fabrication of several interesting microstructures.

An early example was demonstrated by Harris *et al.*, who deposited onto pre-etched microfluidic channels 50 μm wide and 4.5 μm deep (Figure 6.7a) [38]. By carefully aligning the incident vapour plane to the microchannel axis and depositing at a glancing ($\alpha = 85°$), sidewall shadowing is minimized and the channels are filled with a high-surface-area columnar microstructur, as shown in Figure 6.7b. In this image, film material deposited outside the channel was removed by mechanically scraping in order to accommodate cover-slip attachment and the creation of sealed channels. The authors suggest this proof-of-principle device demonstration may be useful for microfluidic lab-on-a-chip devices.

Seto *et al.* investigated GLAD with more complex large-scale substrate topographies by depositing onto lithographically defined arrays of mesa microstructures 17 μm tall and fabricated using deep silicon RIE (Bosch) processing techniques [39]. Depositing onto substrates patterned with such structures defines a nonuniform deposition geometry, illustrated in

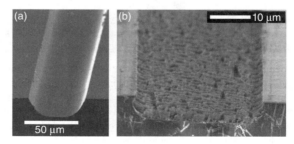

Figure 6.7 A 50 μm wide and 4.5 μm deep microfluidic channel (a) before and (b) after GLAD. The GLAD process can be used to fill the channel with a high-surface-area microstructure, useful for lab-on-a-chip devices. Reproduced by permission of ECS – The Electrochemical Society from [38].

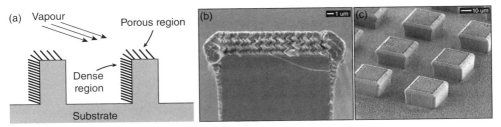

Figure 6.8 (a) Deposition onto raised large-scale surface geometries produces a nonuniform deposition environment. A porous microstructure develops along the top surfaces where the local deposition angle is equal to α. However, the local deposition angle is $90° - \alpha$ along the vertical sidewalls, resulting in the formation of a dense sidewall deposit. (b) Adding φ rotation creates a nonuniform morphology, consisting of a self-sealed porous region with dense sidewalls. (c) Repeating this mesa topography over the substrate creates arrays of self-sealed porous microstructures. Reproduced from [39] with permission of The Royal Society of Chemistry.

Figure 6.8a. Vapour arrives along the flat top-surface of the mesa structure with the intended high-α geometry, thus producing a porous GLAD microstructure as expected. However, the local deposition angle along the mesa sidewall is equal to $90° - \alpha$, and therefore is near normal for large α. Consequently, the sidewall deposit is significantly denser. (This effect has also been discussed in Section 3.2.5, although in the context of single-column seeded growth.) The film density thus varies over the structure, tracking the variation in the local geometry. When combined with φ rotation, a unique self-sealed microstructure is produced where the edges of the porous region are sealed by the dense sidewall deposit (Figure 6.8b, c). Adding a capping layer (Section 6.5) completes the 3D encapsulation of the porous region. The authors envision fabricating arrays of these self-sealed microchamber structures for use as individual microreactors for chemical and biological sensors.

6.4 PHOTOLITHOGRAPHIC PATTERNING OF GLANCING ANGLE DEPOSITION THIN FILMS

Photolithography is a vital process in microdevice fabrication, used to pattern thin-film materials into useful forms. Photolithography involves a few basic steps: application of a photoresist layer, selective UV exposure of the photoresist through a mask, and development and removal of the exposed (positive photoresist) or unexposed (negative photoresist) photoresist areas. Additional low-temperature baking steps are generally included to modify the resist layer; for example, pre-exposure soft bakes to drive off solvent and post-development hard bakes to increase resist stability. The hardened photoresist protects the desired regions of the film from subsequent processing, such as etching, ion implantation, film deposition, and so on.

Standard photolithography process flows can be adapted to micropattern GLAD films, requiring relatively few modifications owing to the porous nature of the GLAD structures. A primary complication in photolithographic processing of GLAD films arises in the photoresist application step. In standard processing, spin-casting is used to prepare high-uniformity photoresist layers of controlled thickness on planar thin-film layers. With the GLAD film,

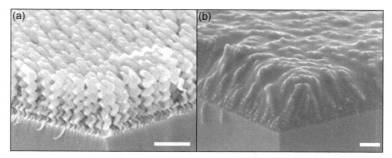

Figure 6.9   An SiO_2 helical film microstructure deposited at $\alpha = 85°$ (a) before and (b) after filling with HPR 504 photoresist. The spin-cast photoresist completely infiltrates the porous film microstructure, protecting the embedded columns from subsequent etch processing. All scale bars indicate 1 µm. Reproduced by permission of ECS – The Electrochemical Society from [38].

however, the photoresist must uniformly fill and encapsulate the intercolumn pore regions, as incomplete pore filling or any gap regions will allow etch penetration during subsequent processing steps. It has been found that, using standard processing conditions, many standard photoresist compounds (including HPR 504 [38,40,41], HPR 506 [41,42], AZ 1518 [43], AZ P4620 [41] and AZ 5214 [40]) provide sufficiently good filling to protect the microstructure against further processing. The photoresist filling is illustrated in Figure 6.9, which shows SiO_2 helical columns deposited at $\alpha = 85°$ before and after photoresist application (HPR 504 photoresist spun onto the film at 5500 rpm for 30 s [38]. The helical columns survive, remaining anchored to the substrate through the spinning process, and the photoresist fully infiltrates and protects the porous microstructure. Mechanical forces associated with surface tension and photoresist shrinkage, discussed further below, pull the embedded columnar microstructures inward at the edges after cleaving the substrate for SEM sample preparation. Typically, the microstructures are also coated with an overburden layer of excess photoresist whose thickness is determined by the rotation speed of the spin-casting process, the resist viscosity and the microstructure porosity. The overburden can be selectively removed using short UV exposures, thereby controllably revealing the column apexes and exposing them to subsequent processing when desired (e.g. Section 6.8).

The spin speed is an important factor during photoresist application to GLAD microstructures. While complete and uniform filling of the porous microstructures is observed over a wide range of rotation speeds, the overburden thickness and mechanical forces acting on the embedded film are sensitive to spin speed. This effect was systematically investigated by Bezuidenhout [41], and Figure 6.10 shows side-view SEM images of HPR 504 resist spun at different speeds into SiO_2 GLAD zigzag microstructures fabricated at (a) $\alpha = 85°$, (b) 84° and (c) 80°. The first column (i) shows the films prior to photoresist application, and the four adjacent columns show the photoresist-infiltrated structures for spin speeds of (ii) 1000, (iii) 2500, (iv) 4000 and (v) 5500 rpm. The overburden layer thickness is reduced as the spin speed is increased, eventually producing a very thin photoresist overcoat at an α-dependent spin-speed threshold. However, increasing the rotation speed past this threshold value does not expose the column apexes but instead yields a compressive stress and a net decrease in the thickness of the film–photoresist composite. The total compression is more significant for films deposited at higher α, presumably due to the increased porosity and

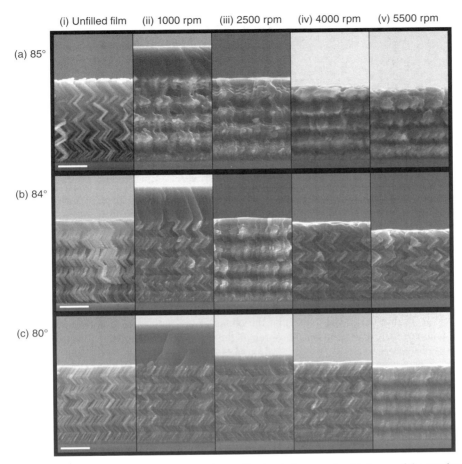

Figure 6.10   The spin speed during photoresist application influences the thickness of the overburden layer and the compressive force applied to the film by the infiltrating photoresist layer. The first column (i) of (a), (b) and (c) shows the unfilled zigzag columnar microstructure, and then after photoresist application at spin speeds of (ii) 1000, (iii) 2500, (iv) 4000 and (v) 5500 rpm. Slower rotation speeds coat the microstructure with a thick overburden layer. Increasing the rotation speed reduces the excess layer and then causes a compression of the microstructured layers for spin speeds beyond a microstructure-dependent threshold. All scale bars indicate 1 μm. Reproduced from [41]. With permission from L. Bezuidenhout.

greater compliance of the microstructure. While the exact origins of this force are unclear, it is likely due to solvent evaporation and shrinkage of the infiltrated photoresist, which adheres strongly to the column surface. Exposure and development of the compressed films releases the microstructure, which then returns to its original thickness, indicating that the film deformation is elastic. To minimize stress build-up at the interface between exposed and unexposed regions of a developed GLAD film, it is recommended that photoresist application be performed at spin speeds near the threshold between overburden formation and film compression. From these preliminary studies, the threshold speed falls between 2500

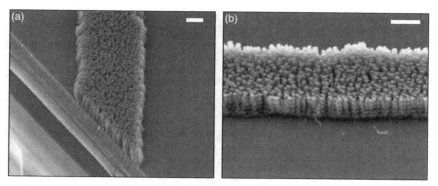

Figure 6.11 Examples of lithographically patterned micrometre-scale lines of helical columnar microstructures. All scale bars indicate 1 μm. Reproduced from [40]. With permission from K.D. Harris.

and 4000 rpm; further process optimization requires experimental determination for specific combinations of resist chemistries, column materials and GLAD film microstructures.

After the exposure and development steps, the patterned GLAD film is ready for subsequent processing; for example, using etching to define lines of column microstructures, as shown in Figure 6.11 [40]. Note that the edges of the patterned regions are disordered, with a characteristic variation on the order of the column diameter. Such disorder is common to photopatterned GLAD films. While standard photolithographic processing can define very sharp features, the columnar microstructure of the GLAD film coarsens the boundary between the exposed and unexposed photoresist regions, thus producing a rougher interface.

6.5 ENCAPSULATION AND REPLANARIZATION OF GLANCING ANGLE DEPOSITION FILMS

While GLAD films obtain much of their functionality from their open, porous microstructure, it is often necessary to cap the porous GLAD film with a dense overlayer (Figure 6.12) when integrating GLAD films into microdevice designs. Adding an encapsulation layer provides a number of benefits. The overlayer can be used to isolate the porous layer from the outside environment and provide mechanical protection to the comparatively delicate nanostructures, thus improving device robustness and handling. The capping layer also replanarizes the thin film, thereby making the surface better suited to subsequent film deposition and microprocessing. Replanarization facilitates post-deposition lithographic processing and integration of common device components, such as electrical contacts and vias, and so on.

6.5.1 Encapsulation layer substrate motions

The general approach to film encapsulation involves fabricating a denser layer at a reduced α on top of the existing porous microstructure, and multiple substrate motion algorithms have been developed to produce capping layer geometries, summarized in Figure 6.13. Note that the capping layer need not be of the same material as the porous layer, and

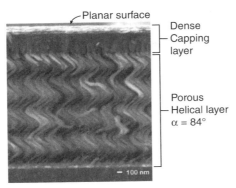

Figure 6.12 A porous, helical GLAD layer encapsulated by dense overlayer. The capping layer provides mechanical protection and replanarizes the film surface for subsequent microprocessing (e.g. multilayer deposition, device patterning, electrode fabrication). Reprinted with permission from [44]. © 1999, American Vacuum Society.

many designs incorporate planar capping electrodes by depositing a metallic overlayer (see Section 6.6.2). The simplest capping method, adopted by several groups [45–47], involves depositing a denser layer of thickness t_{cap} at a fixed deposition angle $\alpha_{cap} < \alpha_p$, where α_p is the deposition angle of the underlying porous microstructure. The capping layer angle is selected to produce a much denser microstructure, and α_{cap} values of 0° [45] and 45° [46,47] have been used in capping-layer studies. This fixed-angle approach has been shown to realize good encapsulation, although stress-related problems arise in thick capping layers. To create

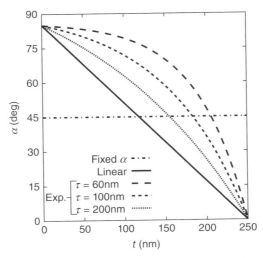

Figure 6.13 Substrate motion algorithms for different capping-layer microstructures, each 250 nm thick. The dot–dash black line shows a fixed-angle capping layer, where α is held at an intermediate value to create the overlayer. The solid black line shows a linear capping-layer motion where α is decreased to 0° linearly with deposited thickness (Equation 6.1). The other lines demonstrate the exponential cap α motion for three different values of the variable τ (Equation 6.2).

thicker overlayers with reduced stresses, different graded-α substrate motions have been developed to realize a smoother transition between the porous and dense microstructures. One such motion decreases α linearly with thickness following

$$\alpha(t) = \alpha_i - (\alpha_i - \alpha_f)\frac{t}{t_{cap}}, \tag{6.1}$$

where α_i is the deposition angle of the porous layer and α_f is the final capping layer angle. The linear motion algorithm has been used in multiple studies [48–50] to cap films with $\alpha_i > 80°$. Stress-fracture-free encapsulation layers were realized with $\alpha_f = 30°$ and $0°$ and t_{cap} ranging from 200 to 500 nm. Compared with the linear capping algorithm, Robbie *et al.* found superior stress reduction in thick overlayers by decreasing α following an exponential function [48]. For an exponential capping layer, the flux angle is reduced according to

$$\alpha(t) = \alpha_i - A \left[\exp\left(\frac{t}{\tau}\right) - 1 \right], \tag{6.2}$$

where the variable τ sets the curvature of the α variation as shown in Figure 6.13, and the constant A is given by

$$A = \frac{\alpha_i - \alpha_f}{\exp(t_{cap}/\tau) - 1}. \tag{6.3}$$

The exponential capping motion has been implemented in multiple studies [44, 48, 51]. When designing GLAD structures involving variable-α substrate motion, as with the linear and exponential capping layer, it is important to recognize that the film growth rate is α dependent. Accurate thickness control using deposition rate monitoring therefore requires deposition ratio calibration (Section 7.3.2) and a corresponding modification of the substrate motion (e.g. Section 5.4.1).

6.5.2 Film stress in encapsulation layers

High-α GLAD films typically exhibit low mechanical stresses due to their open microstructure, and hence stress-induced fracture and de-adhesion problems are not generally a concern. However, stress-related issues often arise when depositing a dense, encapsulating overlayer, leading to delamination and fracture of the film surface (Figure 6.14). It is thus important to be mindful of these effects when encapsulating porous GLAD films. The fixed-angle capping approach is the best studied in terms of stress reduction, and useful design guidelines have been created in order to avoid stress-related failure and improve the quality of capping layers. Moreover, additional process variables (e.g. deposition rate, working gas pressure during sputtering, substrate temperature) can be optimized to further reduce capping layer stress.

Karabacak *et al.* observed a number of interesting results in investigating compressive stresses in capping layers deposited at $\alpha = 0°$ onto porous, slanted column layers fabricated at $\alpha = 85°$ [45]. Figure 6.15a shows measured data of cap-layer stress as a function of cap-layer thickness, when deposited onto a 50 nm porous layer, indicating that the layer stress increases with thickness following a power law. Interestingly, the mechanical stress in a dense

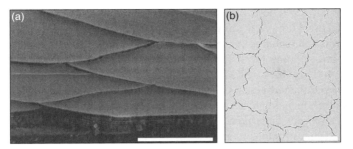

Figure 6.14 Capping layer problems. Stress build-up in the deposited dense layer can cause (a) film delamination and (b) cracking of the encapsulating overlayer. Scale bars indicate 10 μm. (a) Reprinted with permission from [48]. © 1997, American Vacuum Society. (b) Reprinted from [46], with permission from Elsevier.

layer alone (i.e. deposited at $\alpha = 0°$ with no porous layer) is an order of magnitude greater than observed in an equivalent capping layer as the surface roughness and compliance of the porous layer provide a mechanism for stress relief. This mechanism is enhanced by depositing onto thicker porous layers, yielding greater reduction in capping-layer stress. These results indicate that capping-layer stress is influenced by both the capping-layer thickness and the thickness of the underlying porous layer, and both factors must therefore be taken into account when optimizing the encapsulation process.

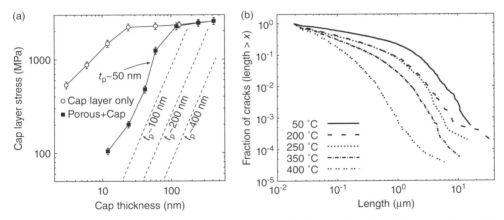

Figure 6.15 (a) Measured (compressive) film stresses in fixed-angle W capping layers deposited onto W vertical columns ($\alpha = 85°$, thickness t_p) as a function of the cap-layer thickness. The film stress increases with cap-layer thickness, scaling following a power law. Note that stress in a capping layer deposited without the porous layer (open circles) is significantly greater, indicating that the underlying microstructured column layer influences stress build-up in the encapsulation layer. The dashed lines indicate theoretical predictions at different t_p. (b) Measured cumulative crack-length distribution found in fixed-angle capping layers (TiO$_2$, 400 nm) deposited onto 2.7 μm tall TiO$_2$ vertical columns at different temperatures. Increasing the substrate temperature reduces (tensile) film stress in the encapsulation layer and suppresses the formation of long cracks. (a) Reprinted with permission from [45]. © 2004, AIP Publishing LLC. (b) Reprinted from [46], with permission from Elsevier.

In another systematic report, Kupsta *et al.* examined the effect of substrate temperature on stress-induced cracking of encapsulation layers deposited at fixed $\alpha = 45°$ onto 2.7 μm vertical post layers fabricated at $\alpha = 81°$. As seen in Figure 6.14b, cracks are observed in the top surface of a capping layer, forming due to the build up of tensile stresses (e-beam deposited TiO_2, 400 nm thick). SEM observations show that these cracks are not related to mechanical failure of the individual GLAD nanostructures, but are caused by the contraction of the capping layer which is enhanced by the compliance of the columns. While the increased compliance of the porous underlayer lowers overall film stress as seen by Karabacak *et al.* [45], the increased nanostructure length (2.7 μm versus 50 nm) allows the columnar structures to bend farther, leading to greater strain and enhanced crack formation. This result suggests that while cap-layer stress is reduced by the compliance of the nanostructured layer, sufficient stiffness is required to prevent fracture in layers under tensile strain. To reduce tensile stress in the capping layer, the authors heated the substrate during cap-layer growth (the porous layer deposition was unheated) and measured the crack dimensions using computerized SEM image analysis. Figure 6.15b shows the cumulative crack-length distribution function measured at different deposition temperatures. As the temperature is increased, the number of long cracks appearing in the cap layer is reduced and replaced by an approximately equal length of shorter cracks. It is suggested that the increased substrate temperature both reduces the tensile strain and increases the tensile strength of the capping layer, preventing the development of long cracks in the microstructure and improving the surface quality of the encapsulating layer.

6.6 INTEGRATING ELECTRICAL CONTACTS WITH GLANCING ANGLE DEPOSITION MICROSTRUCTURES

6.6.1 Planar electrode configurations

Device applications such as gas sensing often use a planar electrode geometry to ensure that the high-surface-area film is fully exposed to the ambient environment and to avoid impeding analyte diffusion into the microstructure. A common planar configuration is the interdigitated electrode (IDE) geometry, schematically illustrated in Figure 6.16a. The IDE consists of separately addressable, interpenetrating, comb-like electrodes, creating a sequence of parallel wires with alternating polarity. The IDE capacitance is determined by the electrode dimensions, including width W, length L and separation s, as well as the dielectric properties of the surrounding medium (substrate and deposited film) [54]. Covering the IDE with a porous GLAD film therefore enables sensing based on the dielectric changes of fluid condensation in the columnar microstructure [52, 55–57].

Depending on the electrode dimensions required, the IDE can be fabricated using standard lithographic processing or using shadow evaporation through an appropriately defined mask. Electrodes of height h ranging from tens to hundreds of nanometres are thus patterned across the surface, creating a raised topography that affects the shadowing environment during GLAD growth, as was discussed previously in Section 6.3. The impact of the electrode on film structure is shown in Figure 6.16b and c. The microstructure can be divided into normal-growth areas, where the surface is flat and the local deposition geometry is uniform, and edge-growth areas, where the step-like electrode topography creates a dense region of

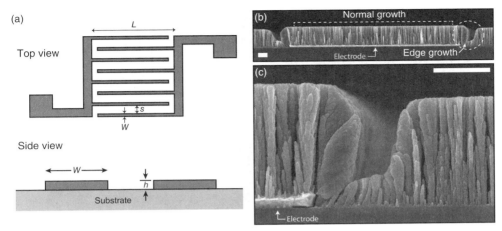

Figure 6.16 (a) An IDE is formed from interpenetrating, comb-like electrodes of thickness h and is a planar electrode geometry useful for capacitive gas sensing [52]. (b) As the electrodes sit above the substrate, they create an initial shadowing topography that affects the subsequent GLAD growth. While growth is normal along the electrode surface, the electrode boundary presents a surface relief that modifies film growth. (c) A close-up view of the edge-growth region reveals the abnormal microstructure that forms along the electrode boundary. Scale bars correspond to 1000 nm. (b), (c) Reproduced from [53]. With permission from J.J. Steele.

low-α growth as indicated. The edge growth has a negative impact on device sensitivity as it has a reduced surface area in comparison with the normal-growth regions. In terms of total device performance, the effect of the edge growth depends on the electrode size (W and s) relative to the shadow length. In larger electrodes, the normal-growth region dominates the microstructure and the influence of the edge region will be small. However, as the electrode dimensions are reduced, the edge effect increases and the edge-growth regions constitute a greater portion of the overall microstructure. The normal-region dominant to edge-region dominant crossover can be inferred from electrode scaling measurements [50]. To mitigate edge-growth effects in small electrode structures, Steele *et al.* used a countersunk IDE geometry, where the electrodes are sunk into an oxide layer to create a planar surface [52]. Although extra microfabrication is required to produce these electrodes, the replanarized surface allows uniform GLAD growth over the entire IDE structure and minimizes the formation of edge-growth regions, as seen in Figure 6.17.

6.6.2 Parallel-plate electrode configurations

Creating a parallel-plate electrode structure requires both pre- and post-deposition processing to create the bottom and top electrodes respectively. The bottom electrode is typically fabricated by depositing an initial conductive layer; for example, Au with a Cr adhesion layer, or a transparent conductor such as ITO. The surface roughness of the bottom electrode layer should be small to minimize its effect on the initial ballistic shadowing and nucleation processes. If the bottom electrode is patterned (e.g. via shadow masking or photolithography), the electrode should cover a sufficiently large area to minimize the contribution of edge

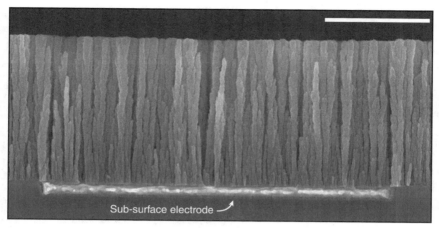

Sub-surface electrode

Figure 6.17 GLAD growth on a countersunk IDE geometry. The buried electrodes are coplanar with the substrate, producing a flat surface and minimizing edge-growth effects at the electrode boundaries. A uniform high-surface-area microstructure is then produced across the entire IDE. The scale bar indicates 1000 nm. Reproduced from [53]. With permission from J.J. Steele.

growth or additional microfabrication steps should be performed to replanarize the surface and produce a uniform shadowing environment over the device area (see Section 6.6.1).

Once the bottom electrode has been defined, the intended GLAD microstructure is then fabricated. When the deposited microstructure is dense ($\alpha < 70$–$80°$, depending on thickness), the top electrode can be deposited and patterned using standard processing techniques, often with little modification. However, at higher α, where the GLAD microstructure is highly porous and/or consists of isolated columnar structures, an encapsulation layer (Section 6.5.1) is generally required to realize a continuous, and therefore conductive, top electrode. Note that by controlling the final deposition angle α_f in the capping algorithm, the porosity, surface roughness and conductivity of the capping layer can be tuned for specific applications. This can be useful for device optimization; for example, allowing top contacts in gas sensor devices to be simultaneously porous and conductive. Device optimization then becomes a balance between increasing porosity to improve gas diffusion into the GLAD microstructure while retaining sufficient electrical conductivity for acceptable electrode operation. Examples of the encapsulation microstructures are shown in Figure 6.18, where the functional GLAD morphology (helices and vertical columns) is connected to the top electrode by a graded-density capping overlayer. The first device (Figure 6.18a and b) is a mechanical resonator based on actuating the Alq_3 helices using applied electric potentials [51]. The second device (Figure 6.18c–e) is a cross-bridge electrode configuration used for four-wire conductance measurements of columnar structures [50]. In the first device, an exponential capping layer was fabricated using the same material as the underlying helical structures, and both the bottom and top electrodes were unpatterned and thus cover the entire substrate. In the second device, a linear capping layer was deposited using Ag to electrically connect the ITO nanocolumns to the upper Ag electrode. Additionally, both the top and bottom electrodes were lithographically patterned to define the microdevice region. The planar top surface provided by the encapsulation process facilitates subsequent photopatterning of the top electrode. In the device presented, the bottom electrodes were not countersunk, thus

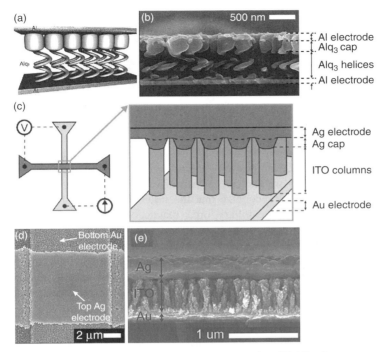

Figure 6.18 Parallel-plate electrode configurations incorporating GLAD microstructures. Connecting the highly porous GLAD microstructures to a continuous, conductive electrode requires adding an encapsulation layer as shown in these devices. (a) Design and (b) realization of a four-layer device [51]: a bottom Al electrode; an Alq_3 helical microstructure deposited at $\alpha = 85°$; an Alq_3 capping layer deposited to replanarize the device; a top Al electrode. Applying a potential between the electrodes allowed electrostatic actuation of the spring-like Alq_3 layers. (c) Design and (d,e) realization of a patterned four-layer device [50]: a lithographically patterned bottom Au electrode; an ITO vertical column layer ($\alpha = 83°$); an Ag capping layer; an Ag top electrode. The top three layers were lithographically patterned to create a cross-bridge device suitable for four-wire electrical characterization of the nanostructured ITO layer. (a) and (b) Reprinted with permission from [51]. © 2007, AIP Publishing LLC. (c)–(e) Reprinted with permission from [50]. © 2013, AIP Publishing LLC.

creating dense growth regions along the perimeter of the bottom electrode (not shown). In small devices, the resulting edge effects produce a measurable impact on the electrical characteristics as quantified through area scaling experiments [50], indicating that countersunk electrode structures may be required for optimized microdevices.

6.7 FILMS IN LIQUID ENVIRONMENTS

Porous GLAD films are often immersed in liquid environments to exploit the increased surface area of the microstructure and enhance device performance. A prominent example is the integration of microstructured GLAD films into electrochemical devices where the GLAD film is infiltrated with a liquid electrolyte solution and used as a high-surface-area electrode or catalyst support to boost reaction efficiencies. Some demonstrated electrochemical applications include oxygen reduction electrocatalysis [59–61], photoelectrochemical water splitting [17, 22, 62, 63], electrochemical capacitors [64], anodes for lithium-ion batteries [65–67], electrochromic cells [68–70] and methanol oxidation catalysis [71]. Other device applications featuring GLAD structures immersed in liquid environments include using GLAD columns as alignment support structures in electrically switchable liquid-crystal cells [72–75], porous electrodes for dye-sensitized solar cells [14, 76], porous media for ultrathin-layer chromatography [16, 77, 78], optical devices based on electrophoretic dye movement [79, 80] and nanostructured substrates for surface-enhanced Raman scattering used in trace detection of chemical and biological agents [81–89]. In addition to being an important part of many device applications, microfabrication techniques frequently rely on immersing films in solvent or wet etchants as part of various processing steps.

However, it is important to recognize that the liquid infiltration and evaporation processes generate significant lateral capillary forces that act on the microstructure as the liquid enters/exits the porous film. These forces become particularly problematic in films deposited at high α as the isolated, high-aspect-ratio columns can experience large bending deformations that lead to irreversible bundling of multiple columns (Figure 6.19a). This nanocolumn bundling behaviour, referred to in the literature as the 'nanocarpet effect', has been systematically investigated in a series of papers examining liquid penetration and evaporation in

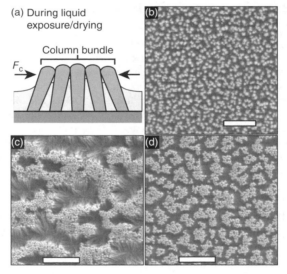

Figure 6.19   (a) The nanocarpet effect: clustering of columns pulled together by lateral capillary forces experienced during infiltration and evaporation of liquids from the porous film. (b) A top-down SEM view of vertical nanocolumns prior to water exposure. (c) and (d) After a water droplet wets into and evaporates from the film, the isolated nanocolumns are deformed by capillary forces and leave irreversibly clustered column bundles. Capillary force variations over the droplet surface lead to nonuniform cluster sizes. (c) Regions near the droplet centre experience large forces and form larger clusters, whereas (d) the capillary forces are weaker towards the droplet periphery, yielding smaller bundles. Scale bars indicate 2 μm. Reprinted with permission from [58]. © 2004, American Chemical Society.

structured GLAD films [33, 58, 90–93]. In sessile water-droplet contact-angle studies, the authors noted that as the droplets spread onto and then evaporated from the surface, they left behind a macroscopic watermark on the surface. SEM examination of the watermark region revealed the deformation and bundling of columnar microstructures, which increases optical scattering in the water-treated regions and creates a visible watermark. Figure 6.19b shows a top-down view of Si vertical columns ($\alpha = 86°$, 890 nm tall) prior to application of the water droplet. The film microstructure consists of individual, well-separated columnar structures as is characteristic of films fabricated at such a large deposition angle. After the water droplet wets the film surface and then evaporates, the high-aspect-ratio columns have been deformed and drawn together by the lateral capillary forces, leading to formation of clusters of bundled columns, as shown in Figure 6.19c and d. The bundling is nonuniform over the watermark, with larger cluster structures (Figure 6.19c) forming towards the watermark centre where the capillary forces are stronger. The forces are weaker towards the droplet edge, producing correspondingly smaller column deformations and yielding smaller cluster structures. To reproduce the observed clustering in arrays of cylindrical columns, finite-element simulations estimated bending forces up to 10 nN in the watermark centre, compared with <1 nN towards the periphery [58]. The susceptibility of a given microstructure to the nanocarpet effect is determined by several parameters, including the mechanical characteristics of the individual columns, the intercolumn separation distance and the surface tension of the liquid. Taller columns promote increased clustering owing to the reduced stiffness of such structures to the lateral capillary force [90, 91]. Experimental studies of Si GLAD columns have shown that the cluster diameter d_c, quantified via numerical image analysis, scales with the column height h according to a power law $d_c \propto h^{1.2 \pm 0.1}$, a scaling similar to the prediction of a simplified energetic model of column bending and cluster formation [91]. The specific column architecture also plays a role, and helical and zigzag microstructures have been shown to be more mechanically stable than vertical columns [92], as expected from moment of inertia considerations.

Multiple strategies have been devised to reduce the impact of the nanocarpet effect and minimize column clustering, based on using an additional processing step to mechanically reinforce the individual columns. Fan *et al.* deposited a 100 nm thick Si layer at near-normal incidence ($\alpha = 3°$) onto a 750 nm tall Si vertical column microstructure [92]. The near-normal deposition fills in the intercolumn gaps at the substrate, thereby increasing the base diameter of the nanocolumns. This structural modification increases the bending stiffness of the structures, rendering them less susceptible to the capillarity-induced deformations. No structural changes were detected before and after water treatment of the reinforced structures, demonstrating the efficacy of the technique in mitigating the nanocarpet effect. A similar reinforcement approach was developed by Kwan and Sit, who used ion milling to modify the columnar structures [29]. As discussed in Section 6.2.4, ion milling increases the base diameter of the nanocolumns via redeposition of sputtered material, thereby stiffening the columns against the lateral capillary forces experienced during immersion and drying. The effect of ion milling is presented in Figure 6.20, which shows pre-water-immersion SEM views of (a) as-deposited vertical columns and the structures after (b) 30 s and (c) 60 s of ion bombardment. The second row of images (Figure 6.20d–f) shows the same samples after water immersion and subsequent drying. The as-deposited sample exhibits substantial column deformation and cluster formation consistent with the previously discussed results. However, the ion-milling process stiffens the column microstructures, making them less susceptible to clustering. After 30 s of ion-milling treatment (Figure 6.20e), the average

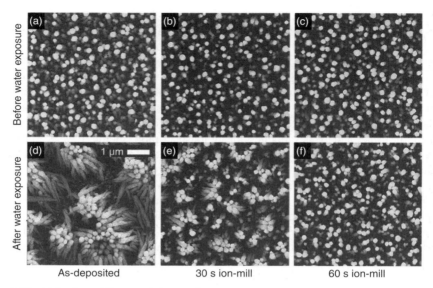

Figure 6.20   Using ion milling to reinforce columnar structures and prevent the nanocarpet effect. The top row of top-down SEM images shows the vertical column microstructures (SiO_2, $\alpha = 86°$, 1.6 μm tall) before water immersion. (a) The as-deposited film; (b) and (c) the film after 30 s and 60 s ion-milling treatment respectively. The bottom row of SEM images shows the same films after water immersion and drying. The ion-milling-induced structural modifications (Section 6.2.4) yield mechanically stiffer columns that are less susceptible to capillary forces and column clustering. Reproduced from [29]. © 2010 IOP Publishing. Reproduced by permission of IOP Publishing. All rights reserved.

cluster size is reduced consistent with a decrease in column deformations. Increasing the ion-milling treatment to 60 s (Figure 6.20f) further strengthens the columnar structures, and the water exposure produced no observable effect on the film microstructure. Both the near-normal deposition and ion-milling reinforcement techniques are straightforward to implement and can be used to realize robust microstructures better suited for stable operation in liquid environments.

Another approach to avoiding the nanocarpet effect during water removal from GLAD films is to use critical-point drying [94], a process commonly used in MEMS fabrication to protect delicate microdevices from the damaging effects of surface tension. The SEM images in Figure 6.21 show the as-deposited film (SiO_2 slanted columnar microstructure) and the film following immersion in ethanol with normal drying and with critical-point drying using liquefied CO_2. The normal drying process induces clustering of the columns consistent with the nanocarpet effect (note that the slanted columnar microstructures cluster differently than the vertical column microstructures discussed previously). During the critical-point drying process, the ethanol is displaced by liquid CO_2 that is heated to a supercritical fluid phase and removed without surface tension effects. The microstructure is consequently unaffected by the drying process, as can be seen qualitatively by comparing Figure 6.21a and c and established quantitatively through image analysis [94].

The critical-point drying results indicate that the majority of the structural deformation and clustering associated with the nanocarpet effect arises during the liquid evaporation

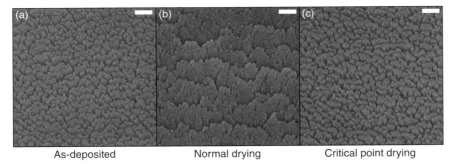

As-deposited   Normal drying   Critical point drying

Figure 6.21   (a) Top-down SEM view of as-deposited thin-film microstructures (SiO_2 slanted columns, $\alpha = 84 - -86°$). The same film after immersion in ethanol and then drying (b) in an ambient environment and (c) using a CO_2-based critical-point drying process. Critical-point drying eliminates surface tension forces during the drying process, allowing liquid removal without deforming the film microstructure. Scale bars indicate 500 nm. Reprinted from [94], with permission from Elsevier.

process, a conclusion also supported by SEM examination of freeze-dried samples [93]. The nanocarpet effect is therefore less of a concern in device applications, where the film is never removed from the liquid environment for the life of the device; for example, in a permanently sealed electrochemical cell. However, in other applications where the microstructured film may undergo immersion and drying cycles (e.g. in a reusable biosensor), significant capillary forces are generated and should be considered when designing the GLAD layer.

6.8 USING GLANCING ANGLE DEPOSITION MICROSTRUCTURES AS REPLICATION TEMPLATES

Porous GLAD films are sufficiently robust to survive extensive post-deposition processing and can be used as microstructured templates for subsequent casting fabrication techniques. The original GLAD structure can then be reproduced in an alternative material with different properties and new functionality. Such templating can also be used to realize GLAD structures in materials less amenable to basic GLAD processing, such as high-mobility evaporants, materials with complex stoichiometry or organic compounds. The following describes the basic templating process and several modifications thereof.

6.8.1 Single- and double-template fabrication processes

The template replication process flow is schematically depicted in Figure 6.22. The process begins with an initial GLAD fabrication step (Figure 6.22a) to produce the initial microstructural template. The porous template is then infiltrated with the desired material (Figure 6.22b). Depending on the infiltration method, an overcoat layer is produced, thus completely burying the columnar structures. The third process step is therefore required to etch back the filling material to expose the apexes of the buried columns (Figure 6.22c). Once exposed, the initial template structures can be removed using an etch process that selectively removes the column material without affecting the filling material (Figure 6.22d). The result is an

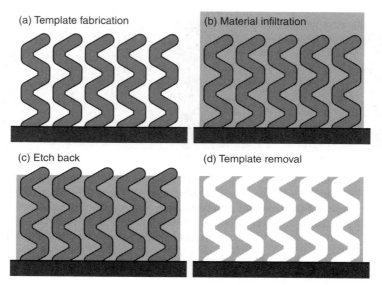

Figure 6.22   The four-step template fabrication process. (a) The initial template master is fabricated using GLAD. (b) The porous master microstructure is filled with a different material. (c) The surface of the filling material is controllably etched back to reveal the column apexes. (d) The column material is removed using a selective-etch process to produce the final, inverted microstructure. Adapted from [95]. With permission from M.A. Summers.

inverted thin-film microstructure: a continuous matrix of the infiltrated material perforated with pore structures defined by the initial column shapes, as shown in Figure 6.23a, b. Note that the etch-back step requires careful control over the removal rate in order to reveal the columns (Figure 6.24).

The first demonstration of the inverted morphology was by Harris *et al.*, who realized perforated thin films of photoresist and spin-on-glass templated from SiO_2 helices [96]. While in this initial work the material was introduced to the film via spin-casting, the templating process has since been applied to other materials and filling techniques. Perforated metal films of Ni and Au were fabricated via electroplating filling of helical microstructures [98, 99]. Organic materials have also been used, including polystyrene [98] and liquid-crystal polymers [100]. LPCVD has been used to in-fill square-spiral microstructures with Si and fabricate inverted square-spiral photonic crystals [95]. The perforated thin films themselves can also be used as templates in another infiltration and selective-etch process. Such a double-templating process can be used to create higher quality helical columnar structures (Figure 6.23c, d) from materials that are challenging or impossible to directly deposit using PVD techniques, such as polymer compounds [43].

6.8.2 Nanotube fabrication via template fabrication

The templating approach described in Section 6.8.1 is based on a complete, uniform infiltration of the porous GLAD film that yields a continuous, inverted microstructure after

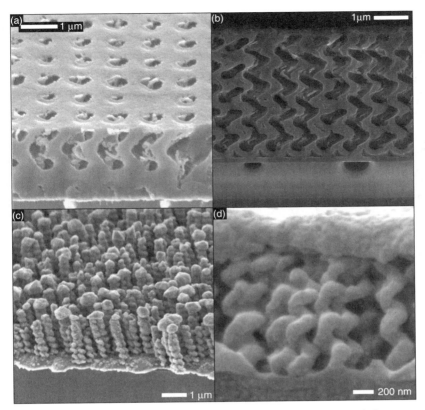

Figure 6.23 Thin-film microstructures realized via templating and double-templating processes: (a) a perforated photoresist thin film templated from a seeded array of SiO_2 helices; (b) inverse Si square-spiral structure created by low-pressure chemical vapour deposition (LPCVD) filling of an SiO_2 square-spiral template; (c) double-template fabrication of Cu helices electrodeposited in a perforated photoresist thin film; and (d) double-template fabrication of polyacrylate helices by filling a perforated photoresist film. (a) Reproduced by permission of ECS – The Electrochemical Society from [96]. (b) Reproduced from [95]. With permission from M.A. Summers. (c) © 2002 IEEE. Reprinted with permission from [97]. (d) Reproduced from [43]. © IOP Publishing. Reproduced by permission of IOP Publishing. All rights reserved.

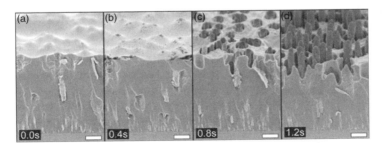

Figure 6.24 The photoresist overburden layer coating the porous GLAD film can be controllably removed using short UV exposures to uncover the column apexes. (a) The original photoresist-filled film, and then the developed film after UV exposures of (b) 0.4 s, (c) 0.8 s and (d) 1.2 s, demonstrating the progressive removal of the overburden layer and exposure of the column apexes for subsequent processing. All scale bars indicate 1 μm. Reproduced from [42] with permission from The Royal Society of Chemistry.

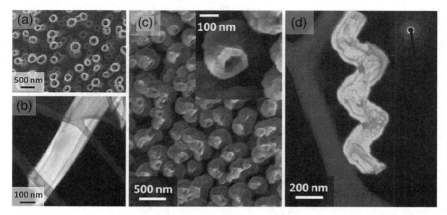

Figure 6.25 Nanotube fabrication by templating GLAD-fabricated columns with thin conformal layers. (a) SEM and (b) TEM images of an Si nanotube fabricated by conformal LPCVD coating of an SiO_2 vertical column. The SiO_2 core material is exposed via ion milling and then removed using an etch process to produce the hollow tube structure. The same procedure applied to helical SiO_2 columns produces the helical nanotube structures shown in (c) and (d). Reproduced with permission from [101]. Copyright © 2009 Wiley–VCH Verlag GmbH & Co. KGaA, Weinheim

processing. However, it is possible to create arrays of hollow nanotube structures by coating the isolated columns with a thin, conformal layer. The core columnar material is then exposed with a similar etch-back step, and then removed with a selective-etch process to produce hollow tubes with a structure determined by the original column, as shown in Figure 6.25. Here, an LPCVD process was used to coat vertically oriented (Figure 6.25a, b) and helical (Figure 6.25c, d) SiO_2 columns with a conformal Si layer on the order of tens of nanometres thick. The Si layer at the column apex was then removed via ion-mill etching to expose the inner SiO_2 core. Buffered-oxide etch was then used to remove core material and produce the final tubular structure shown in Figure 6.25. This nanotube fabrication process could be adapted to other material systems by an appropriate selection of conformal coating techniques (e.g. ALD) and etch chemistries.

REFERENCES

[1] Wang, P.I., Parker, T.C., Karabacak, T. *et al.* (2009) Size control of Cu nanorods through oxygen-mediated growth and low temperature sintering. *Nanotechnology*, **20**, 085605.

[2] Karabacak, T., DeLuca, J.S., Wang, P.I. *et al.* (2006) Low temperature melting of copper nanorod arrays. *Journal of Applied Physics*, **99**, 064304.

[3] Deniz, D., Frankel, D.J., and Lad, R.J. (2010) Nanostructured tungsten and tungsten trioxide films prepared by glancing angle deposition. *Thin Solid Films*, **518**, 4095–4099.

[4] Khare, C., Gerlach, J.W., Höche, T. *et al.* (2012) Effects of annealing on arrays of Ge nanocolumns formed by glancing angle deposition. *Applied Surface Science*, **258**, 9762–9769.

[5] Choi, W.K., Li, L., Chew, H.G., and Zheng, F. (2007) Synthesis and structural characterization of germanium nanowires from glancing angle deposition. *Nanotechnology*, **18**, 385302.

[6] Flaherty, D.W., Dohnálek, Z., Dohnálková, A. *et al.* (2007) Reactive ballistic deposition of porous TiO_2 films: growth and characterization. *Journal of Physical Chemistry C*, **111**, 4765–4773.

[7] Van Popta, A.C., Sit, J.C., and Brett, M.J. (2004) Optical properties of porous helical thin films and the effects of post-deposition annealing, in *Organic Optoelectronics and Photonics* (eds P.L. Heremans, M. Muccini, and H. Hofstraat), vol. 5464 of *Proceedings of SPIE*, SPIE Press, Bellingham, WA, pp. 198–208.

[8] Van Popta, A.C., Cheng, J., Sit, J.C., and Brett, M.J. (2007) Birefringence enhancement in annealed TiO_2 thin films. *Journal of Applied Physics*, **102**, 013517.

[9] Hrudey, P.C.P. (2006) Luminescent chiral thin films fabricted using glancing angle deposition, PhD thesis, University of Alberta.

[10] Kiema, G.K. and Brett, M.J. (2004) Effect of thermal annealing on structural properties and electrochemical performance of carbon films with porous microstructure. *Journal of the Electrochemical Society*, **151**, E194–E198.

[11] Flaherty, D.W., Hahn, N.T., Ferrer, D. *et al.* (2009) Growth and characterization of high surface area titanium carbide. *Journal of Physical Chemistry C*, **113**, 12742–12752.

[12] Schubert, E., Fahlteich, J., Rauschenbach, B. *et al.* (2006) Recrystallization behavior in chiral sculptured thin films from silicon. *Journal of Applied Physics*, **100**, 016107.

[13] Haynes, W.M. (ed.) (2013) *CRC Handbook of Chemistry and Physics*, 94th edn, CRC Press, Oxford, UK.

[14] Kiema, G.K., Colgan, M.J., and Brett, M.J. (2005) Dye sensitized solar cells incorporating obliquely deposited titanium oxide layers. *Solar Energy Materials and Solar Cells*, **85**, 321–331.

[15] Pursel, S.M., Horn, M.W., and Lakhtakia, A. (2006) Blue-shifting of circular Bragg phenomenon by annealing of chiral sculptured thin films. *Optics Express*, **14**, 8001–8012.

[16] Jim, S.R., Oko, A.J., Taschuk, M.T., and Brett, M.J. (2011) Morphological modification of nanostructured ultrathin-layer chromatography stationary phases. *Journal of Chromatography A*, **1218**, 7203–7210.

[17] Hahn, N.T., Ye, H., Flaherty, D.W. *et al.* (2010) Reactive ballistic deposition of α-Fe_2O_3 thin films for photoelectrochemical water oxidation. *ACS Nano*, **4**, 1977–1986.

[18] He, Y.P., Zhang, Z.Y., and Zhao, Y.P. (2008) Optical and photocatalytic properties of oblique angle deposited TiO_2 nanorod array. *Journal of Vacuum Science and Technology B*, **26**, 1350–1358.

[19] Hrudey, P.C.P., Taschuk, M., Tsui, Y.Y. *et al.* (2005) Optical properties of porous nanostructured Y_2O_3:Eu thin films. *Journal of Vacuum Science and Technology A*, **23**, 856–861.

[20] Sanchez-Valencia, J.R., Toudert, J., Borras, A. *et al.* (2011) Selective dichroic patterning by nanosecond laser treatment of Ag nanostripes. *Advanced Materials*, **23**, 848–853.

[21] Filippin, A.N., Borras, A., Rico, V.J. *et al.* (2013) Laser induced enhancement of dichroism in supported silver nanoparticles deposited by evaporation at glancing angles. *Nanotechnology*, **24**, 045301.

[22] Dang, H.X., Hahn, N.T., Park, H.S. *et al.* (2012) Nanostructured Ta_3N_5 films as visible-light active photoanodes for water oxidation. *Journal of Physical Chemistry C*, **116**, 19225–19232.

[23] Robbie, K., Cui, Y., Elliott, C., and Kaminska, K. (2006) Oxidation of evaporated porous silicon rugate filters. *Applied Optics*, **45**, 8298–8303.

[24] Sorge, J.B., Taschuk, M.T., Wakefield, N.G. *et al.* (2012) Metal oxide morphology in argon-assisted glancing angle deposition. *Journal of Vacuum Science and Technology A*, **30**, 021507.

[25] Summers, M.A. and Brett, M.J. (2008) Thermal oxidation of periodically aligned silicon square-spirals. *Microelectronics Engineering*, **85**, 1222–1224.

[26] Kupsta, M.R., Taschuk, M.T., Brett, M.J., and Sit, J.C. (2009) Reactive ion etching of columnar nanostructured TiO_2 thin films for modified relative humidity sensor response time. *IEEE Sensors Journal*, **9**, 1979–1986.

[27] Pursel, S.M., Horn, M.W., and Lakhtakia, A. (2007) Tuning of sculptured-thin-film spectral-hole filters by postdeposition etching. *Optical Engineering*, **46**, 040507.

[28] Huang, Z., Hawkeye, M.M., and Brett, M.J. (2012) Enhancement in broadband and quasi-omnidirectional antireflection of nanopillar arrays by ion milling. *Nanotechnology*, **23** (27), 275703.

[29] Kwan, J.K. and Sit, J.C. (2010) The use of ion-milling to control clustering of nanostructured, columnar thin films. *Nanotechnology*, **21**, 295301.

[30] Colgan, M.J., Vick, D., and Brett, M.J. (2000) Non-lithographic nanocolumn fabrication with application to field emitters, in *Nonlithographic and Lithographic Methods of Nanofabrication – From Ultralarge Scale* (eds A. Karim, L. Merhari, D. Norris, J. Rogers, and Y. Xia), vol. 636 of *MRS Proceedings*, Materials Research Society, doi:10.1557/PROC-636-D9.53.1.

[31] Tsoi, S., Fok, E., Sit, J.C., and Veinot, J.G.C. (2004) Superhydrophobic, high-surface area 3-D SiO_2 nanostructures through siloxane-based surface functionalization. *Langmuir*, **20**, 10771–10774.

[32] Tsoi, S., Fok, E., Sit, J.C., and Veinot, J.G.C. (2006) Surface functionalization of porous nanostructured metal oxide thin films fabricated by glancing angle deposition. *Chemistry of Materials*, **18**, 5260–5266.

[33] Fan, J. and Zhao, Y. (2010) Nanocarpet effect induced superhydrophobicity. *Langmuir*, **26**, 8245–8250.

[34] Van Popta, A.C., Steele, J.J., Tsoi, S. *et al.* (2006) Porous nanostructured optical filters rendered insensitive to humidity by vapor-phase functionalization. *Advanced Functional Materials*, **16**, 1331–1336.

[35] Albrecht, O., Zierold, Patzig, C. *et al.* (2010) Tubular magnetic nanostructures based on glancing angle deposited templates and atomic layer deposition. *Physica Status Solidi (b)*, **247**, 1365–1371.

[36] George, S.M. (2010) Atomic layer deposition: an overview. *Chemical Reviews*, **110**, 111–131.

[37] Taschuk, M.T., Harris, K.D., Smetaniuk, D.P., and Brett, M.J. (2012) Decoupling sensor morphology and material: Atomic layer deposition onto nanocolumn scaffolds. *Sensors and Actuators B*, **162**, 1–6.

[38] Harris, K.D., Brett, M.J., Smy, T., and Backhouse, C. (2000) Microchannel surface area enhancement using porous thin films. *Journal of the Electrochemical Society*, **147** (5), 2002–2006.

[39] Seto, M., Westra, K., and Brett, M.J. (2002) Arrays of self-sealed microchambers and channels. *Journal of Materials Chemistry*, **12**, 2348–2351.

[40] Harris, K.D. (2003) Fabrication and applications of highly porous thin films, PhD thesis, University of Alberta.

[41] Bezuidenhout, L.W. (2011) Molecular separations using nanostructured porous thin films fabricated by glancing angle deposition, PhD thesis, University of Alberta.

[42] Bezuidenhout, L.W., Nazemifard, N., Jemere, A.B. *et al.* (2011) Microchannels filled with diverse micro- and nanostructures fabricated by glancing angle deposition. *Lab on a Chip*, **11**, 1671–1678.

[43] Elias, A.L., Harris, K.D., Bastiaansen, C.W.M. *et al.* (2005) Large-area microfabrication of three-dimensional, helical polymer structures. *Journal of Micromechanics and Microengineering*, **15** (1), 49–54.

[44] Seto, M.W., Robbie, K., Vick, D. *et al.* (1999) Mechanical response of thin films with helical microstructures. *Journal of Vacuum Science and Technology B*, **17**, 2172–2177.

[45] Karabacak, T., Picu, C.R., Senkevich, J.J. *et al.* (2004) Stress reduction in tungsten films using nanostructured compliant layers. *Journal of Applied Physics*, **96**, 5740–5746.

[46] Kupsta, M.R., Taschuk, M.T., Brett, M.J., and Sit, J.C. (2011) Overcoming cap layer cracking for glancing-angle deposited films. *Thin Solid Films*, **519**, 1923–1929.

[47] Hwang, S., Kwon, H., Chhajed, S. *et al.* (2013) A near single crystalline TiO_2 nanohelix array: enhanced gas sensing performance and its application as a monolithically integrated electronic nose. *Analyst*, **138**, 443–450.

[48] Robbie, K. and Brett, M.J. (1997) Sculptured thin films and glancing angle deposition: Growth mechanics and applications. *Journal of Vacuum Science and Technology A*, **15**, 1460–1465.

[49] Hirakata, H., Nishihira, T., Yonezu, A., and Minoshima, K. (2011) Interface strength of structured nanocolumns grown by glancing angle deposition. *Engineering Fracture Mechanics*, **78**, 2800–2808.

[50] Lalany, A., Tucker, R.T., Taschuk, M.T. *et al.* (2013) Axial resistivity measurement of a nanopillar ensemble using a cross-bridge Kelvin architecture. *Journal of Vacuum Science and Technology A*, **31**, 031502.

[51] Dice, G.D., Brett, M.J., Wang, D., and Buriak, J.M. (2007) Fabrication and characterization of an electrically variable, nanospring based interferometer. *Applied Physics Letters*, **90**, 253101.

[52] Steele, J.J., Fitzpatrick, G.A., and Brett, M.J. (2007) Capacitive humidity sensors with high sensitivity and subsecond response times. *IEEE Sensors Journal*, **7** (6), 955–956.

[53] Steele, J.J. (2007) Nanostructured thin films for humidity sensing, PhD thesis, University of Alberta.

[54] Van Gerwen, P., Laureyn, W., Laureys, W. *et al.* (1998) Nanoscaled interdigitated electrode arrays for biochemical sensors. *Sensors and Actuators B*, **49**, 73–80.

[55] Steele, J.J., Gospodyn, J.P., Sit, J.C., and Brett, M.J. (2006) Impact of morphology on high-speed humidity sensor performance. *IEEE Sensors Journal*, **6**, 24–27.

[56] Steele, J.J., Taschuk, M.T., and Brett, M.J. (2008) Nanostructured metal oxide thin films for humidity sensors. *IEEE Sensors Journal*, **8** (8), 1422–1429.

[57] Steele, J.J., Taschuk, M.T., and Brett, M.J. (2009) Response time of nanostructured relative humidity sensors. *Sensors and Actuators B*, **140**, 610–615.

[58] Fan, J.G., Dyer, D., Zhang, G., and Zhao, Y.P. (2004) Nanocarpet effect: pattern formation during the wetting of vertically aligned nanorod arrays. *Nano Letters*, **4**, 2133–2138.

[59] Bonakdarpour, A., Tucker, R.T., Fleischauer, M.D. *et al.* (2012) Nanopillar niobium oxides as support structures for oxygen reduction electrocatalysts. *Electrochimica Acta*, **85**, 492–500.

[60] Khudhayer, W.J., Kariuki, N.N., Wang, X. *et al.* (2011) Oxygen reduction reaction electrocatalytic activity of glancing angle deposited platinum nanorod arrays. *Journal of the Electrochemical Society*, **158**, B1029–B1041.

[61] Khudhayer, W.J., Kariuki, N., Myers, D.J. *et al.* (2012) GLAD Cr nanorods coated with SAD Pt thin film for oxygen reduction reaction. *Journal of the Electrochemical Society*, **159**, B729–B736.

[62] Wolcott, A., Smith, W.A., Kuykendall, T.R. *et al.* (2009) Photoelectrochemical study of nanostructured ZnO thin films for hydrogen generation from water splitting. *Advanced Functional Materials*, **19**, 1849–1856.

[63] Berglund, S.P., Flaherty, D.W., Hahn, N.T. *et al.* (2011) Photoelectrochemical oxidation of water using nanostructured $BiVO_4$ films. *Journal of Physical Chemistry C*, **115**, 3794–3802.

[64] Broughton, J.N. and Brett, M.J. (2002) Electrochemical capacitance in manganese thin films with chevron microstructure. *Electrochemical and Solid-State Letters*, **5**, A279–A282.

[65] Fleischauer, M.D., Li, J., and Brett, M.J. (2009) Columnar thin films for three-dimensional microbatteries. *Journal of the Electrochemical Society*, **156**, A33–A36.

[66] Lin, Y.M., Abel, P.R., Flaherty, D.W. *et al.* (2011) Morphology dependence of the lithium storage capability and rate performance of amorphous TiO_2 electrodes. *Journal of Physical Chemistry C*, **115**, 2585–2591.

[67] Abel, P.R., Chockla, A.M., Lin, Y.M. *et al.* (2013) Nanostructured $Si_{(1-x)}Ge_x$ for tunable thin film lithium-ion battery anodes. *ACS Nano*, **7**, 2249–2257.

[68] Le Bellac, D., Azens, A., and Granqvist, C.G. (1995) Angular selective transmittance through electrochromic tungsten oxide films made by oblique angle sputtering. *Applied Physics Letters*, **66**, 1715–1716.

[69] Gil-Rostra, J., Cano, M., Pedrosa, J.M. *et al.* (2012) Electrochromic behaviour of $W_xSi_yO_z$ thin films prepared by reactive magnetron sputtering at normal and glancing angles. *ACS Applied Materials and Interfaces*, **4**, 628–638.

[70] Garcia-Garcia, F.J., Gil-Rostra, J., Yubero, F., and González-Elipe, A.R. (2013) Electrochromism in WO_x and $W_xSi_yO_z$ thin films prepared by magnetron sputtering at glancing angles. *Nanoscience and Nanotechnology Letters*, **5**, 89–93.

[71] Yoo, S.J., Jeon, T.Y., Kim, K.S. *et al.* (2010) Multilayered Pt/Ru nanorods with controllable bimetallic sites as methanol oxidation catalysts. *Physical Chemistry Chemical Physics*, **12**, 15240–15246.

[72] Robbie, K., Broer, D.J., and Brett, M.J. (1999) Chiral nematic order in liquid crystals imposed by an engineered inorganic nanostructure. *Nature*, **399**, 764–766.

[73] Sit, J.C., Broer, D.J., and Brett, M.J. (2000) Alignment and switching of nematic liquid crystals embedded in porous chiral thin films. *Liquid Crystals*, **27**, 387–391.

[74] Sit, J.C., Broer, D.J., and Brett, M.J. (2000) Liquid crystal alignment and switching in porous chiral thin films. *Advanced Materials*, **12**, 371–373.

[75] Kennedy, S.R., Sit, J.C., Broer, D.J., and Brett, M.J. (2001) Optical activity of chiral thin film and liquid crystal hybrids. *Liquid Crystals*, **28** (12), 1799–1803.

[76] Hsu, S.Y., Tsai, C.H., Lu, C.Y. *et al.* (2012) Nanoporous platinum counter electrodes by glancing angle deposition for dye-sensitized solar cells. *Organic Electronics*, **13**, 856–863.

[77] Jim, S.R., Taschuk, M.T., Morlock, G.E. *et al.* (2010) Engineered anisotropic microstructures for ultrathin-layer chromatography. *Analytical Chemistry*, **82**, 5349–5356.

[78] Oko, A.J., Jim, S.R., Taschuk, M.T., and Brett, M.J. (2012) Time resolved chromatograms in ultra-thin layer chromatography. *Journal of Chromatography A*, **1249**, 226–232.

[79] Hrudey, P.C.P., Martinuk, M.A., Mossman, M.A. *et al.* (2007) Application of transparent nanos-tructured electrodes for modulation of total internal reflection, in *Nanocoatings* (eds G.B. Smith and M.B. Cortie), vol. 6647 of *Proceedings of SPIE*, SPIE Press, Bellingham, WA, p. 66470A.

[80] Krabbe, J.D. and Brett, M.J. (2010) Photonic crystal reflectance switching by dye electrophoresis. *Applied Physics Letters*, **97**, 041117.

[81] Shanmukh, S., Jones, L., Driskell, J. *et al.* (2006) Rapid and sensitive detection of respiratory virus molecular signatures using a silver nanorod array SERS substrate. *Nano Letters*, **6**, 2630–2636.

[82] Chu, H., Huang, Y., and Zhao, Y. (2008) Silver nanorod arrays as a surface-enhanced Raman scattering substrate for foodborne pathogenic bacteria detection. *Applied Spectroscopy*, **62**, 922–931.

[83] Leverette, C.L., Villa-Aleman, E., Jokela, S. *et al.* (2009) Trace detection and differentiation of uranyl(VI) ion cast films utilizing aligned Ag nanorod SERS substrates. *Vibrational Spectroscopy*, **50**, 143–151.

[84] Du, X., Chu, H., Huang, Y., and Zhao, Y. (2010) Qualitative and quantitative determination of melamine by surface-enhanced Raman spectroscopy using silver nanorod array substrates. *Applied Spectroscopy*, **64**, 781–785.

[85] Driskell, J.D., Primera-Pedrozo, O.M., Dluhy, R.A. *et al.* (2009) Quantitative surface-enhanced Raman spectroscopy based analysis of microRNA mixtures. *Applied Spectroscopy*, **63**, 1107–1114.

[86] Driskell, J.D., Zhu, Y., Kirkwood, C.D. *et al.* (2010) Rapid and sensitive detection of rotavirus molecular signatures using surface enhanced Raman spectroscopy. *PLoS One*, **5**, e10222.

[87] Hennigan, S.L., Driskell, J.D., Dluhy, R.A. *et al.* (2010) Detection of *Mycoplasma pneumoniae* in simulated and true clinical throat swab specimens by nanorod array surface-enhanced Raman spectroscopy. *PLoS One*, **5**, e13633.

[88] Abell, J.L., Garren, J.M., Driskell, J.D. *et al.* (2012) Label-free detection of micro-RNA hybridization using surface-enhanced Raman spectroscopy and least-squares analysis. *Journal of the American Chemical Society*, **134**, 12889–12892.

[89] Wu, X., Xu, C., Tripp, R.A. *et al.* (2013) Detection and differentiation of foodborne pathogenic bacteria in mung bean sprouts using field deployable label-free SERS devices. *Analyst*, **138**, 3005–3012.

[90] Fan, J.G. and Zhao, Y.P. (2006) Characterization of watermarks formed in nano-carpet effect. *Langmuir*, **22**, 3662–3671.

[91] Zhao, Y.P. and Fan, J.G. (2006) Clusters of bundled nanorods in nanocarpet effect. *Applied Physics Letters*, **88**, 103123.

[92] Fan, J.G., Fu, J.X., Collins, A., and Zhao, Y.P. (2008) The effect of the shape of nanorod arrays on the nanocarpet effect. *Nanotechnology*, **19**, 045713.

[93] Fan, J.G. and Zhao, Y.P. (2008) Freezing a water droplet on an aligned Si nanorod array substrate. *Nanotechnology*, **19**, 155707.

[94] Shah, P.J., Wu, Z., and Sarangan, A.M. (2013) Effects of CO_2 critical point drying on nanostructured SiO_2 thin films after liquid exposure. *Thin Solid Films*, **527**, 344–348.

[95] Summers, M.A. (2009) Periodic thin films by glancing angle deposition, PhD thesis, University of Alberta.

[96] Harris, K.D., Westra, K.L., and Brett, M.J. (2001) Fabrication of perforated thin films with helical and chevron pore shapes. *Electrochemical and Solid-State Letters*, **4** (6), C39–C42.

[97] Harris, K.D., Sit, J.C., and Brett, M.J. (2002) Fabrication and optical characterization of template-constructed thin films with chiral nanostructure. *IEEE Transactions on Nanotechnology*, **1** (3), 122–128.

[98] Elias, A.L., Harris, K.D., and Brett, M.J. (2004) Fabrication of helically perforated gold, nickel, and polystyrene thin films. *Journal of Microelectromechanical Systems*, **13**, 808–813.

[99] Fernando, S.P., Elias, A.L., and Brett, M.J. (2006) Mechanical properties of helically perforated thin films. *Journal of Materials Research*, **21** (5), 1101–1105.

[100] Elias, A.L., Brett, M.J., Harris, K.D. *et al.* (2007) Three techniques for micropatterning liquid crystalline polymers. *Molecular Crystals and Liquid Crystals*, **477**, 137–151.

[101] Huang, Z., Harris, K.D., and Brett, M.J. (2009) Morphology control of nanotube arrays. *Advanced Materials*, **21**, 2983–2987.

7 Glancing Angle Deposition Systems and Hardware

7.1 INTRODUCTION

GLAD is based on PVD, a vacuum-based deposition technique commonly used in thin-film research and manufacturing. While PVD technology is very mature with widely available hardware and expertise, developing systems for GLAD use requires specialized hardware modifications. This chapter discusses several basic aspects of GLAD system design, beginning with vacuum requirements necessary for vapour collimation during deposition. Accurate film thickness monitoring is crucial for controlled fabrication, and we discuss methods to track deposition rate and calibrate thickness, explaining the geometry- and density-related subtleties in GLAD film fabrication. We also examine uniformity in depth, explaining how to calculate wafer-scale structure and property variations for different GLAD motions. This information is particularly important for systematic design and manufacturing. The chapter also discusses general hardware requirements for the substrate motion, examining tolerances for the α and φ positions, as well as substrate temperature considerations.

7.2 VACUUM CONDITIONS

7.2.1 Vacuum requirements for glancing angle deposition systems

PVD processes such as GLAD are typically performed in controlled, high-vacuum environments designed to both control the vaporization process (evaporation, sputtering) and limit undesirable contamination by residual gas species. Vacuum engineering is fundamental to many industrial applications and research areas, and several specialist guides cover the topic in great depth (e.g. Ref. [1]). The scope here is narrower, and we discuss a few topics of central importance in GLAD technology.

The behaviour of gases in vacuum environments is primarily understood through physical principles derived from kinetic gas theory. Gases are treated as a large number of constantly moving particles (atoms or molecules) within a closed container and the only interactions considered are Newtonian collisions occurring between particles and the container walls and between the particles themselves. Macroscopic physical properties, such as pressure and temperature, emerge via statistical averaging of the microscopic particle kinetics. During PVD processing, vapour species leaving the source travel in straight, linear trajectories until they either collide with another gas particle or with the vacuum chamber wall. For GLAD, an important quantity from kinetic gas theory is the mean-free-path ℓ, which is the average

Glancing Angle Deposition of Thin Films: Engineering the Nanoscale, First Edition.
Matthew M. Hawkeye, Michael T. Taschuk and Michael J. Brett.

distance a gas particle travels before colliding with another particle. The mean-free-path is determined by the gas temperature T and pressure P according to

$$\ell = \frac{kT}{\pi d^2 P \sqrt{2}}, \tag{7.1}$$

where d is the gas particle diameter and k is Boltzmann's constant. A related parameter in vacuum technology is the Knudsen number Kn, equal to the ratio of the gas mean-free-path to the characteristic dimension D of the vacuum chamber:

$$Kn = \frac{\ell}{D}. \tag{7.2}$$

The Knudsen number can be used to determine how gas flows in the deposition chamber. For Kn < 0.01, interparticle collisions dominate and the gas flow is viscous. Vacuum conditions satisfying this requirement are frequently termed rough vacuum. Kn > 1 corresponds to the molecular flow regime where interparticle collisions are rare and particles travel along linear, ballistic trajectories until colliding with the chamber walls or surfaces in the chamber (i.e. the substrate). Vacuum environments in the molecular flow regime are said to be at high vacuum (HV). For typical deposition hardware with dimensions on the order of a metre, the HV regime is achieved at pressures of around 10^{-5} Torr.

Another important quantity is the monolayer formation time τ_{ML}, equal to the characteristic time required to cover an initially bare surface with a monolayer of gas molecules. This is another important quantity in vacuum technology and is given by

$$\tau_{ML} = \frac{a\sqrt{2\pi mkT}}{P}, \tag{7.3}$$

where m is the gas particle mass and a is the monolayer molecular surface density (typically $\sim 10^{19}$ m^{-2}). HV and UHV environments are distinguished based on the monolayer formation time. UHV environments are typically held at pressures $<10^{-9}$ Torr where monolayer formation times can be measured in hours. Surface contamination concerns are therefore significantly reduced compared with experiments in typical HV conditions ($\sim 10^{-6}$ Torr), where a gas monolayer is adsorbed in a matter of seconds. However, UHV conditions are considerably more difficult and expensive to achieve than HV environments are. Reaching UHV requires additional equipment (such as ion pumps), longer pumping times with surface bakeout steps to remove adsorbed gases, and increased maintenance to combat real and virtual leaks. For successful GLAD, HV conditions are nearly always required to achieve ballistic vapour transport. UHV conditions are only used when extreme surface control is required.

Figure 7.1 shows ℓ and τ_{ML} at different pressure values calculated at room temperature ($T = 293$ K) for N$_2$ gas ($d = 3.75$ Å, $m = 4.65 \times 10^{-26}$ kg). From atmospheric conditions down to pressures $<10^{-11}$ Torr, all environments accessible with modern vacuum technology, the mean-free-path varies drastically, ranging from micrometre scales at rough vacuum to intercontinental distances in UHV systems.

GLAD technology operates in the molecular flow regime where the vapour transport is ballistic and the evaporant travels in straight lines from source to substrate. Under these operating conditions the vapour flux arriving at the substrate is highly collimated; that is,

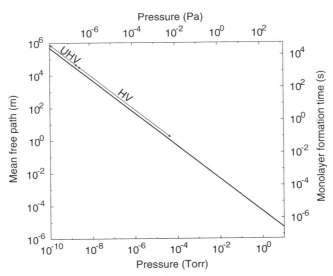

Figure 7.1   The gas mean-free-path ℓ (Equation 7.1) and monolayer formation time t_{ML} (Equation 7.3) as a function of chamber pressure, calculated at room temperature and for N_2 gas. Two pressure ranges of interest are labelled on the graph. The HV range corresponds to $\ell > 1$ m, which will produce molecular flow conditions in an ~1 m sized vacuum chamber. GLAD processes are nearly always performed in HV environments. The ultrahigh vacuum (UHV) range marks the point where $t_{ML} > 1$ h, which provides a practical definition of UHV conditions. In such an environment, requiring pressures $<10^{-9}$ Torr (10^{-7} Pa), surface contamination is greatly reduced, ℓ is extremely long and interparticle collisions are exceedingly rare.

the incident particle trajectories are parallel. Any interparticle collisions occurring between source and substrate randomize the vapour trajectories, thereby increasing the angular divergence of the flux and leading to poor shadow definition. During a GLAD process, the throw distance should therefore be no shorter than the mean-free-path at the deposition pressure. This is a basic operating requirement that should be observed when designing vacuum systems for GLAD use. Improving the vapour collimation by decreasing the operating pressure generally leads to increased microstructural definition. However, while ℓ can be continually increased by reducing the pressure, the vapour-source geometry ultimately limits the degree of collimation (Section 7.3.3).

7.2.2 Physical vapour deposition process gases and higher pressure deposition

Although many PVD processes can be performed at very low pressure where mean-free-path concerns are negligible, there are also many common processes that require higher pressures to function. Prominent examples of such processes include reactive evaporation, where evaporation is performed with a partial pressure of reactive gas in the chamber to promote stoichiometric deposition of compounds at the substrate, and sputter deposition, where the process gas is ionized to form the sputtering plasma. These PVD techniques

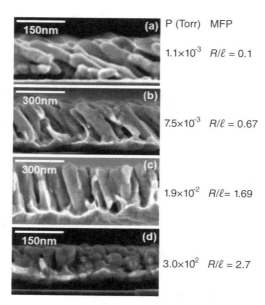

Figure 7.2 Effect of Ar operating pressure on film microstructure in GLAD films prepared by sputtering at $\alpha = 85°$. As the pressure P increases, so does the ratio of the throw distance R to the mean-free-path (MFP; ℓ), producing an increasingly isotropic and nondirectional vapour flux. Microstructural control is lost as this ratio approaches and exceeds unity, producing a typical, normal-incidence column structure. When depositing at raised pressures (e.g. during sputtering or reactive evaporation), care should be taken to ensure sufficient vapour collimation for the GLAD process. Reprinted with permission from [2]. © 2010, AIP Publishing LLC.

require the desired gas species (e.g. Ar, O_2) be controllably introduced to the deposition chamber after pumpdown and maintained at a sufficient operating pressure, which can range from 10^{-4} to 10^{-1} Torr. While the gas pressure is already an important processing variable in these techniques, the pressure takes on added significance when developing GLAD processes as the decreased mean-free-path length influences the resulting film microstructure (and hence properties). Figure 7.2 presents a series of SEM images of Au GLAD films deposited at $\alpha = 85°$ via magnetron sputtering at different Ar pressures [2]. At low pressure (1.1×10^{-3} Torr), ℓ is significantly greater than the target-to-substrate throw distance R ($R/\ell = 0.1$), producing the characteristic tilted-column microstructure. In this low-pressure regime, the degree of collimation of the sputtered vapour is sufficiently high to yield controlled shadowing conditions. The effect of decreased collimation associated with shorter ℓ quickly becomes apparent as the Ar pressure is increased. As R/ℓ increases, a greater portion of the flux is scattered en route to the substrate, leading to an increasingly isotropic vapour. At 7.5×10^{-3} Torr, the column tilt angle is significantly reduced compared with the low-pressure case due to the loss in vapour directionality. Further pressure increases cause the columns to orient along the substrate normal as the lack of vapour directionality effectively reproduces an isotropic deposition geometry.

When developing GLAD-based reactive evaporation or sputtering processes, therefore, it is important to consider the necessary working pressure and the corresponding mean-free-path length and how the resulting film microstructure is affected. In sputtering, low-pressure conditions are generally preferred to minimize flux scattering and maintain flux directionality over a long throw distance [3]. Other strategies can be used to improve GLAD film microstructure, such as adding collimating baffles [4]. In lieu of reactive evaporation at elevated partial pressures of the reactive gas, post-deposition treatments such as oxidation (Section 6.2.2) could be performed to correct or modify the film composition.

7.3 THICKNESS CALIBRATION AND DEPOSITION RATE MONITORING

Accurate thickness control is critical to repeatable fabrication of GLAD nanostructures, being necessary, for example, to realize nanocolumn arms of specified lengths and fabricate helices with precise pitch values. Furthermore, as GLAD substrate motion algorithms are generally specified in terms of thickness, the measured deposition rate can be incorporated into a feedback loop to improve the substrate motion accuracy. Reliable thickness control in GLAD fabrication requires both accurate rate monitoring and proper calibration.

While there are many technologies available for thickness monitoring in thin-film deposition [5], quartz crystal microbalance (QCM)-based devices are most commonly used for real-time deposition monitoring in both PVD and GLAD processes. QCMs are based on piezoeletrically driven resonant oscillation of thin quartz discs. As material deposits on the crystal face, the mass loading shifts the resonance frequency, which is tracked electrically. By calibrating with the material density and acoustic impedance, the data are converted to total thickness and a deposition rate. Optical methods, such as spectroscopic ellipsometry, are also commonly used to monitor film deposition, particularly in optical coating applications. With this approach, the optical properties of the growing film are measured during growth and the corresponding thickness is calculated from optical models. Determining accurate thickness values relies on the accuracy of the optical constants and theoretical models employed. When properly implemented, both QCM and optical approaches can achieve exceptional precision and provide sub-nanometre growth resolution.

7.3.1 Source directionality and tooling factor

Ideally, the rate monitoring element (i.e. the QCM head or the optically probed surface) should be as close as possible to the fabricated samples. This is often difficult to achieve in practice, especially for GLAD, where the substrate is in motion. Off-axis probing, where the rate-monitoring element is placed at a different location in the vacuum chamber, is more easily implemented and thus more commonly seen in GLAD systems.

However, off-axis probing inevitably leads to differences between the deposition rates at the substrate and rate monitor, and this difference must be calibrated. The mass flux M (mass per unit area per unit time) emitted from an evaporation source is often modelled as [6]

$$M = \frac{M_0(n+1)\cos^n \theta \cos \alpha}{2\pi R^2}, \tag{7.4}$$

where θ is the angle measured from the source normal and R is the distance from the source (the throw distance). The $\cos \alpha$ factor represents the geometric flux reduction due to

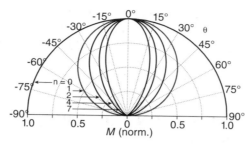

Figure 7.3   Calculated flux distribution for different values of the cosine exponent n and a normally incident flux. For larger values of n, the flux becomes increasingly directional.

the surface being tilted to an angle α; this factor is key to GLAD and is further discussed in Section 7.3.2. The coefficient n determines the directionality of the vapour flux (see Figure 7.3). When $n = 0$, Equation 7.4 describes evaporation from a point source, where the flux is isotropic. When $n = 1$, the vapour flux is directional and follows a cosine distribution typical of a surface source: emission is maximized along $\theta = 0$ and zero at $\theta = 90°$. Larger n values describe increased flux directionality. In practice, n is physically related to the geometry of the source and can be calibrated by measuring the deposition rate at different θ values and then fitting Equation 7.4 to the data via n. The vapour flux distributions associated with sputter deposition can be significantly more complex and are sensitive to many variables, including target and magnet configuration, working pressure and gas composition, material, and so on. Advanced modelling of the deposition environment, using computational tools such as SIMTRA [7], are thus required to deal with specific cases, and we therefore limit further discussions to the analytic model provided by Equation 7.4. However, Equation 7.4 can often be used to approximate long-throw sputtering process, particularly when augmented with empirical calibration.

The ratio between the mass flux rate at the substrate centre M_s and the rate monitor M_{RM} is given by

$$\frac{M_s}{M_{RM}} = \frac{R_{RM}^2 \cos^n \theta_0}{R_0^2 \cos^n \theta_{RM}} \frac{\cos \alpha}{\cos \alpha_{RM}} = F_0 \cos \alpha. \tag{7.5}$$

Here, θ_0 and R_0 describe the location of the substrate centre, θ_{RM} and R_{RM} specify the rate monitor position, and α and α_{RM} respectively define the orientation of the substrate and rate monitor surfaces with respect to the incident vapour. Equation 7.5 also defines the tooling factor F_0, a constant equal to the ratio between the substrate and rate monitor deposition rates when $\alpha = 0°$. Two points should be made clear here. First, variations in the substrate thickness introduce a small correction to θ_0 and R_0. Second, we have assumed point-like locations, whereas in reality the deposition geometry (R and θ) varies over the substrate and rate monitor. Rate monitors are typically made small to minimize these effects. However, substrate uniformity issues are an important topic and are discussed at length in Section 7.4.1; for now, we focus on the centre of the substrate only.

To determine F_0 experimentally, first deposit a film at $\alpha = 0°$ and then measure the thickness at the substrate centre. The ratio of this thickness to the thickness measured by the rate monitor gives F_0. When R_0, θ_0, R_{RM} and θ_{RM} are known, and assuming Equation 7.4

accurately represents the flux distribution, this measurement also yields the vapour directionality n. To ensure accurate and reproducible results, F_0 should be calibrated for specific sets of processing conditions (e.g. material, deposition pressure, substrate temperature) and checked regularly for variation.

7.3.2 Thickness calibration at nonzero α: deposition ratios

With GLAD, depositing onto tilted substrates presents additional subtleties in thickness calibration. The first is simply geometric: when tilted to an angle α, the mass flux deposited on the substrate is reduced by a factor equal to $\cos \alpha$ and the deposited film thickness will be reduced correspondingly. A second complication is related to the reduced, α-dependent density (Section 4.4) of obliquely deposited films. Because the GLAD film density is decreased by a factor ρ_n, the GLAD film thickness is correspondingly greater than expected. This affects both QCM and optical thickness monitoring efforts. The QCM measures the accumulated mass and converts this value to a film thickness and deposition rate by assuming a particular bulk density value. As the substrate α varies and the film density changes, the QCM reading will therefore be systematically incorrect. Similarly, the reduced film density also complicates growth monitoring with optical methods. When indirect monitoring is used and a separate, non-tilted witness substrate is probed, the GLAD film will be thicker than expected. Where direct monitoring is used and the GLAD film itself is probed, the reduced density impacts the optical constants as discussed in Sections 5.1.2 and 5.3. Optical models used to extract film thickness must be correspondingly modified.

The mass fluxes in Equation 7.5 can be expressed relative to their respective densities and thicknesses as $M_{RM} = \rho_b t_{RM}$ and $M_s = \rho_n \rho_b t_s$. Here, ρ_b is the bulk density assumed by the rate monitor, ρ_n is the density of the GLAD film normalized to ρ_b, and t_{RM} and t_s are the film thicknesses at the rate monitor and the substrate. Inserting these into Equation 7.5 and adding the geometric $\cos \alpha$ factor gives

$$D_0 = \frac{t_s}{t_{RM}} = \frac{F_0 \cos \alpha}{\rho_n(\alpha_0)}. \tag{7.6}$$

This equation defines the deposition ratio D_0, an α-dependent calibration factor for converting a thickness measured at the rate monitor to a thickness deposited at the substrate. It is also the ratio between the instantaneous vertical growth rate Γ at the substrate and monitoring element; that is:

$$D_0 = \frac{\Gamma_0}{\Gamma_{RM}}, \tag{7.7}$$

where Γ is in units of deposited thickness per unit time (e.g. ångströms per second).

D_0 is a critical quantity for accurate deposition, combining the geometric reduction in incident flux due to angled deposition, the tooling factor and the reduced density characteristic of oblique deposition. These latter two parameters, represented by F_0 and ρ_n, depend on several processing parameters, including material, substrate temperature, deposition pressure, as well as α and φ. Consequently, calibrating D_0 is an important experimental matter realized by performing a series of depositions at different α, measuring the deposited film

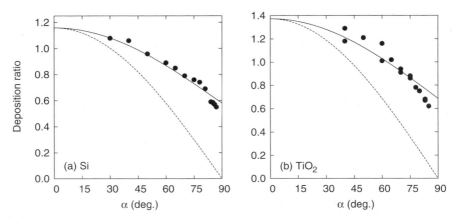

Figure 7.4 Experimentally measured deposition ratios for (a) Si (unpublished) and (b) TiO_2 [8], prepared using different deposition systems. The deposited film thicknesses are greater than the geometric $\cos \alpha$ factor (dashed line) due to the reduced density of the obliquely deposited films. Incorporating a basic density model (Equation 7.6) provides a better prediction of the result, as shown by the solid lines. Better process modelling can be performed by using fittable density expressions to capture material- and deposition-condition dependencies [9].

thickness and comparing the results with the monitoring element thickness. It is also important to periodically recalibrate the deposition ratio to detect any changes caused by drifting experimental conditions. Figure 7.4 shows measured deposition ratios for Si and TiO_2 GLAD films fabricated on different deposition systems. In both materials, the deposition ratio is found to decrease with α, indicating that the film deposition is slower at higher α. However, the deposition ratio remains larger than the $\cos \alpha$ curve (dashed line) because of the reduced film density. The solid line shows Equation 7.6 numerically fit to the data sets, using F_0 as a fitting parameter, and modelling the film density using Equation 4.11. As can be seen, this single-parameter equation provides a reasonable description of the data. However, we emphasize that ρ_n is sensitive to many experimental factors and also changes with film thickness, as discussed in Section 4.4. More complex density expressions, such as Equation 4.12, can be used to improve the data modelling, providing additional parameters to better capture processing differences.

Another effect recently discussed by Buzea *et al.* [10] is an α dependence of the sticking coefficient S, which is the probability that arriving vapour particles remain bound to the substrate. From analysis of molecular dynamics simulations, S has been empirically found to vary with α following

$$S(\alpha) = 1 - \exp\left[-\frac{(\alpha - \alpha_0)^2 + y^2}{\mu^2}\right], \tag{7.8}$$

where y is a nondimensional factor inversely proportional to incident particle energy, μ parameterizes the curvature onset at low α, and α_0 captures the angle where S is minimized (typically near 65–70°). However, this effect only becomes important when the incident particle energy is high (>15 eV), whereas typical particle energies are between 0.1 and

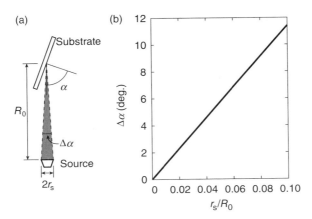

Figure 7.5 (a) Points on the substrate receive vapour flux from a nonzero spread of angles related to the extended source. The collimation of the incoming vapour flux is consequently limited by the deposition system geometry. (b) Reducing the source size r_s and increasing the throw distances R_0 reduces the angular spread $\Delta\alpha$ and improves the resulting vapour collimation.

0.2 eV in evaporation and between 3 and 10 eV in sputtering [6]. The sticking coefficient is therefore effectively equal to unity except in particularly high energy processes.

7.3.3 Extended source: effect on collimation

In addition to affecting the directionality, the source geometry also impacts the collimation of the emitted vapour. Vapour emitted from different points on the source arrives at the substrate along slightly different line-of-sight trajectories (Figure 7.5). The substrate thus receives vapour flux at primary angle α but with a spread $\Delta\alpha$, leading to a source geometry-dependent reduction in the vapour collimation. (The substrate φ angle is similarly affected.) If we consider a circular surface source with radius r_s, the point on the substrate directly above the source receives flux from a cone-shaped volume with an angle

$$\Delta\alpha = 2\arctan\left(\frac{r_s}{R_0}\right). \tag{7.9}$$

This equation is plotted in Figure 7.5, showing how $\Delta\alpha$ scales with r_s/R_0. For context, a typical 7 cm^3 crucible liner for e-beam evaporation has an opening of approximately $r_s = 1.25$ cm (0.49 in.). Assuming the melt spans this opening (in practice it is usually smaller), achieving $\Delta\alpha = 2°$ requires a 71.6 cm throw distance. Note that this is a simple example, whereas in practice the effective source size depends strongly on the PVD method and source configuration (e.g. racetrack geometry in magnetron sputtering). Reducing the source size improves the vapour collimation, generally at the expense of a reduced deposition rate. The collimation of an extended source can also be improved by using baffles or louvres to restrict line-of-sight paths between the source and substrate. Also note that the collimation will vary over the substrate based on the exact geometry and that substrate motion complicates the matter further. This variation will generally be quite small and important only in high-precision work.

7.4 UNIFORMITY CALCULATIONS FOR GLANCING ANGLE DEPOSITION PROCESSES

Uniformity is a key concern in thin-film deposition and can strongly impact experimental measurements, quality of fabricated samples and the manufacturability of specific processes. Variations in film properties arise due to the change in deposition geometry across the substrate, and given the range of α and φ motions employed in GLAD technology the general prediction of uniformity patterns and property distributions becomes a computational task. The following section examines film uniformity calculations for GLAD processes, the results of which can be used to estimate uncertainties in film properties, compare different substrate motions, and develop substrate placement and sampling methodologies in GLAD systems.

7.4.1 Calculating geometry variation over a wafer

We begin by first examining the case of a stationary tilted substrate. Analysing this configuration provides the geometric framework, which is then extended to the more typical GLAD situation of dynamic substrate motion. The following analysis is based heavily on the work of Wakefield and Sit, who performed a detailed, general analysis of deposition uniformity during GLAD that was well supported by experimental data [11].

The deposition geometry is depicted in Figure 7.6, showing a prototypical GLAD arrangement. The substrate is mounted above the source S on a substrate holder with an origin O. (In the following we assume a point-like source, thus neglecting geometry variation due to an extended source.) Two coordinate sets are thus defined: the XYZ reference frame is centred at S with the Z-axis oriented along the source surface-normal, and the xyz reference frame is

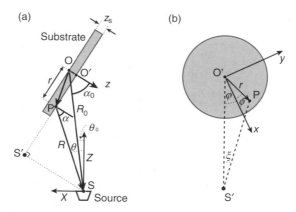

Figure 7.6 Diagram showing the deposition geometry with points and quantities of interest identified. The parameters are defined in the text. Reprinted with permission from [11]. © 2011, AIP Publishing LLC.

centred at O with the z-axis oriented along the substrate surface-normal. In the XYZ frame, O is located at a distance R_0 from the origin, with an azimuthal angle θ_0 and a polar angle Θ_0 (not shown). During GLAD α and φ rotations, only the orientation of the xyz frame is altered; the XYZ and the location of O remain fixed.

The substrate origin O' is offset from O by the substrate thickness z_s. For $z_s \ll R_0$, as is typical, the effect of substrate thickness on the deposition geometry is minimal. However, we include it here for completeness, as it may be useful for analysing certain substrates or mounting geometries. The vector $\mathbf{R}_{OS}$ pointing from O to S defines a number of important parameters: the nominal throw distance R_0 is the length of $\mathbf{R}_{OS}$; the nominal deposition angle α_0 is the angle between $\mathbf{R}_{OS}$ and the substrate normal ($\hat{\mathbf{z}}$); and the nominal vapour plume angle θ_0 is the angle between $-\mathbf{R}_{OS}$ and the source normal. Together, these nominal parameters define the deposition geometry at the substrate origin. Point P corresponds to a position of interest on the substrate, located at a distance r from the substrate origin. The substrate coordinates are indicated in Figure 7.6b, showing the view of the substrate looking down the substrate normal (z-axis). The point P can be specified by the (x, y) or (r, ϕ) coordinates. The GLAD rotation angle φ rotates the x- and y-axes. A second vector $\mathbf{R}_{PS}$ pointing from P to S is thus defined, creating a set of local parameters R, α and θ that depend on the location of P.

When calculating the uniformity, we are interested in how these local parameters vary across the substrate, particularly in reference to the nominal parameters. The geometry is specified by the vector equation (expressed in the substrate xyz-coordinate frame)

$$\mathbf{R}_{PS} = \mathbf{R}_{OS} - \mathbf{R}_{OP}$$

$$= \begin{pmatrix} -r \cos \alpha \cos(\varphi + \phi) + z_s \sin \alpha \\ -r \sin(\varphi + \phi) \\ R_0 - r \sin \alpha \cos(\varphi + \phi) - z_s \cos \alpha \end{pmatrix}, \tag{7.10}$$

where $\mathbf{R}_{OP}$ is the vector pointing from O to P. The local throw distance is given by the magnitude $|\mathbf{R}_{PS}| = R$, found from

$$\frac{R}{R_0} = \sqrt{1 + \gamma_1^2 + \gamma_2^2 - 2\gamma_1 \sin \alpha_0 \cos(\varphi + \phi) - 2\gamma_2 \cos \alpha_0}, \tag{7.11}$$

where $\gamma_1 = r/R_0$ and $\gamma_2 = z_s/R_0$. The local deposition angle α is found from $\mathbf{R}_{PS} \cdot \hat{\mathbf{z}}$, yielding

$$\cos \alpha = \frac{\cos \alpha_0 - \gamma_2}{R/R_0}. \tag{7.12}$$

The vapour plume angle is determined by $-\mathbf{R}_{PS} \cdot \hat{\mathbf{Z}}$, leading to

$$\cos \theta = \frac{\cos \Theta_0}{R/R_0} [\gamma_1 \cos(\varphi + \phi) \sin(\alpha_0 + \theta_0) + \gamma_1 \sin(\varphi + \phi) \tan \Theta_0$$
$$+ \gamma_2 \cos(\alpha_0 + \theta_0) - \cos \theta_0]. \tag{7.13}$$

When O is centred above the source (i.e. lies on the Z-axis), $\Theta_0 = 0$ and Equation 7.13 simplifies to

$$\cos\theta = \frac{1}{R/R_0}[\gamma_1\cos(\varphi + \phi)\sin\alpha_0 + \gamma_2\cos\alpha_0 - 1]. \tag{7.14}$$

Because the column growth direction lies in the deposition plane (the plane containing the substrate normal $\hat{z}$ and the incident flux trajectory defined by $\mathbf{R}_{PS}$), it is important to know how the orientation of this plane varies over the substrate. This information allows deviations in the column orientation across the wafer to be calculated. As above, the nominal and local deposition planes refer to the deposition planes at O' and at P, which are indicated by the dashed lines $\overline{O'S'}$ and $\overline{S'P}$ respectively (Figure 7.6b). The angle ξ corresponds to the angle between the nominal and local deposition planes, given by

$$\sin\xi = -\frac{\gamma_1\sin(\varphi + \phi)}{\sqrt{\sin^2\alpha_0 + \gamma_1^2 - 2\gamma_1\sin\alpha_0\cos(\varphi + \phi)}}. \tag{7.15}$$

The sign of ξ depends on $\varphi + \phi$, with $0 \leq \varphi + \phi < 180$ producing negative ξ values and $180 \leq \varphi + \phi < 360$ producing positive ξ values.

As an example, Figure 7.7 shows the uniformity of the four geometric parameters R/R_0, α, θ and ξ calculated for $\alpha_0 = 80°$, with the substrate centred over the source ($\Theta_0 = \theta_0 = 0°$). Note that the x- and y-axes are presented as shown in Figure 7.6b with $\varphi = 0$ and that the x and y values are normalized to R_0. For context, the dashed line on each plot marks the outline for a circular wafer with a radius equal to one-tenth the nominal throw distance. This would correspond to a standard 100 mm (4 in.) diameter wafer mounted in a deposition system with a 50 cm throw distance, a representative situation in a typical GLAD vacuum system. The local deposition geometry varies significantly over such a wafer. The throw distance varies by as much 20% along the substrate x-axis yet varies minimally along the y-axis. The deposition angle shows a similar pattern, changing by 2° along the x-axis and negligibly in the y-direction. The x and y variations are reversed for the vapour plume angle and the deposition plane angle. The vapour plume angle is zero at the substrate centre and quickly increases when moving along the y-axis, nearly reaching 6° at the border of the indicated substrate. The deposition plane angle also varies significantly along the y-axis, ranging from $-6°$ to $+6°$ across the wafer. The variation in deposition geometry produces a dramatic effect on film thickness uniformity, especially when the substrate is in motion. These effects are considered in the following sections.

7.4.2 Mapping out thickness variation

The geometric parameters can be combined with the deposition ratio (Section 7.3.2) to calculate the thickness variation over the substrate. As above, we distinguish between the nominal deposition ratio D_0, the deposition ratio at O' (given by Equation 7.6), and the local deposition ratio D given by

$$D = \frac{F\cos\alpha}{\rho_n(\alpha)}, \tag{7.16}$$

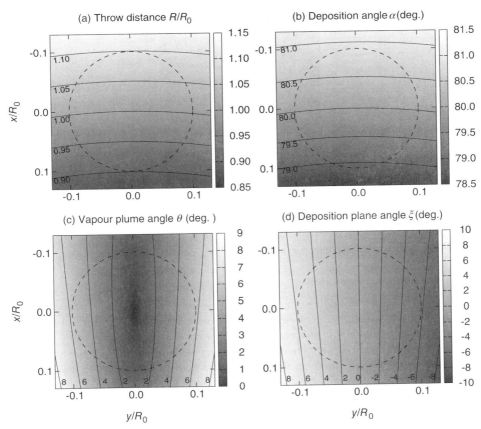

Figure 7.7 Variation in the deposition geometry over the substrate, calculated for a nominal deposition angle $\alpha_0 = 80°$. The x- and y-coordinates are normalized to R_0 and the black dashed line marks the size of a 100 mm diameter wafer in a system placed 50 cm from the source.

where F is the local tooling factor:

$$F = \frac{R_{RM}^2 \cos^n \theta}{R^2 \cos^n \theta_{RM}}. \tag{7.17}$$

When possible, other models of the vapour flux distribution, based on simulation or experiment, can be used to construct a more appropriate expression describing specific system geometries and PVD methods (e.g. sputtering). The spatial variation of D is determined by the variation in R, α and θ as calculated from Equations 7.11, 7.12 and 7.13. However, determining D also requires specifying $\rho_n(\alpha)$. This parameter is primarily determined by α, but is also sensitive to many other experimental variables, including material, substrate temperature, substrate rotation speeds, and so on. Accurate prediction of thickness distributions, therefore, requires experimental calibration of density (Section 4.4). For the purposes of this discussion we model $\rho_n(\alpha)$ via Equation 4.11.

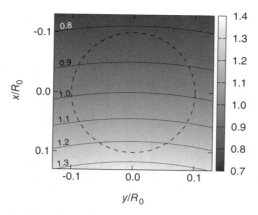

Figure 7.8 Variation in the deposition ratio (normalized to D_0) over the substrate, calculated for a nominal deposition angle $\alpha_0 = 80°$. The x- and y-coordinates are normalized to R_0 and the black dashed line marks the size of a 100 mm diameter wafer in a system placed 50 cm from the source.

Figure 7.8 shows Equation 7.16, after D has been normalized to D_0, plotted for a stationary substrate tilted to $\alpha_0 = 80°$ and assuming a typical cosine flux distrubution ($n = 1$). Over a typical wafer size (dashed line denoting 100 mm diameter wafer at $R_0 = 50$ cm, as before), a significant thickness gradient is produced under these conditions, exceeding 40% along the x-direction due to the large variation in the throw distance across the wafer. In the y-direction, the thickness is significantly more uniform, with a less than 2% variation over the indicated wafer. The thickness uniformity pattern depends strongly on the deposition geometry and substrate motion algorithm, as investigated in the following sections.

7.4.3 Calculating parameter variations for moving substrates

The previous discussions have examined how the deposition geometry varies over a stationary, tilted substrate. However, this is clearly a limiting case, as during the majority of GLAD algorithms the substrate undergoes a set of complex motions, specified by $\alpha(t_0)$ and $\varphi(t_0)$, where t_0 is the nominal thickness deposited at O', related to t_{RM} via Equation 7.6. The set of parameters R, α, θ, ξ and D can thus be viewed as instantaneous quantities describing the local deposition geometry at a particular growth interval dt_0. Given a specific substrate motion, Equations 7.11–7.15 can be used to calculate the deposition geometry over the wafer at any point in the deposition.

Figure 7.9 illustrates this for the case of constant φ rotation at a fixed $\alpha_0 = 85°$. Figure 7.9a shows the substrate with four points of interest highlighted. The origin is at O', as before. Points A, B and C lie along the x-axis at distances of $r = 0.05R_0$, $0.1R_0$ and $0.15R_0$ respectively. When the substrate is rotated in φ during the deposition, A, B and C rotate around the origin following the indicated paths. Points away from the x-axis will trace similar trajectories, although with a different phase. Figure 7.9a and b shows the specific $\alpha(t_0)$ and $\varphi(t_0)$ motion algorithm, with the deposition angle fixed at $\alpha_0 = 85°$ while φ is increased linearly, completing one revolution every 500 nm. As described in Section 2.5.1, such a substrate motion produces helical columnar structures with a 500 nm pitch. Figure 7.9d,e

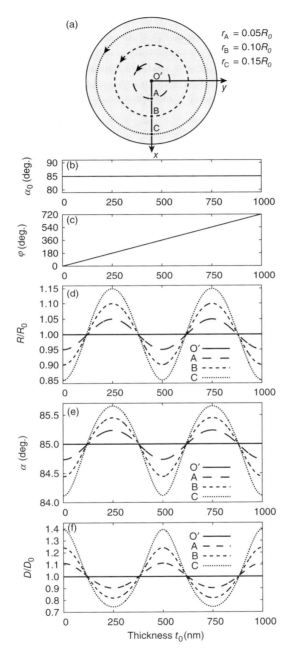

Figure 7.9   Variation in deposition geometry during substrate rotation. (a) An illustration of the substrate geometry with four points labelled: O′ is the substrate origin, while A, B and C are points on the x-axis at a distance $r = 0.05R_0$, $0.1R_0$ and $0.15R_0$ from O′. For the (b) α_0 and (c) φ substrate motions presented, the variation in (d) throw distance R/R_0 (Equation 7.11), (e) deposition angle α (Equation 7.12) and (f) deposition ratio (Equation 7.16 with $n = 1$) are calculated at the indicated substrate positions. During the φ rotation, each position traces different spatial trajectories and experiences different variations of the deposition geometry, leading to complex nonuniformity in, for example, deposited thickness and α-dependent properties.

and f respectively presents the variation in the R, α and D parameters traced out over the course of the substrate motion at the four substrate points, O$'$, A, B and C. As expected, the three parameters are constant at O$'$ because of its central location, whereas points A, B and C map out a complex nonuniformity over each φ rotation cycle. The variation in R, α and D is cyclical, with the magnitude of the parameter variation increasing with r. However, the variations are not sinusoidal, and several distortions can be observed:

- The parameter variation is asymmetric, with the minimum and maximum parameters values not equidistant from the O$'$ values. For example, the minimum and maximum α values along the point C trajectories are 84.1° and 85.7° respectively. This asymmetry is most pronounced for the deposition ratio, where the point C minimum and maximum values are $D/D_0 = 0.74$ and 1.39 respectively.
- The curvature of the contour lines in Figure 7.7a and b and Figure 7.8 slightly skew the duty cycle of the parameter variation about the O$'$ point. The A, B and C points spend a greater portion of every cycle at $R/R_0 > 1$, $\alpha > 85°$ and $D/D_0 < 1$.
- The magnitude of these distortions becomes more pronounced for points farther from O$'$, generally with a nonlinear r dependence.

The deposition geometry variation has several implications on the properties of the film. Away from the origin, there will be a progressively worse deviation between the nominal structure design and the actual fabricated structure. This is particularly important in the extreme shadowing regime as small α perturbations create large changes in the shadowing length and the resulting film properties. Furthermore, variations in R, α and θ cause complex deposition rate fluctuations, as shown in Figure 7.9f. Although not shown, ξ changes during the φ rotation lead to column orientations that vary with thickness. The exact parameter variations and the corresponding effects on microstructure depend strongly on the specific $\alpha(t_0)$ and $\varphi(t_0)$ motions used. The preceding discussion provides methods for quantitative evaluation of deposition nonuniformity over the course of a specific motion algorithm.

7.4.4 Calculating thickness uniformity for moving substrates

Determining the total deposited thickness t_{total} requires integrating the instantaneous quantities at each point of the wafer over the course of the deposition. The instantaneous thickness deposited at point P is equal to the thickness deposited at the substrate centre dt_0 multiplied by the ratio of the deposition rates at the two points; that is, $dt = \Gamma/\Gamma_0\, dt_0$ or equivalently $dt = D/D_0\, dt_0$. An expression for D/D_0 is obtained by combining Equations 7.16 and 7.17 with Equations 7.5 and 7.6, giving

$$\frac{D}{D_0} = \frac{R_0^2 \rho_{\text{n}}(\alpha_0) \cos^n \theta \cos \alpha}{R^2 \rho_{\text{n}}(\alpha) \cos^n \theta_0 \cos \alpha_0}. \tag{7.18}$$

Several of the parameters in this expression vary with $\alpha_0(t_0)$ and $\varphi(t_0)$. The variations in R, α and θ are calculated using Equations 7.11, 7.12 and 7.13 respectively. The two density parameters $\rho_{\text{n}}(\alpha)$ and $\rho_{\text{n}}(\alpha_0)$ must also be determined, either through empirical calibration or using a theoretical model. In the following, Equation 4.11 is used to calculate the α-dependent

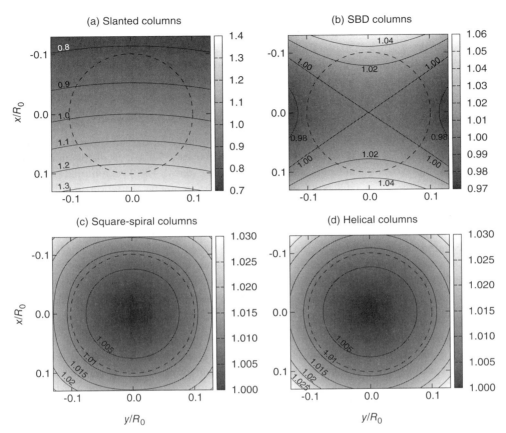

Figure 7.10   Thickness uniformity patterns calculated for GLAD film microstructures consisting of (a) slanted columns (fixed φ), (b) SBD columns (discrete $180°$ φ rotations), (c) square-spiral columns (discrete $90°$ φ rotations) and (d) helical columns (continuous φ rotation). The specific thickness uniformity pattern is strongly determined by the particular φ motion, with the uniformity improving as the symmetry of the φ motion increases. In all four cases α was fixed at $85°$ and the film thickness is normalized to the substrate centre thickness. The x- and y-coordinates are normalized to R_0 and the black dashed line marks the size of a 100 mm diameter wafer in a system placed 50 cm from the source.

density, and a cosine vapour flux distribution ($n = 1$) is assumed. The final thickness at P is calculated by integrating

$$t_f = \int_0^{t_f} dt = \int_0^{t_{0,f}} \frac{D(t_0)}{D_0(t_0)} dt_0 \tag{7.19}$$

for the specified $\alpha_0(t_0)$ and $\varphi(t_0)$ motions, where $t_{0,f}$ is the final thickness at O'.

Figure 7.10 shows the calculated thickness distributions (normalized to the thickness at $x = 0$, $y = 0$) for four different substrate motions. Figure 7.10a is for a slanted post microstructure, where α_0 is fixed at $85°$ and $\varphi = 0$ for the entire deposition. Figure 7.10b is for an SBD process (Figure 5.6) where α_0 is fixed at $85°$ and φ is rotated between $0°$

and 180° after every growth interval $p = 5$ nm. Figure 7.10c is for a square-spiral column motion algorithm (Figure 2.14), wherein α_0 is fixed at 85° and φ is rotated by 90° after every growth interval $p = 250$ nm. Finally, Figure 7.10d presents results for a helical column microstructure (Figure 2.12), with α_0 fixed at 85° and φ increased linearly to complete one revolution every 500 nm (i.e. a 500 nm helical growth pitch). In each motion, the final thickness at O′ was 1000 nm. As in previous graphics, the x and y coordinates are normalized to R_0 and the dashed black line denotes a 100 mm diameter wafer at $R_0 = 50$ cm for context. Note that the greyscale in each plot is different.

The spatial distribution of the thickness depends strongly on the substrate motion. As discussed Section 7.4.3, the slanted column motion (Figure 7.10a) produces significant thickness variation over the indicated substrate along the x-axis (>40%) yet minimal variation along the y-axis (<2%). This is a consequence of the geometry variation in this static configuration, and introducing φ rotation dramatically changes the uniformity properties. By symmetrically depositing along the positive and negative x-directions with the SBD algorithm (Figure 7.10b), much of the geometry variation is averaged out and the thickness uniformity is greatly improved. Over the indicated substrate, the x- and y thickness variations are 1.5% and 3.3% respectively. Furthermore, the uniformity pattern is significantly changed, forming a saddle-shaped surface: moving away from the centre point, the thickness decreases along the x-axis and increases along the y-axis. These changes can all be reconciled by averaging the geometry distributions in a $\varphi = 0°$ and 180° configuration. The uniformity pattern changes again using the square-spiral motion (Figure 7.10c). The thickness uniformity is better than 1% across the indicated substrate, a direct consequence of the extra rotation-induced geometry averaging. Note that the contour lines are not perfectly circular but are actually squared-off due to the fourfold symmetry of the square-spiral motion algorithm. Circular contour lines are produced by moving to the helical column motion (Figure 7.10d), where the incident flux arrives equally from all φ directions. The rotational averaging also produces high thickness uniformity with a less than 1% thickness variation over the indicated wafer. Thickness uniformity maps such as these can be useful in sample positioning during fabrication and the development of characterization methodologies. To improve sample-to-sample uniformity, it is recommended to arrange samples following the contour lines of the particular motion algorithm employed. Post-deposition characterizations should compare samples from similar contour levels in order to minimize thickness differences and reduce unintended experimental variation.

7.4.5 Calculating column orientation uniformity

Besides the thickness variation, another important parameter is the column growth direction, which is biased towards the position of the source. Across the substrate, this direction is determined by the local α, which determines the column tilt β, and ξ, which quantifies the local orientation of the deposition plane. The instantaneous column growth direction depends on these two parameters and the substrate φ rotation and is specified by the vector (in xyz-coordinates)

$$\mathbf{q} = \begin{pmatrix} \sin\beta\cos(\xi - \phi) \\ \sin\beta\sin(\xi - \phi) \\ \cos\beta \end{pmatrix}. \tag{7.20}$$

Evaluating this expression at a given substrate point and for specific α and φ parameters requires knowing the relationship between β and α. This relationship is process specific and can either be calibrated experimentally or estimated from the models discussed in Section 2.4.2. The following discussion uses Equation 2.2 to provide β estimates, whereas experimental data and/or empirical $\alpha–\beta$ relationships could be used in practice. The ξ parameter varies over the wafer following Equation 7.15. For depositions with stationary substrates (fixed α and φ), the net column orientation is given by Equation 7.20. In the dynamic substrate case, the net column orientation is calculated by integrating the instantaneous orientation over the entire substrate motion. As with the thickness calculation, the nonuniform growth rate must be accounted for by adding the D/D_0 term. Integrating Equation 7.20 over the deposited thickness as before leads to

$$\mathbf{q}_{net} = \int_0^{t_f} \mathbf{q}\, dt = \int_0^{t_{0,f}} \frac{D(t_0)}{D_0(t_0)} \mathbf{q}\, dt_0. \tag{7.21}$$

After calculation, $\mathbf{q}_{net}$ can be written as

$$\mathbf{q}_{net} = \begin{pmatrix} q_x \\ q_y \\ q_z \end{pmatrix} = \begin{pmatrix} \tan \beta_{net} \cos \xi_{net} \\ \tan \beta_{net} \sin \xi_{net} \\ 1 \end{pmatrix} t_f. \tag{7.22}$$

This expression describes a column of vertical length component t_f and having a column tilt angle β_{net} given by

$$\cos \beta_{net} = \frac{\mathbf{q}_{net} \cdot \hat{\mathbf{z}}}{|\mathbf{q}_{net}|} = \frac{q_z}{\sqrt{q_x^2 + q_y^2 + q_z^2}} \tag{7.23}$$

and a net orientation angle ξ_{net} given by

$$\sin \xi_{net} = \pm \frac{|(\mathbf{q}_{net} \times \hat{\mathbf{z}}) \times \hat{\mathbf{y}}|}{\mathbf{q}_{net} \times \hat{\mathbf{z}}}, \tag{7.24}$$

where the $\pm$ sign is determined as for Equation 7.15. Figure 7.11 shows a series of vector plots describing the orientation of the growth direction in the xy-plane across the substrate. The length of each vector is proportional to $(q_x^2 + q_y^2)^{1/2}$; shorter arrows thus indicate more vertical columns with a small β_{net}, while long arrows correspond to more inclined columns and larger β_{net} values. The proportionality constant varies between the graphs in order to emphasize the uniformity patterns associated with each substrate motion. The dashed line indicates the same wafer outline as in previous examples. Calculations for four different substrate motions are considered, producing slanted columns (Figure 7.11a), SBD columns (Figure 7.11b), square-spiral columns (Figure 7.11c), and helical columns (Figure 7.11d). These are the same substrate motions examined in Figure 7.10 and the previous discussions.

Keeping the substrate stationary during deposition produces a slanted column microstructure where the columns are predominantly aligned along the x-axis (Figure 7.11a). Away from the substrate centre along the y-axis, the orientation acquires a nonzero y component and the columns tilt inwards towards the x-axis, yielding a macroscale variation in column alignment. As was observed with the thickness calculations, adding φ rotation changes the

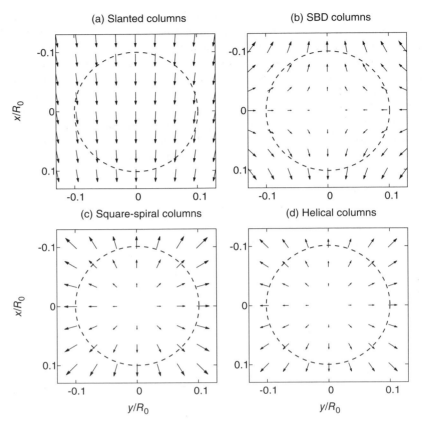

Figure 7.11 Calculations of the net column orientation (Equation 7.22) for GLAD film microstructures consisting of (a) slanted columns (fixed φ), (b) SBD columns (discrete 180° φ rotations), (c) square-spiral columns (discrete 90° φ rotations) and (d) helical columns (continuous φ rotation). ($\alpha = 85°$ in each case.) The x- and y-coordinates are normalized to R_0 and the black dashed line marks the size of a 100 mm diameter wafer in a system placed 50 cm from the source.

column orientation distribution. For SBD-fabricated columns, the rapid φ alternation produces a more complex pattern (Figure 7.11b). At the substrate centre ($x = 0, y = 0$), the deposition is balanced and the net column growth is oriented along the substrate normal. Moving along the x-axis the columns tilt away from the substrate centre, whereas along the y-axis the columns tilt towards the centre point. Elsewhere the column orientation is mixed, containing different x and y components and producing the diagonal patterns. The square-spiral (Figure 7.11c) and helical (Figure 7.11d) column motions produce similar column orientation maps. As with the SBD columns, the deposition is balanced at the substrate centre, yielding net-vertical column growth. Moving away from the centre point, the columns grow radially outwards and the amount of tilt increases for points farther from the centre. As with the thickness uniformity information, knowing the distribution of column orientation for a specific substrate motion is useful in substrate positioning/alignment and for accurate characterization.

7.5 SUBSTRATE MOTION HARDWARE

The complexity of GLAD systems spans a very wide range. Simple, stationary, oblique deposition can be performed in standard PVD tools by mounting substrates on machined wedges or hinged holders. φ rotation can be added to such approaches by repositioning the substrates or holders between deposition steps; one can envision achieving many φ positions with multifaceted wedges or manually rotatable stages. At the other end of the complexity range lie purpose-built, dedicated GLAD systems with computer-controlled, closed-loop substrate motion. Such systems are highly customizable and available from a number of manufacturers around the world, including:

- Kurt J. Lesker Company, Ltd (AXXIS series);
- Mantis Deposition, Ltd (M600);
- Advanced Technological Solutions Ltd (TriAxis Series);
- Polyteknik AS (Cryofox Explorer 500 GLAD);
- Pascal Technologies, Inc. (GLAD substrate holder feedthrough);
- VACOM Vakuum Komponenten and Messtechnik GmbH (GLAD stage); and
- MDC Vacuum Products LLC (multi-axis positioning system).

The latter three entries in the list are fully integrated GLAD systems built into a standard feedthrough or stage design, ready to be mounted onto existing vacuum systems. It is also common to retrofit basic PVD systems with custom-built GLAD substrate motion apparatus, which can take on many forms. Vacuum-compatible motors are available from several manufacturers, and electrical feedthroughs can be used to establish computer connections to motors mounted inside the vacuum system. Alternatively, motors can be mounted outside the system, mechanically linked to internal substrate stages via rotary feedthroughs.

It is difficult to define the motion performance requirements for a general purpose GLAD system precisely. Tolerances on α and φ motions for specific applications are largely unexplored, and experimental optimization is required to properly gauge the impact on film microstructure and properties. In a research setting, where the emphasis is on system capability and flexibility, it is often best to design using large margins and overbuild the motion capabilities. With specific application and property tolerances in mind, experiments can be quickly performed to establish tailored performance requirements for custom systems. The following discussion examines several general considerations that are helpful for the design and analysis of GLAD hardware capabilities.

7.5.1 α motion accuracy and precision

The microstructure is highly sensitive to α, and it follows that controlling the α positional uncertainty is an important factor in GLAD control and film reproducibility. The necessary tolerance levels must be established based on the specific property and intended application. However, as a gateway to quantitative sensitivity analysis, consider the normalized film density ρ_n, which is directly related to many structural and physical properties. As previously

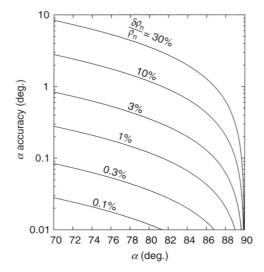

Figure 7.12 Deposition angle accuracy levels required to achieve different relative density tolerances $(\delta\rho_n/\rho_n)$ as a function of α. As α increases, the density becomes very sensitive to small α changes and motor positioning should be highly accurate to ensure sufficient reproducibility in the high-α/extreme-shadowing regime.

discussed in Section 4.4, ρ_n is intimately related to the ballistic shadow length and the α dependence can be approximated using a geometrical shadowing model as

$$\rho_n(\alpha) = \frac{2\cos\alpha}{1 + \cos\alpha}. \tag{7.25}$$

This equation was first presented as Equation 4.11 and is reproduced here for clarity. The effect of uncertainty in α, given by $\delta\alpha$, on ρ_n can be estimated following standard uncertainty propagation calculations, leading to

$$\delta\rho_n = \left|\frac{d\rho_n}{d\alpha}\right|\delta\alpha. \tag{7.26}$$

Figure 7.12 presents the $\delta\alpha$ tolerances required to achieve different relative density uncertainties $\delta\rho_n/\rho_n$ as a function of α. The α and ρ_n uncertainties are proportional; a ten-times improvement in $\delta\alpha$ produces a corresponding ten-times improvement in $\delta\rho_n$. However, note that $\delta\rho_n$ exhibits a clear, nonlinear α dependence. As α increases and ballistic shadow effects are amplified, the α sensitivity increases and the α position tolerances become more stringent. For example, achieving a 1% density uncertainty requires controlling α to better than 0.3° at $\alpha = 70°$. However, increasing α to 80° or 85° imposes α tolerances of nearly 0.1° or 0.05° respectively. The curves diverge as $\alpha \to 90°$, a consequence of the extreme ballistic shadowing lengths. The analysis leads to some key conclusions. As a basic guideline, achieving a specific level of control over density in a GLAD process requires roughly ten times that level of control in the α positioning. For example, a 10% density tolerance requires better than 1° α control, whereas a 1% density tolerance level necessitates better than 0.1° α positioning control. Note that these considerations are irrespective of other parameters that provide additional uncertainty in α, such as extended source effects or flux scattering. Stringent density control at very high α can therefore demand high levels of accuracy and precision in the α positioning, as well as consideration of vapour collimation issues. Given the strong link between the microstructural density and many resulting properties, these conclusions are general and should be found in many examples.

Calibration and monitoring of the α position is an essential part of improving growth control. Digital levels with sub-degree absolute accuracy are readily available and useful for general calibration work. Motors with position encoders enable feedback loops for improved motor control, and limit switches (electrical, optical or magnetic) are also helpful in establishing absolute motor positions. Gear boxes can be used to improve motor resolution, although at the expense of adding backlash problems.

7.5.2 φ motion requirements

Substrate φ rotation is a key feature of GLAD with nearly all motion algorithms implementing φ rotations in some manner. The motion hardware, which includes the φ motor, gearing systems and motion-control feedback system, should be able to track φ changes with minimal error, lag/overshoot or other distortion. The tolerance levels are dictated by the substrate motion, microstructure and application. It is particularly important to consider φ errors in motion algorithms that rely heavily on many discrete φ rotations, such as phisweep (Figure 2.33), SBD (Figure 5.6) or polygonal helices (Figure 2.14), as accumulated errors can produce significant drift from intended positions. For example, long SBD processes can easily contain hundreds of discrete, 180° φ rotations, and sub-degree errors can be quickly compounded to produce large errors. Film structures requiring precise alignment of column axes – for example, square-spirals for photonic crystals and SBD columns for waveplates – also place an emphasis on accurate φ positions. Accurate φ rotation speeds are important to obtain designed pitch values in helical or vertical column microstructures.

The rotation speed requirements vary significantly between the different motion algorithms and are also dictated by the deposition rate, since the φ motions are typically specified relative to the deposited thickness (e.g. degrees per nanometre). For example, fabricating helical column microstructures requires a slow, continuous φ rotation at a rate defined by the helical pitch (Figure 2.12), which is typically >100 nm. The SBD algorithm, in contrast, requires rapid, 180° φ rotations between each sublayer (Figure 5.6). Because the SBD sublayers may be only a few nanometres in thickness, the substrate rotation must be completed over a sub-nanometre thickness to minimize deviation from the intended algorithm. Figure 7.13 presents these considerations, plotting the achievable degrees per nanometre for a given deposition rate and substrate rotation speed. The four contours on the plot, labelled A, B, C and D, denote the necessary rotation speeds for four different substrate motion algorithms. Contour A corresponds to a helical motion with a 150 nm pitch, which requires rotation speed of 360°/150 nm = 2.4°/nm. Contour B denotes the rotation speed required for the slow-corner algorithm (Figure 3.13), used to suppress defect formation at the corners of square-spiral columns. In this example, φ is rotated 90° over a thickness $L/7$, where $L = 100$ nm is the length of the square-spiral arm segment, corresponding to a rotation speed of 6.3°/nm. Contour C shows the 36°/nm rotation speed necessary to fabricate vertical column structures with a 10 nm growth pitch (Figure 2.12). Finally, contour D corresponds to

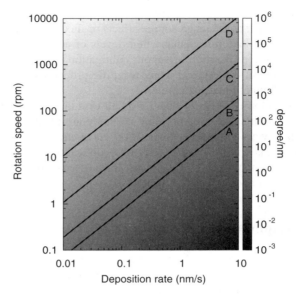

Figure 7.13   To perform the necessary φ motions, typically specified in degrees-rotated per nanometre deposited, sufficient substrate rotation speed is required at the given deposition rate. The four contours show the necessary φ rotation speeds as a function of deposition rate for (A) a helical-column motion (2.4°/nm), (B) the slow-corner algorithm (6.3°/nm), (C) vertically oriented columns (36°/nm) and (D) an SBD motion (360°/nm). A general-purpose GLAD system should be capable of rotation speeds spanning several orders of magnitude. During sensitive portions of a rapid φ rotation algorithm, the deposition rate can be reduced to improve motor performance.

a typical SBD motion algorithm where φ is rotated by 180° over a 0.5 nm growth interval, equating to a 360°/nm rotation speed.

In a general-purpose GLAD system, the φ motion system should provide accurate motion at rotation speeds spanning several orders of magnitude, as well as be able to modify rotation speeds according to the measured deposition rate. An erratic deposition rate can be problematic, and control systems should be sufficiently responsive to compensate. During particularly sensitive portions of a substrate motion algorithm, the deposition rate can be temporarily reduced or the vapour blocked with a shutter to give the φ motors more time to accurately complete the desired motion.

7.5.3 Additional factors to consider

The motion systems (motors, transmission, and gearing) should not only be able to perform the necessary α and ϕ motions at the desired accuracy levels, but multiple secondary issues should also be considered in practical design. Many of these issues are present in any vacuum system with extensive tooling and fixturing apparatus [12]. Friction between moving surfaces and mechanical vibrations can increase particulate generation during substrate motions, leading to substrate contamination and pinhole defects in the films. Moving parts can be shielded from the vapour flux with removable liners to prevent film build-up and reduce flaking. The extra surface area introduced by the substrate motion apparatus increases the surface gas load and can affect pumping times. Heating is a particular problem for motors and gearing mounted in vacuum due to lack of air-cooling. Even in processes without intentional substrate heating the substrate temperature rises due to source radiation and latent heat of condensation, especially in high deposition-rate processing and thick film growth. This heat is transferred through the gearing system and mechanical linkage to the motor, and can accelerate wear of mechanical components or even cause motor failure. Mechanical systems should therefore be designed to accommodate the expected temperature swings in heated or unheated deposition processes. For example, the thermal mass of the motor assembly can be increased with heat sink structures or low thermal-conductivity mechanical components can be used to isolate sensitive areas. Water-cooling could be required in high-temperature processes.

7.5.4 Substrate heating and cooling approaches

A number of groups have examined GLAD deposition with substrate temperature control. Substrate heating is relatively straightforward and has been accomplished using radiative lamp heating [13, 14] or resistive heating approaches [15–18]. Substrate cooling can be achieved using Peltier cells and electrical feedthroughs for smaller wafers, or by flowing water or liquid N_2 [15, 19] through a cooling loop contacting the substrate mounts. The primary challenge associated with implementing heating/cooling equipment in a GLAD system is coping with the α and φ motions of the substrate. Electrical and cooling-fluid connections must have sufficient slack to accommodate the full range of rotation. To provide consistent heating at all α, lamp heaters should rotate with the substrate assembly, multiple lamps should be used or lamp power should be varied with α to compensate. Implementing heating/cooling systems in a manner that does not interfere with substrate φ rotation is a notable issue, as the hot/cold surfaces should remain in contact with the substrate as much as possible for effective energy transfer. This requirement leads to a balance between enhancing the thermal conductance between the surfaces by increasing the contact area/pressure while limiting friction forces that resist motion and increase wearing of the parts. To ease the heating and cooling requirements, the thermal isolation of the substrate-holder/cooling assembly from the surrounding system can be improved by using low thermal conductivity ceramic components.

It is also important to recognize that, even without implementing a direct heating method, the substrate temperature will generally rise during processing due to a number of effects, including radiative heating from the deposition source, latent heat release during vapour condensation, exothermic surface reactions and energetic particle bombardment. The exact temperature rise associated with a specific process and deposition chamber is difficult to predict as it depends on the magnitude of each heating mechanism relative to the substrate energy loss through re-radiation and heat conduction through the substrate stage. The indirect heating should therefore be calibrated experimentally. For example, Seto bonded a resistance temperature device to the substrate and monitored the temperature rise during SiO thermal evaporation [20]. These measurements showed that the substrate temperature reached 150°C after a 90 min deposition. While this temperature rise is small compared with the SiO melting point (1702 °C), it would be a concern for lower melting point materials, and indirect heating effects should be considered in such cases. Several strategies can be pursued for reducing indirect heating, such as depositing at decreased rates, reducing the process time, increasing the substrate thermal mass and increasing the substrate stage thermal conductivity.

7.6 SCALABILITY TO MANUFACTURING

To date, GLAD is primarily a research endeavour with only a few examples of publicized work developing the process for larger scale production [21–24]. However, future commercial applications using GLAD technology will require a greater emphasis on cost-effective fabrication. In anticipation of future developments, some general considerations are discussed in this section.

The size of the substrate motion assembly is a key factor in manufacturability, determining the size and number of wafers that can be processed at a time. Gross die-per-wafer calculations can then be used to estimate total production volumes for a specific GLAD system. However, increasing the wafer size will eventually lead to unacceptable uniformity in one or more key parameters; for example, in thickness, α or column orientation. The uniformity of each parameter depends on the specific substrate motions employed, which can be measured experimentally or calculated using the uniformity framework developed in Section 7.4. Certain substrate motions produce higher uniformity conditions and may be better suited to economical fabrication with general-purpose GLAD systems. It is also possible to improve uniformity and deposit onto larger wafers by increasing the throw distance R_0. Aside from the increased capital expense, increasing the chamber size also leads to longer pump times and increases the overall process cycle time, thus offsetting some of the benefit provided by moving to larger wafers. Large deposition systems should move to load-locked vacuum chambers to improve the cycle time.

An alternative approach is roll-to-roll processing, where a rolled, flexible substrate is unwound in the deposition chamber, coated with the desired layer and then re-rolled. Large substrate areas can then be coated in a single fabrication cycle, leading to greatly increased fabrication throughput. The roll-to-roll approach has long been used to perform oblique deposition (with no φ rotation) for magnetic storage tapes [21–23]. Implementations generally use strategically placed masks or shrouds to restrict the deposition area and appropriately define the α geometry. More recently, Krause et al. have extended the concept to realize complex and dynamic α and φ conditions in a roll-to-roll configuration [24], providing a route to high-throughput GLAD fabrication.

REFERENCES

[1] O'Hanlon, J.F. (2003) *A User's Guide to Vacuum Technology*, 3rd edn, Wiley, Hoboken, NJ.
[2] García-Martín, J.M., Alvarez, R., Romero-Gómez, P. *et al.* (2010) Tilt angle control of nanocolumns grown by glancing angle sputtering at variable argon pressures. *Applied Physics Letters*, **97**, 173103.
[3] Sit, J.C., Vick, D., Robbie, K., and Brett, M.J. (1999) Thin film microstructure control using glancing angle deposition by sputtering. *Journal of Materials Research*, **14**, 1197–1199.
[4] Rossnagel, S.M., Mikalsen, D., Kinoshita, H., and Cuomo, J.J. (1991) Collimated magnetron sputter deposition. *Journal of Vacuum Science and Technology A*, **9**, 261–265.

[5] Buzea, C. and Robbie, K. (2005) State of the art in thin film thickness and deposition rate monitoring sensors. *Reports on Progress in Physics*, **68**, 358–409.

[6] Ohring, M. (2002) *Materials Science of Thin Films*, 2nd edn, Academic Press, San Diego, CA.

[7] Depla, D. and Leroy, W.P. (2012) Magnetron sputter deposition as visualized by Monte Carlo modeling. *Thin Solid Films*, **520**, 6337–6354.

[8] Hawkeye, M.M. (2011) Engineering optical nanomaterials using glancing angle deposition, PhD thesis, University of Alberta.

[9] Poxson, D.J., Mont, F.W., Schubert, M.F. *et al.* (2008) Quantification of porosity and deposition rate of nanoporous films grown by oblique-angle deposition. *Applied Physics Letters*, **93**, 101914.

[10] Buzea, C., Kaminska, K., Beydaghyan, G. *et al.* (2005) Thickness and density evaluation for nanostructured thin films by glancing angle deposition. *Journal of Vacuum Science and Technology B*, **23**, 2545–2552.

[11] Wakefield, N.G. and Sit, J.C. (2011) On the uniformity of films fabricated by glancing angle deposition. *Journal of Applied Physics*, **109**, 084332.

[12] Mattox, D.M. (2010) *Handbook of Physical Vapor Deposition (PVD) Processing*, 2nd edn, William Andrew, Oxford, UK.

[13] Malac, M. and Egerton, R.F. (2001) Observations of the microscope growth mechanism of pillars and helices formed by glancing-angle thin-film deposition. *Journal of Vacuum Science and Technology A*, **19**, 158–166.

[14] Kupsta, M.R., Taschuk, M.T., Brett, M.J., and Sit, J.C. (2011) Overcoming cap layer cracking for glancing-angle deposited films. *Thin Solid Films*, **519**, 1923–1929.

[15] Robbie, K., Beydaghyan, G., Brown, T. *et al.* (2004) Ultrahigh vacuum glancing angle deposition system for thin films with controlled three-dimensional nanoscale structure. *Review of Scientific Instruments*, **75**, 1089–1097.

[16] Flaherty, D.W., Dohnálek, Z., Dohnálková, A. *et al.* (2007) Reactive ballistic deposition of porous TiO_2 films: growth and characterization. *Journal of Physical Chemistry C*, **111**, 4765–4773.

[17] Patzig, C. and Rauschenbach, B. (2008) Temperature effect on the glancing angle deposition of Si sculptured thin films. *Journal of Vacuum Science and Technology A*, **26**, 881–886.

[18] Flaherty, D.W., Hahn, N.T., Ferrer, D. *et al.* (2009) Growth and characterization of high surface area titanium carbide. *Journal of Physical Chemistry C*, **113**, 12742–12752.

[19] Kupsta, M.R. (2010) Advanced methods for GLAD thin films, Master's thesis, University of Alberta.

[20] Seto, M.W.M. (2004) Mechanical response of microspring thin films, PhD thesis, University of Alberta.

[21] Nakamura, K., Ohta, Y., Itoh, A., and Hayashi, C. (1982) Magnetic properties of thin films prepared by continuous vapor deposition. *IEEE Transactions on Magnetics*, **18**, 1077–1079.

[22] Feuerstein, A. and Mayr, M. (1984) High vacuum evaporation of ferromagnetic materials – a new production technology for magnetic tapes. *IEEE Transactions on Magnetics*, **20**, 51–56.

[23] Bijker, M.D., Visser, E.M., and Lodder, J.C. (1998) Oblique metal deposited thin films for magnetic recording. *Tribology International*, **31**, 553–560.

[24] Krause, K.M., Taschuk, M.T., and Brett, M.J. (2013) Glancing angle deposition on a roll: towards high-throughput nanostructured thin films. *Journal of Vacuum Science and Technology A*, **31**, 031507.

A Selected Patents

Date filed	Number	Title	Inventors	Comments
Jul 29, 1998	EP1007754 B1	Glancing angle deposition of thin films	M.J. Brett, K.J. Robbie	GLAD patent covering variety of structures and deposition algorithms
Jul 23, 1996	US 5866204 A	Method of depositing shadow sculpted thin films	K.J. Robbie, M.J. Brett	GLAD patent covering seeding and computer control
Jul 30, 1997	US6206065	Glancing angle deposition of thin films	K.J. Robbie, M.J. Brett	GLAD patent covering variety of structures and deposition algorithms, computer control
Apr 1, 2012	US20120276549	Photonic biosensors incorporated into tubing, methods of	C.J. Choi	Uses GLAD to fabricate Ag nanoparticles on SiO_2 posts for SERS
Apr 4, 2012	US2012/0309080	Surface enhanced Raman spectroscopy nanodome biosensors and methods of manufacturing the same	B.T. Cunningham *et al.*	Similar to US20120276549, except with Au nanoparticles
Apr 10, 2007	US8040591	Ionic electrophoresis in TIR-modulated reflective image displays	L.A. Whitehead	Cites GLAD as a method to improve reflective display performance through coating one electrode
Apr 18, 2006	US 6777770 B2, US7109563, US7485942	Films deposited at glancing incidence for multilevel metallization	K.Y. Ahn, L. Forbes	Produces low-k dielectric materials with GLAD to reduce signal propagation delays in integrated circuits

(*continued*)

Glancing Angle Deposition of Thin Films: Engineering the Nanoscale, First Edition.
Matthew M. Hawkeye, Michael T. Taschuk and Michael J. Brett.
© 2014 John Wiley & Sons, Ltd. Published 2014 by John Wiley & Sons, Ltd.

Date filed	Number	Title	Inventors	Comments
Apr 24, 2007	US7211351, US7344804	Lithium/air batteries with LiPON as separator and protective barrier and method	J.J. Klaassen	GLAD is cited as one method of providing porous thin films to enhance reactant transport into a structured electrode
Apr 24, 2009	US8287937 B2	Endoprosthese	R. Radhakrishnan, S.R. Schewe, V. Schoenle	GLAD is cited as a method for coating stents
Apr 25, 2008	EP2152206A1, US20090123517 A1	Medical devices for releasing therapeutic agent and methods of making the same	A. Flanagan *et al.*	A device with engineered pore structures to controllably release medical compounds; GLAD is one method to produce the structures
Apr 27, 2011	EP2388817 A2, US20110266665 A1	Press-pack module with power overlay interconnection	A. Elasser *et al.*	GLAD is used to engineer a compliant layer between a press-pack module and semiconductor packages, controlling the applied force
Apr 29, 2011	US2011/0270434 A1	Magnetic nanostructured propellers	P. Fischer *et al.*	GLAD is used to fabricate nanostructured helices that can be manipulated by magnetic fields, rotating to propel themselves in fluids
Apr 5, 2010	US8337951	Superhydrophobic surface and method of forming same using high-aspect ratio nano-texture	A.S. Biris *et al.*	Fabrication method for superhydrophobic surfaces by GLAD
Apr 5, 2011	EP2508469 A1	Three-dimensional nano-sculptured structures of high surface-mobility materials and methods of making same	J.M. Caridad Hermandez, V. Krstic	Method to produce GLAD structures of high-mobility materials by providing in-situ heat dissapation with a layer of high thermal conductivity material

Date filed	Number	Title	Inventors	Comments
Apr 9, 2009	US20100259823 A1, EP 2241909 A2	Nanostructured anti-reflection coatings and associated methods and devices	Y.A. Xi *et al.*	Nanostructured anti-reflection coatings that can be produced by GLAD
Aug 4, 2007	US20100079863, EP2084567A1	Optical element, method for production thereof, and usage thereof	M. Thiel *et al.*	GLAD is one proposed method for fabricating optical isolators
Aug 6, 2007	US7656525	Fiber optic SERS sensor systems and SERS probes	Y. Zhao, Y. Liu	SERS substrates can achieve high performance using GLAD nanorods. This patent discloses some fibre-based and instrumental configurations
Aug 9, 2006	US20090056917	Nanostructured micro heat pipes	A. Majumdar	GLAD is used to enhance capillary flow for two-phase cooling of high-performance integrated circuits
Dec 22, 2006	US20100260946	Nanostructure arrays and fabrication methods therefor	D. Jia *et al.*	The use of nanosphere lithography to provide a seeding layer for subsequently deposited GLAD structures
Dec 28, 2007	US20090168136	Solid-state optical modulator	S.J. Jacobs *et al.*	GLAD is one method for producing a chiral material for use in an optical modulator
Feb 10, 2009	EP 2243172 A1, US20090211632	Photovoltaic device based on conformal coating of columnar structures	M.J. Brett *et al.*	GLAD used to control interface shape in photovoltaic devices
Feb 14, 2006	US8075752, EP1861702A	Method and apparatus providing an electrochemical sensor at	A. Dalmia *et al.*	Electrochemical sensor using an electrolytic GLAD film as a support layer

(continued)

Date filed	Number	Title	Inventors	Comments
May 9, 2005	US20050241939	Method and apparatus for providing and electrochemical gas sensor	A. Dalmia *et al.*	Similar to US8075752
Oct 22, 2001	US6908538	Electrochemical gas sensor having a porous electrolyte	A. Dalmia *et al.*	Similar to US8075752
Feb 19, 2010	EP2408487A2	Polymeric/inorganic composite materials for use in medical devices	L. Atanasoska *et al.*	GLAD is one method for producing coatings on implantable medical devices
Feb 23, 2007	US7576905	Electrostatically-controlled diffraction gratings using ionic	L.A. Whitehead, J.S. Huizinga	GLAD is an option for improving index modification by structuring electrodes
Jan 13, 2003	US6770353	Co-deposited films with nano-columnar structures and formation process	P. Mardilovich *et al.*	Co-deposition using multiple GLAD sources to produce composite materials
Jul 12, 2004	US7149076	Capacitor anode formed of metallic columns on a substrate	Shi Yuan	GLAD is the cited method for producing the columnar structures discussed in this patent
Jul 13, 2010	EP 2275842 A1, US 20110012086 A1	Nanostructured functional coatings and devices	L. Tsakalakos *et al.*	An application describing a nanostructured thin film combining antireflection and optical downconversion properties. GLAD is cited as one possible method to fabricate the structures
Jul 27, 2009	EP2460201 A1	Bulk heterojunction organic photovoltaic cells made by glancing angle deposition	R.S. Forrest, N. Li	GLAD is used to manufacture organic photovoltaic cells
Jul 2, 2008	US7880876	Methods of use for surface enhanced Raman spectroscopy (SERS) systems for the detection of bacteria	Y. Zhao *et al.*	GLAD nanorods used as SERS substrates to detect bacteria, or differentiate between different strains of *Escherichia coli*

Date filed	Number	Title	Inventors	Comments
Jun 16, 2008	US7889334	Surface enhanced Raman spectroscopy (SERS) systems for the detection of bacteria and methods of use thereof	D.C. Krause *et al.*	Similar to US7880876, targeting different bacterial and viral strains
Mar 15, 2006	US7738096	Surface enhanced Raman spectroscopy (SERS) systems, substrates, fabrication thereof, and methods of use thereof	Y. Zhao *et al.*	Similar to US7880876, covering a broad range of potential analytes
Jul 28, 2006	US7583379	Surface enhanced raman spectroscopy (SERS) systems and methods of use thereof	Y. Zhao *et al.*	GLAD is one method to fabricate a nanocolumn array for SERS substrates
Jun 9, 2008	US7940387	Surface enhanced Raman spectroscopy (SERS) systems for the detection of viruses and methods of use thereof	Richard A. Dluhy *et al.*	Similar to US7880876, targeting different bacterial and viral strains
Oct 21, 2005	US7658991	Structures having aligned nanorods and methods of making	Y. Zhao *et al.*	Fabrication of aligned GLAD nanocolumns on cylindrically symmetric surfaces
Jul 7, 2003	US6864571, EP1642333A2	Electronic devices and methods for making same using nanotube regions to assist in thermal heat-sinking	M. Arik *et al.*	Application of nanotubes as a heat sink layer for LEDs and integrated circuitry. GLAD is cited as a method of forming suitable nanocolumns
Jul 8, 2008	US7842554	VLSI hot-spot minimization using nanotubes	C.D. Dimitrakopoulos	Application of nanotubes to ameliorate hotspots on integrated circuitry. GLAD is cited as one method for fabricating suitable nanostructures

(*continued*)

Date filed	Number	Title	Inventors	Comments
Oct 18, 2007	US8063483	On-chip temperature gradient minimization using carbon nanotube cooling structures with variable cooling capacity	C.D. Dimitrakopoulos	Similar to US7842554
Jun 12, 2007	US 2008/0003778 A1	Low-temperature welding with nano structues	G. Eyck *et al.*	Uses GLAD films as a low-temperature layer for thermal bonding of two substrates
Jun 19, 2007	US20070242250	Objective with crystal lenses	A. Goehnermeier *et al.*	Describes high-performance objective for lithography. GLAD is cited as one method for producing anisotropic optical coatings to compensate for other optical materials
Mar 12, 2003	EP1483614A2	Objective lens consisting of crystal lenses	A. Goehnermeier *et al.*	Similar to US20070242250
Jun 29, 2010	US20100328896 A1	Article including thermal interface element and method of preparation	D.M. Shaddock *et al.*	Describes a compliant heat transfer layer for high-power semiconductor devices. GLAD helical nanocolumns are one method to fabricate the device
Jun 4, 2010	US20100307238	Humidity sensor and method of manufacturing the same	A.C. van Popta, M.J. Brett	Describes a nanostructured humidity sensor composed of photocatalytic material for optical cleaning, fabricated using GLAD
Mar 13, 2007	US20080224327	Microelectronic substrate including bumping sites with nanostructures	D. Suh	Describes bumping sites enhanced with columnar nanostructures; GLAD is cited as one method for producing the nanocolumns

Date filed	Number	Title	Inventors	Comments
Mar 18, 2009	US20100241071	Polymeric/inorganic composite materials for use in medical devices	L. Atanasoska *et al.*	Cites GLAD as one step in coating implantable devices
Mar 18, 2011	US2011/0238149 A1	Endoprosthesis	L. Atanasoska *et al.*	GLAD is used to provide high-adhesion-strength coatings for implantable stents
Mar 23, 2010	US2010/0245820 A1	Mass sensor	M.M. Schubert *et al.*	Uses GLAD as a high-surface-area substrate for mass sensing, interrogated optically
Mar 24, 2004	US7211881	Structure for containing desiccant	J. McKinnell	GLAD is one cited PVD method for fabricating the multilayer structure described here
Mar 24, 2008	US7869032	Biosensors with porous dielectric surface for fluorescence	W. Zhang *et al.*	A porous GLAD thin film is used to enhance the sensor's surface area
Mar 27, 2009	US20100073678	Blast injury dosimeter	D.H. Smith *et al.*	GLAD is cited as one fabrication method for producing the photonic crystals described
Mar 9, 2007	EP1995794A1, US8049233 B2	Light-emitting device	H. Fukushima	GLAD photonic crystals are one of the possibilities cited for improving optical performance
May 20, 2010	US20100295635 A1	Terahertz resonator	E. Schubert *et al.*	Helical GLAD nanocolumns are used to provide inductance in a tunable resonant circuit for the THz range
Nov 19, 2001	US6589334	Photonic band gap materials based on spiral posts in a lattice	S. John, O. Toader	Patents the square-spiral photonic crystal structure produced with GLAD
Nov 29, 2006	US8094270, EP1796400A1	Thin-film optical retarders	K.D. Hendrix	GLAD is one possible method discussed to achieve form birefringence

(*continued*)

Date filed	Number	Title	Inventors	Comments
Nov 29, 2011	US20120132944	Light-emitting device, light mixing device and manufacturing methods thereof	M.-H. Hsieh *et al.*	GLAD is used to provide an engineered refractive index layer to guide emitted light
Nov 30, 2006	US20080259976	Organic columnar thin films	P.C.P. Hrudey, M.J. Brett	GLAD patent on organic materials for the nanocolumns
Oct 15, 2010	US20120305061 A1/EP2489046A1	Transparent conductive porous nanocomposites and methods of fabrication thereof	P.G. O'Brien *et al.*	GLAD is one method cited for producing transparent conducting oxide thin films with engineered porosity, producing designed optical effects
Oct 17, 2008	US20110056561, EP2201587 A2	Branched materials for photovoltaic devices	B.-K. An *et al.*	The electron acceptor material described may be deposited using GLAD
Oct 18, 2007	US20100313875 A1	High temperature solar selective coatings	C. Kennedy	Uses GLAD to engineer emissivity of high-temperature materials to produce solar selective coatings for solar thermal applications
Oct 22, 2003	US7445814	Methods of making porous cermet and ceramic films	P. Mardilovich	GLAD is one method for producing the porous film described
Oct 9, 2009	US20100125319	Cell-repelling polymeric electrode having a structured surface	T. Scheuermann	GLAD is cited as one method for providing the required textured surface
Sep 20, 2005	US7450227 B2	Surface enhanced Raman spectroscopy (SERS) substrates exhibiting uniform high enhancement and stability	D.W. Dwight *et al.*	GLAD is cited as a method for producing SERS substrates
Sep 22, 2009	US2010/0085566 A1	Surface enhanced Raman spectroscopy on optical resonator	B.T. Cunningham *et al.*	GLAD is used to add metal nanoparticles to optical resonators to produce high-quality SERS substrates

Index